# LIFE 4.0

This book explains what artificial intelligence is, explores its use cases and applications, and helps readers understand key AI concepts and terms such as machine learning, deep learning, and neural networks. It also examines various issues and concerns surrounding AI, including ethics, bias, and the impact on jobs. In addition, it guides readers on learning AI and starting a career in the field. It is designed for both technical and non-technical audiences, and it introduces the fundamentals of AI in an accessible and practical way.

Key Features:

- Explains what AI is, its applications and use cases, and how it is transforming our lives.
- Clarifies common AI terminology, including neural networks, machine learning, deep learning, and data science.
- Navigates ethical and societal discussions surrounding AI.
- Helps readers build machine learning and data science projects.
- Supports professionals in working with AI teams and developing AI strategies within their organizations.

This book is intended for professionals in applied data analytics, artificial intelligence, and related academic fields.

# LIFE 4.0

## Human Life in the Age of Artificial Intelligence

Ali Soofastaei

CRC Press
Taylor & Francis Group
Boca Raton London New York

CRC Press is an imprint of the
Taylor & Francis Group, an **informa** business

Cover design: © Ali Soofastaei

First edition published 2027
by CRC Press
2385 NW Executive Center Drive, Suite 320, Boca Raton FL 33431

and by CRC Press
4 Park Square, Milton Park, Abingdon, Oxon, OX14 4RN

*CRC Press is an imprint of Taylor & Francis Group, LLC*



ISBN: 9781032533469 (hbk)
ISBN: 9781032595740 (pbk)
ISBN: 9781003455219 (ebk)

DOI: 10.1201/9781003455219

Typeset in Times
by codeMantra

*To my son, SAM*

*May you grow up with a fearless curiosity, a steady moral compass, and a heart that remains unmistakably human. In a world where intelligence is everywhere, may your kindness, courage, and imagination be what truly sets you apart.*

*This book is for you and for the future you will help shape*

# Contents

# About the Author

**Ali Soofastaei** is a seasoned technology leader, data strategist, and applied AI practitioner whose career sits at the intersection of engineering, analytics, and large-scale industrial transformation. With a foundation in mechanical engineering and systems thinking and a doctorate focused on information technology within the context of mechanical and mining engineering, Ali has spent two decades helping complex organizations convert raw operational complexity into measurable performance improvement. His professional journey spans global operators and high-impact environments, where the problems are rarely "data problems" alone; they are socio-technical challenges involving people, processes, assets, constraints, and competing priorities. In that space, Ali has built a reputation for connecting the dots, turning scattered signals into coherent decision systems, and translating ambitious digital visions into practical, field-ready execution.

Ali's work is grounded in a simple idea: technology is only valuable when it changes decisions and outcomes. Across global roles in mining and energy, he has led end-to-end programs that connect data, architecture, and advanced analytics to real operational levers of productivity, safety, energy efficiency, reliability, and sustainability. His approach blends rigorous measurement with pragmatic delivery: define the value hypothesis, establish trusted baselines, engineer reliable data pipelines, deploy models that operators can use, and embed the solution into workflows so the improvement survives beyond the pilot. He is particularly known for designing scalable frameworks, governance models, KPI structures, value-driver trees, and operating rhythms that allow analytics to move from isolated use cases into an organizational capability.

A consistent theme in Ali's leadership is his focus on trust: trust in data, trust in models, and trust across teams. He has worked extensively with multi-disciplinary stakeholders, executives, site leadership, engineers, maintenance teams, operators, IT teams, and external partners to align on a shared definition of value and a realistic path to delivery. Whether shaping data governance, modernizing analytics platforms, or deploying AI-enabled optimization, he emphasizes clarity of ownership, transparency of assumptions, and measurable impact. This ability to translate technical depth into executive-level clarity while still earning credibility with frontline teams has made him a natural bridge between strategy and execution.

Ali is also recognized as a data storyteller and change catalyst. He believes successful digital transformation is as much about culture as it is about tooling: building curiosity, improving decision literacy, and ensuring that technology augments human expertise rather than replacing it. He advocates for "human-centered analytics" systems that are interpretable, actionable, and designed to support real decision-makers under real operational pressure. Over time, that mindset has shaped his broader perspective on AI: not as a standalone innovation, but as an evolving layer in society's operating system, influential, transformative, and deserving of thoughtful stewardship.

# Preface

We are living through a hinge moment in human history, not because machines have suddenly become "alive," but because artificial intelligence is rapidly becoming everywhere: woven into the systems that mediate how we learn, earn, heal, govern, communicate, create, and even how we understand ourselves. *LIFE 4.0* is my attempt to name this shift clearly: a new human epoch in which AI is not just a tool we occasionally use but also a persistent collaborator, an embedded advisor, coordinator, and invisible infrastructure inside everyday life.

I chose this topic for one reason: the future isn't arriving "later." It's being negotiated now, quietly, continuously, through millions of design choices, policies, business models, and personal habits. In every previous technological era, the most significant consequences were not only technical but also social. The steam engine reshaped labor, the internet rewired attention, and mobile devices changed the texture of relationships. AI will do all of that faster and across more domains at once. That acceleration creates a real risk: that we drift into a future we didn't mean to build. This book exists to slow the drift, to offer a structured way to think, talk, and decide while the clay is still wet.

You'll notice a repeated pattern throughout the chapters: a simple "1.0 → 4.0" evolution for each domain. This is deliberate. It's easy to get hypnotized by the newest model, the flashiest demo, or the loudest headline. But history teaches something calmer and more useful: every transformation has roots, phases, and predictable tensions; innovation versus stability; personalization versus privacy; efficiency versus equity; and convenience versus agency. By tracing domains from their foundations through industrial and digital phases into AI-native futures, we can spot what's genuinely new, what's repeating, and where the actual trade-offs live.

The scope of *LIFE 4.0* is broad because AI's impact is broad. In finance, we move from manual bookkeeping to algorithmic markets and toward a world of AI-guided wealth, automated compliance, hyper-personalized financial services, and new questions about fairness, access, and systemic risk. In entertainment, we shift from passive consumption to interactive, generative, immersive experiences that can delight us and manipulate attention at an unprecedented scale. In education, AI opens doors to personalized learning and always-on tutoring, while raising hard questions about development, inequality, and what we want "learning" to mean in the first place. Even food evolves from traditional practices to industrial supply chains to smart kitchens and AI-driven nutrition, changing not only what we eat but also how culture and identity express themselves through cuisine.

Some chapters are intentionally uncomfortable because the future will be, at times, uncomfortable. Warfare is being reshaped by autonomy, surveillance, cyber operations, and the speed of decision-making, where seconds can matter, and accountability can blur. Governance faces its own inflection: AI can not only strengthen public services and civic engagement but also intensify surveillance, deepen polarization, and erode trust if deployed without legitimacy and safeguards. Healthcare and wellness may become more predictive, preventive, and personalized; yet, privacy, consent, and equity become existential issues when our bodies and behaviors are continuously measured. Relationships, too, are changing; connections can expand beyond geography and language, while virtual companions and algorithmic social dynamics reshape intimacy, identity, and belonging.

And then, there's the quiet revolution: the AI that doesn't announce itself. Digital assistants and agentic tools are beginning to coordinate schedules, summarize life, manage tasks, filter information, and "help" us make decisions, sometimes so smoothly that we forget we are delegating agency. Sports and spectacle are being transformed by analytics, innovative equipment, AI-augmented training, and immersive broadcasting, redefining performance, fandom, and fairness. Across all of it runs a single thread: AI not only expands human capability but also amplifies whatever incentives and values we embed inside it. That is why ethics is not a side chapter; it's the spine.

For future generations, the stakes are straightforward. They won't just inherit technologies; they'll inherit the norms those technologies enforce: what privacy means, how truth is verified, who gets opportunity, how work is valued, how health is managed, how conflict is constrained, how childhood is shaped, and what kind of relationships feel "normal." If we build LIFE 4.0 with care, they may inherit a world where AI amplifies creativity, expands access to education and healthcare, reduces waste, and helps societies govern more intelligently. If we build it poorly or passively, they may inherit a world of brittle institutions, widened inequality, automated manipulation, and decisions made about them rather than with them.

This preface is also an invitation. You don't need to agree with every forecast in these pages to benefit from them. The purpose is not prediction as performance; it's foresight as responsibility. Read these chapters as thought experiments grounded in trajectories we can already see. Use them to ask better questions in your work, your community, your family, and your own life. Because the most important outcome of the AI epoch isn't the sophistication of our machines, it's the quality of the civilization we choose to build around them.

**Dr. Ali Soofastaei**
*Brisbane, Australia, 2026*

# Acknowledgments

I want to express my sincere gratitude to everyone who contributed, directly and indirectly, to the ideas, conversations, and experiences that shaped *LIFE 4.0*. I am especially thankful to the colleagues, collaborators, and leaders I have worked with across industry and academia, whose openness, challenging questions, and real-world perspectives helped ground this book in practical reality rather than abstraction. My most profound appreciation goes to my wife for her patience, encouragement, and steady support throughout the long hours of thinking, writing, and revising. Your belief in this work made it possible. Finally, to every reader who chooses to engage thoughtfully with the questions raised here: thank you for being part of the conversation that will define how we build an AI-enabled future with responsibility, wisdom, and care.

# Book Structure

*LIFE 4.0* is designed to be read in two ways: straight through as a single argument or selectively as a practical field guide. To make that possible, every significant domain chapter follows the same logic and the same internal "map," so you always know where you are and what comes next.

## THE FLOW OF THE BOOK

### Chapter 1 Sets the Lens

The opening chapter defines what I mean by LIFE 4.0: a new human epoch in which AI shifts from a tool we occasionally use into an embedded layer of everyday life, quietly shaping decisions, services, relationships, and institutions. This section also introduces the core framework that recurs throughout the book: the progression from 1.0 to 4.0, which shows how each domain evolved and why AI changes the direction and speed of that evolution.

### Chapters 2–12 Are Domain Deep-Dives

Each of the remaining chapters explores one domain where AI is transforming "how life works," including the following:

- *Finance and wealth* (Financial Singularity)
- *Entertainment and culture* (Digital Revelries)
- *Education and learning systems* (Education 4.0)
- *Ethics and moral decision-making* (Moral Code 4.0)
- *Warfare and security* (Battle Bytes)
- *Government, institutions, and civic life* (Polity Pulse 4.0)
- *Food systems and everyday consumption* (Culinary Code)
- *Health, wellness, and medicine* (Healing Bytes)
- *Sports, performance, and spectacle* (Digital Olympiad)
- *Relationships and intimacy* (Digital Heartbeats)
- *Daily life coordination through assistants and agents* (Digital Assistant)

These chapters can be read in the published order or approached as "modules" depending on your interests.

## THE STRUCTURE INSIDE EACH CHAPTER

To keep the book coherent and to make comparisons across domains easy, each domain chapter follows a consistent structure:

1. **Abstract**
   A concise snapshot of what's changing in the domain and why it matters.
2. **Introduction**
   The domain's role in human life today and the core pressures driving transformation.
3. **Domain 1.0–4.0 (The Evolutionary Arc)**
   A structured timeline:
   - *1.0: Foundations* (how it began, what it was built to do)
   - *2.0: Digitization* (automation, systems, early data)
   - *3.0: Data and Connectivity* (platforms, networks, scale effects)
   - *4.0: AI-Augmented Intelligence* (prediction, autonomy, personalization, agents)

   This is the backbone of the book: it shows continuity and discontinuity (what AI genuinely breaks open).
4. **Opportunities and Transformational Potential**
   The "why this could be better" section includes efficiency, access, personalization, new business models, and gains in human capability.
5. **Challenges, Risks, and Ethical Considerations**
   The "what could go wrong" section includes privacy, bias, security, transparency, labor transitions, power concentration, and social trust.
6. **The Future Outlook (LIFE 4.0 Scenario)**
   A forward-looking scenario that brings the chapter together in a concrete, lived-in way: what daily life might feel like if AI becomes deeply embedded in that domain.
7. **Conclusion**
   The chapter's key takeaways are what matters most, what to watch, and what choices are likely to be irreversible.
8. **References**
   Curated sources for readers who want to go deeper or validate claims and trends.

## HOW TO READ IT, DEPENDING ON YOUR GOAL

- If you want the whole narrative, start at chapter 1 and continue in order. You'll see how the same forces, data, automation, incentives, governance, and ethics repeat across domains in different forms.
- If you're using this as a practical guide, read chapter 1, then jump to the domains you touch professionally (finance, health, education, governance, etc.). The repeated structure will keep you oriented.
- If you care most about consequences and values, read Moral Code 4.0 early, then return to it as a reference lens while moving through the other chapters.

## A NOTE ON THE BOOK'S INTENT

This is not a book about worshipping technology, and it's not a book about fearing it. It's a book about choice because future generations will not simply inherit "AI."

They will inherit the rules, defaults, and norms we set today: what privacy means, what fairness looks like, how truth is verified, how work is valued, and how human agency is protected (or quietly traded away for convenience).

The structure of *LIFE 4.0* exists to make those choices visible, domain by domain, while there is still time to deliberate about them.

# 1 LIFE 4.0

## 1.1 LIFE 4.0

LIFE 4.0 describes the emergence of a new human epoch shaped by artificial intelligence (AI), intelligent systems, and deep human–machine integration across everyday life. Building on earlier phases of social and technological evolution, this book conceptualizes LIFE 4.0 as the convergence of digital intelligence with core human domains, including health and medicine, relationships, food and consumption, sports and entertainment, governance, and daily tasks. Unlike earlier technological revolutions that primarily enhanced productivity or connectivity, LIFE 4.0 represents a qualitative shift in how humans live, decide, relate, and experience meaning, as AI increasingly operates not only as a tool but as a collaborator, advisor, and embedded presence in social systems.

The book traces a common evolutionary framework across domains, moving from analog, human-centered systems (1.0) through industrial and institutional scaling (2.0) to digitally networked platforms (3.0), and finally toward AI-mediated, adaptive, and personalized ecosystems (4.0). In LIFE 4.0, intelligence becomes ambient: healthcare is predictive and personalized, relationships are augmented by virtual and AI companions, food systems are optimized through data and automation, sports are enhanced by analytics and immersive technologies, and everyday tasks are increasingly delegated to digital assistants. Across these domains, AI reshapes decision-making, efficiency, accessibility, and experience, while simultaneously raising profound ethical, cultural, and societal questions.

A central argument of LIFE 4.0 is that technological advancement alone does not define progress. The integration of AI into intimate and collective aspects of life introduces tensions between efficiency and humanity, personalization and privacy, and augmentation and dependency. The book therefore places strong emphasis on ethical governance, human-centered design, cultural diversity, and inclusivity, arguing that the success of LIFE 4.0 depends on how societies choose to guide, regulate, and live with intelligent systems. Case studies, global perspectives, and interdisciplinary research are used throughout to illustrate both the transformative potential and the risks of AI-driven life systems.

Ultimately, LIFE 4.0 envisions a future in which AI enhances human well-being, creativity, and resilience rather than replacing human agency or social connection. It calls for a deliberate, values-driven transition toward a world in which intelligent technologies serve as partners in human flourishing. By framing AI not merely as a technological force but as a civilizational one, LIFE 4.0 offers a comprehensive lens for understanding and shaping the next chapter of human life in the AI epoch.

DOI: 10.1201/9781003455219-1

## 1.2 FINANCIAL SINGULARITY: NAVIGATING WEALTH IN THE AGE OF AI

Finance has undergone a dramatic transformation from manual, trust-based systems (Finance 1.0) to digitized banking (Finance 2.0) and globally connected, data-driven financial ecosystems (Finance 3.0). This chapter presents Finance 4.0 as the AI-augmented intelligence era, where machine learning, automation, blockchain, and advanced analytics are deeply embedded across financial services. Finance becomes faster, more predictive, and increasingly autonomous. At the core of Finance 4.0 is AI-driven decision-making. Algorithms now assess credit risk, detect fraud, manage portfolios, and execute trades at speeds and scales beyond human capability. Personalized financial services, once reserved for wealthy clients, are increasingly accessible through robo-advisors, AI-powered budgeting tools, and digital platforms. Financial systems become embedded in daily life, operating seamlessly in the background of economic activity.

The opportunities are transformative. AI can dramatically improve efficiency, reduce costs, and expand financial inclusion by serving populations previously excluded from traditional banking. Alternative data and digital platforms enable new forms of credit access, particularly in emerging economies. Predictive analytics improve risk management and stability, while automation reshapes financial institutions into more agile, data-centric organizations. Yet these benefits come with significant risks. Algorithmic opacity, systemic bias, cybersecurity threats, and the concentration of financial power in large technology-driven institutions pose serious challenges. The increasing autonomy of financial systems also raises questions about accountability, transparency, and the potential for AI-driven market instability or cascading failures.

In the LIFE 4.0 vision, finance becomes an invisible yet intelligent infrastructure that supports human well-being rather than dominates it. The chapter argues that responsible governance, ethical AI design, and robust regulation are essential to ensure that Finance 4.0 enhances prosperity, resilience, and trust rather than amplifying inequality or systemic risk.

## 1.3 DIGITAL REVELRIES: THE NEW AGE OF ENTERTAINMENT IN LIFE 4.0

Entertainment has always mirrored the dominant technologies and social structures of its time, evolving alongside human civilization. From communal storytelling, ritual performances, and live spectacles (Entertainment 1.0), through mass media such as print, radio, cinema, and television (Entertainment 2.0), and into the era of digital platforms, streaming, and interactive media (Entertainment 3.0), each phase redefined how people create, consume, and share leisure experiences. This chapter situates Entertainment 4.0 within this historical trajectory, highlighting how AI, immersive technologies, and data-driven personalization are transforming entertainment into a deeply interactive, adaptive, and participatory ecosystem.

Entertainment 4.0 is characterized by the convergence of AI, virtual and augmented reality, real-time analytics, and platform economies. Content is no longer static or one-directional; instead, it is dynamically generated, personalized, and often co-created with audiences. Recommendation algorithms curate music, films, and games; AI-generated characters and narratives adapt to user behavior; and immersive environments blur the boundary between physical and digital experience. Entertainment becomes less about passive consumption and more about engagement, presence, and emotional resonance. This transformation brings significant opportunities. AI-driven entertainment expands creative possibilities, lowers barriers to entry for creators, and enables global audiences to access tailored experiences at unprecedented scale. Independent artists can reach niche communities, interactive storytelling can respond to individual preferences, and immersive platforms can foster new forms of social connection. Entertainment 4.0 thus democratizes creativity while redefining cultural production as a continuous, evolving dialogue between humans and intelligent systems. At the same time, profound challenges emerge. Algorithmic curation risks reinforcing echo chambers, homogenizing culture, or prioritizing engagement over well-being. Issues of data privacy, intellectual property, deepfakes, and the psychological impacts of hyper-personalized and immersive media raise ethical concerns. The monetization of attention through AI-driven platforms also intensifies debates around digital addiction, manipulation, and the commodification of human emotion.

Looking ahead, the LIFE 4.0 vision of entertainment imagines a future where creative expression, technology, and human values are carefully balanced. Entertainment becomes not only a source of pleasure but also a tool for education, empathy, and social cohesion. The chapter ultimately argues that while AI will continue to reshape how entertainment is produced and experienced, human creativity, ethical stewardship, and cultural diversity must remain at the core of this new digital revelry.

## 1.4 EDUCATION 4.0: CLASSROOM CHRONICLES IN THE AI EPOCH

Education has long reflected the economic and technological demands of society, evolving from teacher-centered, rote-based instruction (Education 1.0) to early digitization (Education 2.0) and internet-enabled, learner-centered models (Education 3.0). This chapter frames Education 4.0 as the latest phase in this evolution, aligned with the Fourth Industrial Revolution and defined by AI, data analytics, and adaptive learning environments. Education is no longer confined to classrooms or standardized curricula but is increasingly personalized, continuous, and globally connected.

Education 4.0 leverages AI to transform how learning is delivered, assessed, and experienced. Intelligent tutoring systems, learning analytics, and adaptive platforms tailor content to individual learners' needs, pacing, and learning styles. Teachers are augmented, not replaced, by AI tools that automate routine tasks, provide real-time insights, and free educators to focus on mentorship, creativity, and socio-emotional development. Learning becomes more flexible, experiential, and aligned with real-world problem-solving.

The opportunities for Education 4.0 are substantial. Personalized learning can improve outcomes and engagement, AI-driven analytics can identify learning gaps early, and digital platforms can expand access to education across geographic and socio-economic boundaries. Lifelong learning becomes a practical reality as individuals continuously reskill in response to rapid technological change. Education shifts from a one-time phase of life to an ongoing process embedded throughout the human lifespan. However, Education 4.0 also raises critical challenges. Digital divides risk exacerbating inequality, algorithmic bias can influence assessment and opportunity, and data privacy concerns loom large when student data is continuously collected and analyzed. There are also cultural and institutional challenges, as traditional education systems struggle to adapt curricula, teacher training, and governance structures to AI-enabled learning environments.

In the LIFE 4.0 outlook, education becomes a cornerstone of human–AI collaboration. Schools and universities evolve into hubs of creativity, ethics, and critical thinking rather than mere knowledge transmission. The chapter emphasizes that the success of Education 4.0 depends not only on technological innovation but on inclusive policies, ethical frameworks, and a renewed commitment to human-centered learning.

## 1.5 MORAL CODE 4.0: ETHICAL COMPASSES IN THE AI EPOCH

Human morality has historically evolved through tradition, religion, philosophy, and law, shaping collective norms and ethical behavior (Moral Code 1.0–3.0). This chapter frames Moral Code 4.0 as the ethical response to the integration of AI and autonomous systems into human decision-making. Ethics must now account not only for human actions but also for machine-mediated and machine-initiated outcomes.

Moral Code 4.0 confronts new ethical questions: How should AI systems make decisions that affect human lives? Who is accountable for algorithmic harm? How do we encode values such as fairness, dignity, and responsibility into machines? As AI systems increasingly influence healthcare, finance, governance, and warfare, ethics shifts from abstract philosophy to practical system design and policy.

The chapter highlights opportunities to use AI itself as a tool for ethical enhancement, supporting fairer decision-making, identifying bias, and improving transparency. Ethical-by-design approaches, value-sensitive engineering, and global AI governance frameworks offer pathways to align technological progress with human values.

At the same time, risks are substantial. Moral outsourcing to machines, erosion of human agency, cultural homogenization of values, and the misuse of AI for manipulation or control challenge existing ethical foundations. Without deliberate intervention, AI may reflect and amplify the biases and power structures of its creators.

In the LIFE 4.0 outlook, Moral Code 4.0 serves as the guiding compass for all other domains. The chapter concludes that ethical stewardship is not optional but foundational in the AI epoch. Human wisdom, pluralistic values, and moral responsibility must guide the development and deployment of intelligent systems to ensure that LIFE 4.0 remains humane and meaningful.

## 1.6 BATTLE BYTES: WARFARE IN THE AGE OF ARTIFICIAL INTELLIGENCE

Warfare has evolved from human-centric combat in ancient and feudal societies (Warfare 1.0) through industrialized mass warfare (Warfare 2.0) and digitally networked, precision-based operations (Warfare 3.0). This chapter introduces Warfare 4.0 as the era of AI-enabled conflict, characterized by autonomous systems, intelligent cyber operations, and algorithmically accelerated decision-making.

In Warfare 4.0, AI transforms the battlespace across all domains: land, sea, air, cyber, and space. Autonomous drones, robotic platforms, AI-assisted command systems, and machine-speed cyber weapons redefine how wars are fought. Decision cycles accelerate dramatically, and the role of human soldiers shifts toward oversight, strategy, and ethical judgment rather than direct combat.

The opportunities are both strategic and humanitarian. AI has the potential to reduce risks to human soldiers, enhance defensive capabilities, and improve situational awareness. Autonomous systems can perform dangerous tasks, while AI-driven analytics can support more precise and informed military decision-making.

Yet the risks are severe. Autonomous weapons raise concerns about loss of human control, accidental escalation, and accountability for machine actions. AI-driven cyber warfare lowers the threshold for conflict and increases the risk of rapid, opaque escalation. An AI arms race could destabilize global security and undermine existing norms of warfare.

In the LIFE 4.0 perspective, the chapter stresses that while warfare will inevitably adapt to AI, human values must remain central. Ethical frameworks, international norms, and meaningful human control are essential to prevent technology from outpacing moral responsibility. The future of warfare, the chapter argues, depends as much on governance and ethics as on technological capability.

## 1.7 POLITY PULSE 4.0: GOVERNING IN THE AGE OF ARTIFICIAL INTELLIGENCE

Governance has evolved from centralized, authority-based rule (Governance 1.0) through constitutional democracy and bureaucratic institutions (Governance 2.0) to the digital era of e-governance and globalization (Governance 3.0). This chapter defines Governance 4.0 as the integration of AI, big data, and predictive analytics into public decision-making and administration. Governance becomes increasingly adaptive, data-driven, and anticipatory.

In Governance 4.0, AI supports policy design, service delivery, and regulatory enforcement. Governments deploy machine learning to analyze complex societal data, predict policy outcomes, optimize resource allocation, and respond to citizen needs in real time. Smart cities, AI-driven public services, and automated administrative processes exemplify this shift toward intelligent governance ecosystems.

The potential benefits are significant. AI can enhance efficiency, transparency, and responsiveness in government, reduce bureaucratic friction, and enable more evidence-based policymaking. Data-driven insights can improve public services, anticipate social challenges, and foster more participatory forms of civic engagement through digital platforms. However, Governance 4.0 also presents profound ethical and political risks. Algorithmic bias, surveillance, erosion of privacy, and the potential for techno-authoritarianism challenge democratic values. The delegation of decision-making to algorithms raises questions about accountability, legitimacy, and human agency in governance systems.

The LIFE 4.0 outlook envisions governance that harnesses AI while preserving human judgment, democratic oversight, and ethical accountability. The chapter emphasizes that AI must augment, not replace, human governance, and that strong institutional safeguards are essential to ensure that Governance 4.0 serves public good, equity, and democratic resilience.

## 1.8 CULINARY CODE: SAVORING THE AI-INFUSED FUTURE OF FOOD

This chapter examines how food systems have evolved from traditional, fire-based cooking practices (Culinary 1.0) through industrialized mass production (Culinary 2.0) and digital food platforms and smart kitchens (Culinary 3.0), toward Culinary 4.0, an era defined by AI-driven agriculture, robotics, and personalized nutrition. Food is presented as both a biological necessity and a cultural cornerstone now undergoing intelligent transformation.

Culinary 4.0 integrates AI across the entire "farm-to-fork" chain. Precision agriculture uses machine learning, sensors, and predictive analytics to improve yields, reduce resource use, and adapt to climate change. In kitchens, robotics and smart appliances automate repetitive tasks while enhancing consistency, safety, and efficiency.

A significant focus of the chapter is personalized nutrition. AI systems analyze dietary habits, health data, and even genetic or metabolic indicators to provide tailored nutritional guidance at scale. This shift positions food as a key lever for preventive health, linking culinary practices directly to wellness and longevity.

The chapter also confronts challenges and risks. Labor displacement, data privacy, cultural homogenization, and unequal access to AI-driven food technologies are explored. There is concern that culinary heritage and communal eating traditions could be undermined if personalization and automation are not thoughtfully integrated.

In its LIFE 4.0 vision, the chapter argues that AI should augment, not erase, human creativity and cultural diversity in food. The future of cuisine is framed as a collaborative system where intelligent tools support sustainability, health, and innovation while preserving the social and cultural meanings of food.

## 1.9 DIGITAL OLYMPIAD: SPORTS AND SPECTACLE IN THE AGE OF AI

This chapter traces the evolution of sports from ritualized, community-based competition (Sports 1.0) through professionalization and mass spectatorship (Sports 2.0) to the data-rich digital era (Sports 3.0), culminating in Sports 4.0, where AI reshapes athletic performance, training, and fan engagement. Sports are presented as a mirror of broader technological and social change.

Sports 4.0 is defined by AI-augmented training, smart equipment, and predictive analytics. Athletes use machine learning to optimize biomechanics, manage workloads, and reduce injury risk, while coaches rely on data-driven insights for strategy and talent development. Performance becomes increasingly precise, personalized, and preventative.

For spectators, AI enables immersive and interactive experiences. Smart stadiums, augmented and virtual reality, real-time analytics, and customized content transform fans from passive viewers into active participants. The boundaries between physical venues and digital arenas continue to blur, expanding global access to sports.

The chapter also highlights ethical and competitive concerns. Athlete data ownership, privacy, algorithmic bias, and competitive balance are significant issues. There is an ongoing debate about how much technological augmentation is acceptable before it undermines the spirit of fair competition.

In the LIFE 4.0 outlook, the chapter envisions a future where AI enhances athletic excellence and inclusivity without eroding human values. It concludes that sports must remain fundamentally human endeavors, with technology serving as an enabler of health, fairness, and shared global experience.

## 1.10 HEALING BYTES: WELLNESS AND MEDICINE IN THE AI EPOCH

This chapter situates modern healthcare within a long historical arc, tracing the evolution from holistic, community-based healing traditions (Wellness 1.0) through industrial biomedicine (Wellness 2.0) and digital health systems (Wellness 3.0), culminating in Wellness 4.0, an era defined by AI, genomics, and intelligent health ecosystems. It emphasizes that medicine has always reflected prevailing technologies and philosophies, and that AI now represents a structural transformation rather than a simple technological upgrade.

Wellness 4.0 is characterized by data-driven, predictive, and personalized care. AI-powered diagnostics, precision medicine, and continuous monitoring via wearable and embedded devices enable healthcare systems to shift from reactive treatment to proactive prevention. Machine learning models analyze imaging, genomics, and patient data at scale, augmenting clinicians' decision-making and improving accuracy, speed, and consistency in diagnosis and treatment planning.

The chapter highlights significant opportunities arising from this transformation. AI-driven medicine promises improved access to care, earlier disease detection, reduced costs, and better population-level health outcomes. Telemedicine, digital therapeutics, and remote monitoring extend healthcare beyond hospitals into daily life, making wellness an ongoing, adaptive process rather than an episodic intervention.

At the same time, the chapter critically examines risks and ethical challenges. Issues of data privacy, algorithmic bias, unequal access, and the potential depersonalization of care are explored in depth. The increasing role of automated systems raises questions about trust, accountability, and the balance between clinical judgment and machine recommendations. Looking ahead, the LIFE 4.0 outlook envisions a hybrid healthcare model in which AI augments, rather than replaces, human caregivers. The chapter concludes that the success of Wellness 4.0 depends on ethical governance, inclusive design, and a continued emphasis on empathy, dignity, and human-centered care as medicine enters the AI epoch.

## 1.11 DIGITAL HEARTBEATS: NAVIGATING RELATIONSHIPS IN THE AI EPOCH

This chapter explores how human relationships have evolved from kinship-based bonds (Relationships 1.0) through institutionally mediated social structures (Relationships 2.0) and digitally networked interactions (Relationships 3.0), into an emerging Relationships 4.0 paradigm shaped by AI, social robots, and immersive virtual environments. It frames relationships as a foundational element of social life that is now being reconfigured by intelligent systems.

Relationships 4.0 introduces artificial agents as active participants in social ecosystems. Conversational AI, companion robots, and virtual avatars provide new forms of interaction, care, and emotional support. These systems increasingly simulate empathy, responsiveness, and continuity, enabling people to form meaningful attachments beyond exclusively human relationships.

The chapter emphasizes the transformative potential of these developments. AI-mediated relationships can reduce loneliness, support aging populations, strengthen long-distance bonds, and enable new forms of inclusion for individuals facing social, physical, or psychological barriers. Immersive technologies further allow shared experiences across distance, enhancing intimacy and social presence. However, the chapter also addresses serious concerns. Emotional dependency, authenticity, erosion of human social skills, and privacy risks pose significant challenges. Cultural differences in the acceptance of AI companionship and the potential commercialization of intimacy are explored, highlighting the need for clear ethical boundaries and safeguards.

In its LIFE 4.0 outlook, the chapter proposes a "dual-partnership" future in which human relationships remain central while AI companions play complementary roles. It concludes that sustaining meaningful human connection in the AI epoch requires ethical design, cultural sensitivity, and policies that protect dignity, autonomy, and emotional well-being.

## 1.12 DIGITAL ASSISTANT: NAVIGATING EVERYDAY TASKS IN THE AI EPOCH

This chapter explores how daily life has evolved from manual, routine-based living (Everyday LIFE 1.0) through mechanization and early automation (2.0) to digital assistance (3.0), and into Everyday LIFE 4.0, an era in which AI-powered digital assistants orchestrate tasks, decisions, and interactions across personal and professional domains.

Everyday LIFE 4.0 is characterized by intelligent systems that anticipate needs rather than merely respond to commands. AI assistants manage schedules, optimize household energy use, coordinate transportation, filter information, and support decision-making in real time. Life becomes increasingly adaptive, personalized, and data-informed.

The chapter underscores the benefits of this shift. Digital assistants reduce cognitive load, increase efficiency, and improve accessibility for individuals with disabilities or time constraints. By automating routine tasks, AI frees human attention for creativity, relationships, and higher-order thinking. However, significant risks are also examined. Over-reliance on AI may erode autonomy, situational awareness, and critical thinking. Privacy, surveillance, and algorithmic nudging raise concerns about manipulation, dependency, and loss of agency in everyday decisions.

The LIFE 4.0 outlook presented in the chapter emphasizes balance. The future of everyday life is envisioned as a partnership between humans and intelligent systems, where AI remains transparent, accountable, and subordinate to human values. The chapter concludes that Everyday LIFE 4.0 must enhance human capability without diminishing independence, responsibility, or meaning.

# 2 Financial 4.0

**Abstract**

Finance is undergoing a profound transformation in the era of artificial intelligence (AI). This chapter explores Finance 4.0, the latest evolutionary phase of financial systems, characterized by AI-augmented decision-making, big data analytics, blockchain, and automation. We trace the journey from Finance 1.0, rooted in traditional banking foundations and manual processes, through the Finance 2.0 digitization of banking and Finance 3.0 era of data and connectivity, to the current Finance 4.0 era of AI-driven financial intelligence. A global perspective is maintained throughout, illustrating how these shifts impact both advanced and emerging economies in terms of financial access, efficiency, and innovation. We discuss the transformational opportunities AI brings to wealth management, banking, and financial markets, including greater efficiency, personalization, and inclusion, as well as the challenges and ethical considerations, such as security, bias, job displacement, and regulatory oversight. A forward-looking LIFE 4.0 scenario envisions the future of wealth in an AI-pervasive world, where finance is seamlessly embedded in daily life. The chapter aims to educate and inspire by grounding predictions in clear explanations and real-world examples, demonstrating how AI is reshaping the financial world and what that means for individuals, institutions, and society at large.

## 2.1 INTRODUCTION

The convergence of advanced technologies is driving an unprecedented evolution in the financial industry, often referred to as Finance 4.0. This concept denotes "the latest phase of evolution in the finance industry, driven by technological advancements and digital transformation" [1]. Just as Industry 4.0 heralded a fourth industrial revolution, blending the physical and digital realms, Finance 4.0 marks a new era in which artificial intelligence (AI), machine learning, blockchain, cloud computing, and the Internet of Things (IoT) coalesce to redefine financial services [1, 2]. In essence, Finance 4.0 is AI-augmented finance, leveraging intelligent algorithms and vast data to enhance or even automate financial decision-making and services.

This chapter examines how we arrived at Finance 4.0 by first reviewing the historical foundations (Finance 1.0 through 3.0) that set the stage. Over centuries, finance has evolved from manual, paper-based systems to highly digital, connected networks. Each stage, from traditional banking to early digitization and the data-driven connectivity of recent decades, built capabilities and exposed limitations that the next stage sought to address. Today's leap into AI-driven finance is both a culmination of those trends and a paradigm shift promising to "reimagine the entire financial ecosystem, not just digitize existing processes" [1]. Financial institutions worldwide are embracing cutting-edge technologies to enhance customer experience, improve efficiency, and drive innovation [1].


DOI: 10.1201/9781003455219-2

Importantly, Finance 4.0 is a global phenomenon. While its roots lie in technologies developed mainly in advanced economies, its impact and applications span the globe. For example, AI-powered mobile finance platforms are extending credit and services to populations in Africa and Asia that were previously excluded from formal banking [2, 3]. At the same time, major banks on Wall Street or in London are deploying AI for algorithmic trading and risk management. A 2025 industry survey found that over 85 percent of financial firms worldwide are actively using AI in areas such as fraud detection, risk modeling, and customer service, reflecting how rapidly these tools have become essential [4]. The rise of FinTech (financial technology firms) and Big Tech's entry into finance (e.g., tech giants offering payment or lending services) means competition and innovation in financial services are now truly borderless.

Despite its promise, Finance 4.0 also raises critical challenges and questions. How do we ensure equitable access to AI-enhanced financial services and avoid widening the gap between those with and without digital connectivity? [2] How can privacy and security be protected when finance is increasingly data-driven and cyber threats loom large? What ethical frameworks are needed to govern AI decisions in finance, decisions that affect people's livelihoods and wealth, to ensure fairness and transparency? [3, 4] Regulators and policymakers are grappling with these issues, as seen by growing attention to AI's systemic risks by bodies like the U.S. Financial Stability Oversight Council [4] and the development of "responsible AI" guidelines in financial institutions [3].

The goal of the chapter is to educate and inspire readers about how AI is reshaping the financial world, grounded in factual, clear explanations. In the sections that follow, we review the four stages of modern finance (1.0 through 4.0) to provide historical context. We then delve into the opportunities and transformational potential of AI in finance, from efficiency gains to greater financial inclusion, as well as the challenges, risks, and ethical considerations that must be navigated. Finally, we present an outlook in a LIFE 4.0 scenario, painting a picture of how wealth might be managed and experienced in an AI-pervasive society. By understanding both past and present, and by considering future possibilities, readers across disciplines and regions can better appreciate the profound shifts underway and actively participate in shaping a financial future that benefits all.

## 2.2 FINANCE 1.0–4.0

### 2.2.1 Finance 1.0: Foundations

Finance 1.0 refers to the foundational era of finance, the age of traditional banking and analog processes that defined financial services up until the late 20th century. In Finance 1.0, transactions and record-keeping were predominantly manual, paper-based, and reliant on face-to-face interactions [1]. This era's hallmarks included hand-maintained ledgers, physical cash and coins as the primary medium of exchange, and human judgment as the cornerstone of credit decisions and risk assessment. Financial institutions such as banks emerged and expanded during this period. Still, their operations were grounded in processes that would have been very familiar to a banker from a century earlier: balancing books by hand, verifying identities in person, and managing risk through personal relationships and experience.

Historically, the foundations of modern finance were built over centuries during this Finance 1.0 stage. The establishment of early banks and fiat currencies dates back to the Renaissance and early modern period; for instance, the first central bank (Sweden's Riksbank) was founded in 1668 [2]. Over the 17th–19th centuries, instruments like checks, bills of exchange, and government bonds came into use, but these, too, were paper instruments cleared through manual processes [5]. Double-entry bookkeeping, invented in medieval Europe and popularized by the Italian Medici bank, provided a systematic method for recording transactions, significantly improving financial accounting [5]. Trust was paramount: customers dealt directly with bank tellers and managers, and the banking adage "know your customer" was taken literally; lending decisions often depended on personal knowledge or community reputation.

During the Finance 1.0 era, speed and scale were limited. International transactions could take days or weeks via correspondence. As late as the mid-20th century, moving money or information relied on physical means (cash, checks, mailed documents) or on rudimentary electronic communications such as telegraph transfers. For example, the process of settling an international stock trade in the 1960s might involve mailing stock certificates and waiting for checks to clear. Financial data processing, to the extent it existed, was done on paper or with basic adding machines. Human labor was essential at every step, from filing loan applications to calculating interest by hand or with simple calculators.

Despite these constraints, Finance 1.0 built the trusted frameworks of banking, insurance, and capital markets. Legal and regulatory structures emerged (such as central banks, stock exchanges, and accounting standards) to bring stability and protect consumers. These frameworks would later support the transition to more technologically driven finance. The key takeaway is that Finance 1.0 provided the foundational infrastructure of trust, institutions, and practices, but operated with minimal technology by today's standards. This set the stage for the next era, as burgeoning transaction volumes and globalization in the post-war period pushed the limits of manual finance and opened the door to digitization (Table 2.1).

**TABLE 2.1**
**Comparing Finance 1.0 versus 4.0**

| Category | Finance 1.0 | Finance 4.0 |
|---|---|---|
| Process | Manual | Automated |
| Medium | (Gold, silver, paper cash) | Digital currency |
| Decision-making | Human judgment | AI analytics |
| Customer interaction | In-person branch | Mobile application |

### 2.2.2 Finance 2.0: Digitization Era

Finance 2.0 represents the Digitization Era of finance, when analog processes gave way to electronic systems and the first wave of financial information technology (IT) transformed how financial services are delivered. Roughly spanning the late 1960s through the 1990s, this era introduced computers into banking and capital markets, enabling greater speed, efficiency, and convenience. If ledgers and face-to-face service defined Finance 1.0, Finance 2.0 was represented by the advent of computers, electronic payments, and early online networks.

A pivotal moment often cited as the start of this digital finance revolution was the installation of the first Automated Teller Machine (ATM) in 1967 by Barclays Bank in London [6]. The ATM exemplified Finance 2.0 by allowing customers 24/7 access to cash without a human teller – a radical convenience at the time. Soon, ATMs spread globally, and along with credit cards (which had first appeared in the late 1950s), they digitized everyday banking transactions. As one historical review noted, the advent of the ATM in 1967 marked the start of the second wave of technology in financial services, moving finance "from analogue to digital" [6].

Throughout the 1970s and 1980s, banks and financial institutions invested heavily in mainframe computers and electronic databases. This automation of back-office processes meant that tasks such as clearing payments, calculating interest, and updating accounts could be completed much faster and with fewer errors. For example, in 1973, the Society for Worldwide Interbank Financial Telecommunication (SWIFT) network was created, providing a standardized electronic messaging system for international money transfers. SWIFT replaced slower telex and mail communications, dramatically speeding up cross-border payments. Stock exchanges also began to adopt electronic trading systems: the NASDAQ market launched in 1971 as the world's first electronic stock market, using computers to match buy and sell orders. This transformed the stock market by vastly increasing trading speed and efficiency [5].

By the 1990s, Finance 2.0 reached consumers directly through the emergence of online banking. As internet access spread, some pioneering banks offered services to check balances or transfer funds via dial-up internet or proprietary networks. Though limited at first, by the late 1990s, online banking portals were increasingly common, allowing customers to conduct routine banking from their personal computers. In 1994, Stanford Credit Union became one of the first institutions to let customers access accounts on the World Wide Web, and by the end of the decade, major banks had followed. This period also saw the emergence of electronic brokerage (e.g., E*Trade, founded in 1982, offering online stock trading for retail investors by the 1990s) and electronic bill payment systems.

The Digitization Era's key characteristics can be summarized as follows:

- **Electronic Payment Systems**: The rise of credit and debit cards, ATMs, and electronic funds transfer meant cash was no longer king. By the 1980s, payroll checks were being directly deposited into bank accounts via Automated Clearing House (ACH) systems, and consumers could pay bills electronically. These innovations made transactions faster and more convenient.
- **Back-Office Automation**: Banks implemented core banking software to maintain account ledgers, perform batch processing of transactions overnight, and generate electronic statements. This reduced manual errors and

allowed banks to significantly scale up their customer base without a corresponding increase in staff.

- **Capital Markets Technology**: Stock and bond markets began using digital systems. Besides NASDAQ, many traditional exchanges (NYSE, London, Tokyo) hybridized their floor trading with electronic order systems in the 1980s–1990s. Settlement and custodial functions are also computerized, reducing settlement times and risks.
- **Telecommunications Infrastructure**: The expansion of global telecommunications and networks like SWIFT meant information could flow rapidly between financial centers. This fostered the globalization of finance, as seen in the growth of currency trading and international banking in the late 20th century.

The impact of Finance 2.0 was transformative: financial services became faster, more efficient, and more accessible to many. A bank customer in 1995, for instance, could use an ATM in any major city to withdraw cash from their account back home, a convenience unimaginable in 1960. However, Finance 2.0's digitization was still largely transactional and static by today's standards. Online banking might show you your balance, but it wouldn't analyze your spending or automatically invest your savings. The user interfaces were rudimentary, and most decisions (loans, investments) were still made by humans, albeit aided by computer reports. The stage was set for the next era, which would leverage the internet, data, and, eventually, AI to revolutionize finance further.

### 2.2.3 Finance 3.0: Data and Connectivity Era

Finance 3.0, the Data and Connectivity Era, builds upon digitization by exploiting the internet, mobile technology, and data analytics to radically enhance the reach and complexity of financial services. Roughly spanning the early 2000s through the 2010s (with overlap in the late 1990s), this era is defined by connectivity: ubiquitous internet access, smartphones in billions of pockets, and real-time data flows across global networks. Finance 3.0 saw the rise of FinTech startups and new platforms, the proliferation of digital payment systems, and a shift toward data-driven decision-making in finance. In this era, financial services became more democratized and user-centric, but also more complex as massive volumes of data were generated and harnessed.

Several key developments characterize Finance 3.0:

- **Internet Banking and Mobile Finance**: While online banking began in Finance 2.0, it truly flourished in 3.0 with high-speed internet and mobile devices. By the 2010s, most banks offered fully featured mobile banking apps, and new entrants provided "mobile-first" banking services for everything from payments to budgeting. For example, Kenya's M-Pesa (launched in 2007) enabled mobile phone-based money transfer and financial services for millions without traditional bank accounts, exemplifying the leapfrogging of technology in emerging markets. This connectivity meant consumers could manage finances anytime, anywhere, check balances, pay bills, and apply for loans, all digitally.

- **Rise of FinTech Platforms**: Post-2008, an explosion of fintech companies disrupted traditional financial sectors. Peer-to-peer lending platforms (e.g., LendingClub, founded in 2006) allowed individuals to borrow and lend directly online. Crowdfunding sites (such as Kickstarter, founded in 2009) created new ways to raise capital. Robo-advisors (such as Betterment and Wealthfront, circa 2008–2010) applied algorithms to automate investment advice and portfolio management for retail investors. Payment innovators like PayPal (founded 1998), Stripe (2010), and Alipay (China) facilitated online commerce and digital payments on a global scale. Collectively, these FinTech innovations offered consumers more choices and often lower costs than traditional bank products.
- **Big Data Analytics and Automation**: With more transactions moving online, the volume of financial data has grown exponentially. Financial institutions began using advanced data analytics to derive insights and improve decisions. For instance, banks applied data mining to transaction records to detect fraud patterns and assess credit risk with greater accuracy, moving beyond simple rules to statistical models [1]. Hedge funds and trading firms began using algorithmic trading extensively by analyzing market data in real time, enabling algorithms to execute trades in milliseconds to exploit minor price differences. Automation extended to customer service as well: rule-based chatbots appeared in this era to answer basic customer queries via bank websites or SMS. Overall, many financial processes (loan underwriting, compliance checks, portfolio rebalancing) became partially or fully automated using software, improving speed and consistency.
- **Global Connectivity and Inclusion**: Finance 3.0 dramatically expanded the global financial ecosystem. By 2019, 64 percent of consumers worldwide had adopted at least one fintech service, according to an EY survey (up from 16 percent in 2015), a testament to how quickly these services went mainstream globally. Notably, emerging economies often led in the adoption of mobile finance. China, for example, saw an explosion of mobile payments in the 2010s: apps like WeChat Pay and Alipay became the default way to pay for goods, from street vendors to department stores, virtually eliminating the need for cash in urban China. India too undertook a digital finance revolution (e.g., with the UPI payment system and fintech adoption over 80 percent by 2020) [7]. This era contributed significantly to financial inclusion, bringing millions of "unbanked" individuals into the formal financial system through digital wallets and microfinance apps.
- **New Digital Assets and Decentralization**: A significant late Finance 3.0 development was the birth of cryptocurrency and blockchain technology. Bitcoin's creation in 2009 introduced the idea of a decentralized digital currency not controlled by any government or bank. By the mid-2010s, thousands of cryptocurrencies existed, and blockchain, the underlying distributed ledger technology, was being explored for various financial uses (smart contracts, remittances, trade finance). While crypto markets

remained niche and volatile, they demonstrated a new model of finance (sometimes dubbed "DeFi," for decentralized finance) that, in theory, could operate without traditional intermediaries. Blockchain's promise of secure, peer-to-peer transactions influenced thinking in mainstream finance, prompting pilot programs to integrate distributed ledgers to improve efficiency (e.g., cross-border interbank payments and clearing and settlement).

In summary, Finance 3.0 was a period of unprecedented connectivity and data-driven innovation in finance. The financial system became far more dynamic: information moved instantly, markets operated nearly 24/7 across time zones, and products could be tailored to user behavior. Consumers benefited from greater convenience and choice, including mobile payments and instant personal loans via an app, and many previously excluded people gained access to financial tools. At the same time, the system's complexity grew. The sheer volume of transactions and the speed of markets introduced new risks (e.g., the 2010 "Flash Crash," in which algorithmic trading was implicated in a rapid market plunge). Data privacy became a concern as fintech apps and social media data were mixed (e.g., some lenders began assessing creditworthiness using alternative data such as phone records or social network activity). Regulators worldwide had to play catch-up with innovations like peer-to-peer lending and crypto-assets that didn't fit neatly into existing rules.

By the end of the Finance 3.0 era (late 2010s), the stage was set for AI. Many building blocks were in place: vast datasets, high connectivity, cloud computing infrastructure, and evolving algorithms. Indeed, basic forms of AI (such as machine learning credit models or robo-advisor algorithms) were already in use. Still, Finance 4.0 would elevate this to a whole new level, integrating AI deeply into the financial system's core.

### 2.2.4 Finance 4.0: AI-Augmented Intelligence Era

Finance 4.0 is the AI-augmented intelligence era in finance, the cutting edge where AI, machine learning, and automation are embedded across financial services, augmenting (and in some cases replacing) human decision-making. In this current era (2020s and beyond), we see financial institutions leveraging AI at scale for everything from customer interactions to complex trading strategies. Finance 4.0 is distinguished not just by the presence of new technologies, but by a paradigm shift: finance is no longer just digitally connected; it is, to a significant degree, intelligently autonomous. Systems can learn from data and make decisions with minimal human intervention, heralding efficiencies and capabilities previously unattainable.

Key features and developments of the Finance 4.0 era include:

- **AI-Powered Decision-Making**: Financial firms now use AI algorithms to analyze vast datasets and make decisions in milliseconds. For example, AI-driven trading algorithms ingest news, market data, and even satellite images or social media sentiment to inform investment decisions far beyond what any team of human analysts could process. In credit markets, banks deploy machine learning models to assess loan applications; these models can consider hundreds of variables (including alternative data sources) to predict default risk than traditional scoring models more accurately [3]. Notably, AI credit scoring is expanding access in emerging markets by evaluating applicants with little or no formal credit history using novel data sources (such as mobile phone usage or e-commerce transactions) [3]. AI enables more precise risk assessment and decision-making.

- **Automation and Autonomous Processes**: Building on the automation trend of prior eras, Finance 4.0 sees the advent of autonomous finance or "self-driving" finance. AI increasingly handles routine processes from fraud monitoring to regulatory compliance checks without manual input. Banks employ Robotic Process Automation (RPA) bots and intelligent agents to handle repetitive back-office tasks (like reconciling transactions or validating documents) 24/7 with high accuracy. Some wealth management firms now offer "set-and-forget" investment portfolios where algorithms automatically rebalance and optimize asset allocations for clients. In theory, even complex tasks such as portfolio optimization or market making can be automated mainly by AI that continuously adapts to market conditions. The end goal is often described as touchless or straight-through processing, in which financial transactions and workflows execute automatically from start to finish [8]. Deloitte, for instance, predicted that by 2025, many finance "transactions will be touchless as automation and blockchain reach deeper into finance operations," freeing up humans for higher-level analysis [8].
- **Conversational AI and Customer Service**: One of the most visible changes for consumers in Finance 4.0 is the widespread use of AI-driven chatbots and virtual assistants. Banks and financial apps now commonly deploy chatbots to handle customer inquiries, conduct transactions, or provide financial advice through natural language conversations. Examples include Bank of America's "Erica" virtual assistant and chatbots used by fintech services in messaging apps. These AI assistants can handle millions of customer interactions simultaneously, providing instant responses and personalized suggestions. They are also learning to understand context and sentiment better, making interactions more human-like. The result is improved client engagement and cost savings on customer support. According to industry reports, such AI tools have significantly improved customer satisfaction by providing quick, 24/7 service, while also freeing human staff to focus on complex cases [1]. In short, AI is enhancing both the efficiency and quality of customer service in finance.
- **Big Data and Predictive Analytics at Scale**: While big data analytics started in Finance 3.0, in the AI era, it reached new heights. Predictive analytics driven by machine learning can identify subtle patterns and signals in massive datasets, including transaction histories, market prices, economic indicators, and text from financial news, to forecast trends or detect anomalies. For example, AI systems scan real-time payments data to flag suspicious transactions that might indicate fraud or money laundering, often catching issues humans might miss. Insurers use AI to analyze telematics

data (such as car sensors or fitness trackers) to dynamically price policies based on behavior. Asset managers deploy AI to predict market movements or identify undervalued securities by crunching terabytes of information (so-called quanta mental investing, blending quantitative and fundamental analysis). The ability to "analyze vast amounts of data to detect patterns, identify anomalies, and make predictions" is a hallmark of Finance 4.0 [1]. Crucially, these AI systems continuously learn and improve as more data becomes available, leading to self-optimizing processes.

- **Blockchain, Digital Assets, and Smart Contracts**: Another technology pillar of Finance 4.0 is the maturation of blockchain and its integration with AI. Decentralized finance (DeFi) platforms on blockchains like Ethereum now allow automated financial products (loans, trading, insurance) governed by smart contracts, self-executing agreements in code. Some foresee AI-driven smart contracts that could negotiate and execute financial agreements autonomously. Additionally, central banks are exploring Central Bank Digital Currencies (CBDCs), essentially digital cash, which could transform payments and monetary policy with real-time data on money flows. Blockchain's role in Finance 4.0 is to provide a trusted, transparent ledger for transactions, which, when combined with AI, can lead to robust new systems. For example, AI might dynamically manage a blockchain-based investment fund with minimal human input, or adjust interest rates in a digital currency system in response to economic data. While still emerging, these combinations hint at autonomous, decentralized financial networks in the future. As one industry view noted, blockchain and AI together help reimagine finance by eliminating intermediaries, increasing transparency, and enabling autonomous financial products [1].
- **Integration of IoT in Finance**: The proliferation of IoT devices from smart appliances to connected cars is creating new data streams that Finance 4.0 can leverage. For instance, usage-based insurance uses vehicle sensors to set premiums in real time based on driving behavior. Supply chain financing might use IoT trackers to verify goods in transit before releasing funds. Even smart home devices might one day integrate with financial planning (imagine a fridge that detects a product recall and automatically triggers an insurance claim or refund). These are early stage ideas, but they illustrate how AI in finance can intersect with the physical world through IoT, blurring the lines between financial and non-financial services.

The Finance 4.0 era has already arrived. By the mid-2020s, AI was no longer experimental in finance but essential infrastructure: global banks each have hundreds of AI projects in deployment, and regulators are issuing guidelines for AI governance. A 2025 global outlook noted that "in 2025, over 85% of financial firms are actively applying AI" across various functions [4]. The benefits realized include faster service (e.g., loan approvals in minutes), cost reductions (through workflow automation), and better outcomes (fewer fraudulent transactions and more accurate risk forecasts). At the same time, the profound changes of Finance 4.0 bring challenges, which we examine in the following sections. But there is little doubt that AI-augmented finance is here to stay, evolving rapidly. Financial institutions that harness AI wisely can deliver faster, smarter, and more tailored services, positioning themselves as leaders [4]. Whereas those that lag may struggle to meet rising customer expectations and competitive pressures in this new era.

## 2.3 OPPORTUNITIES AND TRANSFORMATIONAL POTENTIAL

The advent of Finance 4.0 and AI-driven finance brings a wealth of opportunities and transformational potential for the financial sector and society at large. If harnessed effectively, these technologies can make financial services more inclusive, efficient, and innovative than ever before. This section highlights the significant opportunities that Finance 4.0 presents.

### 2.3.1 Improved Efficiency and Cost Reduction

AI and automation allow financial institutions to process transactions and operations at a fraction of the time and cost of manual handling. Mundane tasks from data entry to compliance reporting can be executed by algorithms with near-perfect accuracy and speed. This efficiency gain not only lowers operational costs but also reduces errors and frees employees to focus on higher-value work. For instance, AI-powered underwriting can approve standard loan applications in seconds rather than days, and "touchless" processing means many back-office workflows require no human touch [8]. According to consulting analyses, widespread digital automation in finance could significantly cut administrative costs (some estimates suggest by 20–30 percent) while improving throughput. These savings can translate into cheaper services for consumers (e.g., lower fees) and improve profitability for firms.

### 2.3.2 Enhanced Decision-Making and Risk Management

One of the most potent aspects of AI is its ability to analyze complex data and detect patterns invisible to human analysts. This leads to better-informed decision-making in areas like investment, risk assessment, and financial forecasting. For example, machine learning models can identify early warning signals of credit default or market volatility, enabling proactive risk mitigation. Banks now use AI for real-time fraud detection, flagging anomalous transactions among millions with greater accuracy than rule-based systems [4]. In trading, AI systems can back-test strategies on decades of data in moments and optimize portfolios under myriad scenarios, potentially yielding more stable returns. Overall, Finance 4.0 tools help institutions anticipate and manage risks more effectively – evidence shows that companies using AI in finance report improved forecasting accuracy and more robust risk controls (e.g., AI-based credit models significantly outperform traditional logistic regression models in predicting loan defaults) [7]. By reducing human bias and emotional decision-making, AI can also lead to more rational, data-driven outcomes.

### 2.3.3 Personalization of Financial Services

Finance 4.0 enables an unprecedented degree of personalization in financial services. AI algorithms can tailor financial advice, product offerings, and user experiences to individual customer needs and behaviors. For instance, robo-advisors create custom investment portfolios aligned with an individual's risk tolerance and goals, something historically only wealthier clients could get from personal financial advisors. AI-powered banking apps can analyze a customer's spending patterns and provide personalized budgeting tips or savings plans. We also see insurers using AI to customize insurance pricing to each user's usage (e.g., safe drivers pay less), and lenders dynamically adjusting loan terms. This personalization improves customer satisfaction and engagement, makes products feel "made for you," and can lead to better financial outcomes (such as people saving more due to tailored nudges). It also helps providers differentiate their services in a competitive market and build loyalty through relevant recommendations. As one blog noted, "AI enables financial institutions to offer tailored recommendations and services, improving customer satisfaction and loyalty" [1].

### 2.3.4 Greater Financial Inclusion

A profound opportunity of Finance 4.0 is to expand financial services to underserved populations globally. Digital and AI-driven finance can drastically lower the cost of serving low-income customers, making it viable for banks and fintech to offer accounts, credit, insurance, and investment to people who were previously excluded. For example, AI-based credit scoring using alternative data (such as mobile phone payment history or utility bills) can give lenders insight into the creditworthiness of "thin-file" customers who lack traditional credit scores [3]. In emerging markets, this has already enabled millions of people to access microloans or buy-now-pay-later services for the first time. Mobile banking and e-wallet platforms provide basic financial accounts to anyone with a phone, even in remote areas, eliminating the need for physical bank branches. According to the McKinsey Global Institute, widespread adoption of digital finance could provide access to financial services for 1.6 billion unbanked people and increase the GDP of emerging economies by 6 percent (an estimated $3.7 trillion) by 2025, by empowering individuals and small businesses [9]. Such gains come from increased investment, greater entrepreneurship, and better resource allocation enabled by financial inclusion. In short, AI and digital tech can help "bank the unbanked" and level the economic playing field across regions and socio-economic groups.

### 2.3.5 New Business Models and Revenue Streams

Finance 4.0 blurs industry boundaries and enables innovative business models. Traditional financial institutions are partnering with fintech or tech companies to create hybrid solutions (e.g., banks providing banking-as-a-service via APIs to technology platforms). AI can unlock new revenue streams by analyzing customer data to cross-sell relevant services (with customer consent and proper data ethics, of course). We also see entirely new products born of technology, such as cryptocurrency and digital asset services, and platforms for tokenized assets (where tangible assets like real estate or art are split into digital shares that people can invest in with minimal capital). Additionally, embedded finance is a growing model in which non-financial companies integrate financial services into their offerings (for instance, a ride-sharing app providing drivers with bank accounts and instant payments via AI-assessed cash advances). These new models are transforming how financial services are distributed and monetized, potentially reaching customers in contexts and channels that have never been used for finance. The overall effect is a more dynamic, competitive market spurring continuous innovation.

### 2.3.6 Elevating the Strategic Role of Finance Professionals

Within organizations, the deployment of AI and automation in the finance function (such as corporate finance or the CFO's department) can elevate finance's role from number-crunching to strategic advisory. As routine accounting and reporting tasks become automated, finance teams can focus more on analysis, strategy, and decision support for the business. A study on finance functions found that a large majority expect Finance 4.0 technologies to increase the value contribution and performance of finance departments [10]. CFOs can leverage AI to run predictive forecasts and scenario analyses, providing better insight for strategy. This shift can make finance professionals key partners in innovation and planning, rather than merely seen as back-office record-keepers. In essence, human talent in finance is freed to concentrate on what humans do best: creative problem-solving, relationship management, and strategic thinking, with AI handling the heavy data lifting. This transformation of work is an opportunity to pursue more fulfilling roles and improve business outcomes, as decisions are informed by both human judgment and AI analysis.

### 2.3.7 Macro-economic Benefit

On a macro level, Finance 4.0 has the potential to boost economic growth and stability. By improving credit and capital, AI-driven finance can fuel entrepreneurship and consumption. Enhancing risk detection can help regulators and institutions identify asset bubbles or frauds earlier, possibly averting financial crises or minimizing their impact. The efficiency gains in payments and transactions reduce friction in the economy. For instance, instant payments can improve cash flow for small businesses, and smart contracts can reduce delays in trade finance. One optimistic scenario, estimated by McKinsey, projected that the broad adoption of digital finance could create up to 95 million new jobs in emerging economies through increased business activity and productivity [9]. While such estimates have uncertainty, they underscore the scale of possible positive impact. Even environmental and social goals could benefit: fintech and AI are being used to direct investment to sustainable projects and to improve the distribution of aid or public benefits through digital platforms, thereby contributing to development goals.

In summary, Finance 4.0 offers a transformative toolkit to reimagine wealth management and financial services in ways that can benefit individuals, businesses, and society. It promises a world where financial services are more accessible (available to anyone with a mobile device), more responsive (tailored and on-demand), and more efficient (low-cost and high-speed). The realization of these opportunities, however, is not automatic. It depends on forward-thinking adoption by institutions, supportive public policy, and vigilant management of the accompanying risks. The following section will delve into those challenges and risks, because maximizing the upside of Finance 4.0 goes hand in hand with mitigating its potential downsides.

## 2.4 CHALLENGES, RISKS, AND ETHICAL CONSIDERATIONS

While Finance 4.0 and AI-driven finance hold great promise, they also introduce significant challenges, risks, and ethical dilemmas that must be addressed. As financial systems become more complex and autonomous, stakeholders from bank executives to regulators and consumers need to navigate these issues carefully to ensure technology serves the public interest and financial stability. Below, we discuss the most salient challenges and ethical considerations in the age of Finance 4.0.

### 2.4.1 Data Privacy and Security

Financial services in Finance 4.0 rely heavily on personal and sensitive data (transaction histories, credit records, even social media or geolocation data). This raises privacy concerns: how is this data collected, used, and shared? Consumers may not fully understand or consent to the ways AI models use their data to make decisions. Strict data protection practices and regulations (such as GDPR in Europe) are critical to preventing the misuse of personal financial data. At the same time, cybersecurity threats are a significant risk. As finance becomes digitally interconnected, the system is a target for cyberattacks. Breaches can lead to theft of funds or personal information on a massive scale. A vivid example was the 2017 Equifax hack, which exposed the financial records of over 140 million people. In the AI era, attackers might even try to manipulate AI models (through adversarial attacks or feeding false data) to exploit the system. Thus, ensuring robust cybersecurity and encryption, and building AI systems that are resilient to tampering, is paramount. Financial institutions must invest in cutting-edge security and often work jointly with governments to combat cyber threats, as the consequences of a successful attack on critical financial infrastructure could be systemic.

### 2.4.2 Algorithmic Bias and Fairness

AI models learn from historical data, which may contain biases that lead to unfair or discriminatory outcomes. In finance, this is a pressing ethical issue. For instance, an AI credit-scoring model might inadvertently learn to penalize certain demographic groups if the training data reflects historical biases (such as lower loan approval rates in minority neighborhoods). This could reinforce existing inequalities, the exact opposite of AI's potential for inclusion. Ensuring fairness and avoiding bias in AI decisions is exceptionally challenging. It requires careful choice of model inputs, debiasing techniques, and continuous monitoring. As practitioners have noted, if not carefully designed, AI systems "can reinforce the exact inequalities they intend to alleviate" [3]. Financial firms must adopt responsible AI principles: test models for disparate impacts, adjust, and include human oversight, especially for high-stakes decisions. Some are implementing fairness guidelines (e.g., excluding protected attributes like race/gender from models, or using AI explainability tools to identify proxy variables that cause bias). Regulators are also beginning to demand algorithmic accountability. For example, the EU's proposed AI Act would regulate "high-risk AI," including credit scoring, and mandate audits for bias. The ethical mandate is clear: as finance automates decisions, we must prevent digital redlining or discriminatory pricing hidden behind complex algorithms.

### 2.4.3 Lack of Transparency (The "Black Box" Problem)

Many AI algorithms, especially complex machine learning or deep learning models, operate as a "black box"; their internal logic is not easily interpretable by humans. In finance, this lack of transparency can be problematic. If an AI denies someone a loan or flags a transaction as fraudulent, both the customer and the institution might want to know why. Explainability is not just nice-to-have; in some cases, it's legally required (e.g., regulators often require lenders to explain adverse credit decisions). Black-box AI poses a challenge to compliance and customer trust. Furthermore, even risk managers might struggle to understand their firm's AI-driven trading strategies, making it hard to predict how they'll behave in unusual market conditions. This opacity can become a systemic risk if many institutions use similar inscrutable models and those models react adversely to a scenario, amplifying instability. The field of Explainable AI (XAI) is working on solutions, and some financial institutions are adopting simpler, more interpretable models for specific use cases or using techniques to approximate explanations for complex models. Still, balancing performance and interpretability remains a key challenge. Addressing this will require continuous innovation and possibly regulatory standards for AI transparency in finance.

### 2.4.4 Job Displacement and Workforce Transition

Automation in Finance 4.0 inevitably raises concerns about job losses in the financial sector. Roles that involve routine processing (transaction clerks, reconciliation accountants, etc.) are being supplanted by AI and RPA. For example, a bank that deploys an AI document processing system might need far fewer loan analysts to review applications. A 2020 PwC analysis projected that by 2030, up to 30 percent of jobs in banking and insurance could be affected by automation to varying degrees. While many of these workers will be reskilled or move to more analytical roles, there will be transitional pain. The challenge is ensuring a just transition for the workforce: retraining employees for new roles (such as supervising AI systems or focusing on customer relationships and complex problem-solving that AI cannot handle). Historically, technology has often created new jobs even as it destroys some old ones. For instance, demand for data scientists, AI specialists, and cybersecurity experts is rising in Finance 4.0. But these new jobs require different skills. This creates an ethical and practical imperative to invest in education and training. Financial institutions have started partnering with universities and online platforms to upskill staff in analytics and tech. Governments may also need to expand support (through grants or tax incentives) for workforce development in AI skills to avoid a scenario of structural unemployment or a widening gap between tech-savvy and other workers.

### 2.4.5 Regulatory and Compliance Challenges

Regulators around the world face the difficult task of keeping up with technology. Finance 4.0 innovations often don't fit neatly into existing regulations. Consider cryptocurrency and DeFi, which operate outside traditional banking oversight, raise concerns about money laundering, consumer protection, and even the efficacy of monetary policy. AI decision-making in credit or insurance might challenge existing fair lending or anti-discrimination laws. Additionally, AI models can evolve (through machine learning) in ways regulators didn't explicitly approve, potentially deviating from compliance requirements. Regulators worry about what's been termed the "black box audit problem." How do you audit an AI? In response, we see emerging frameworks: for example, some jurisdictions propose requiring financial firms to obtain certification for their AI models or, at least, to maintain documentation of model development and testing (an "algorithmic accountability report"). International coordination is another challenge: technology crosses borders faster than regulatory regimes can adapt, potentially leading to regulatory arbitrage (in which firms operate in jurisdictions with lax rules when one country imposes strict rules). Bodies such as the Financial Stability Board and the BIS are studying AI in finance to develop best practices. The key is to strike a balance, avoiding stifling innovation with heavy-handed rules, yet preventing risks to consumers and markets. Collaborative efforts, such as "regulatory sandboxes," where fintech and regulators work together on pilot projects, are one way to address this. Nonetheless, regulatory adaptation will be an ongoing effort, with likely new laws concerning AI usage, data privacy, and capital requirements for tech-driven activities.

### 2.4.6 Systemic Risk and Model Risk

As AI becomes entrenched in finance, there's a risk that systemic stability could be affected by technology failures or model errors. If many banks rely on similar AI risk models (perhaps all trained on similar data), they might all underestimate a particular type of risk, leading to a synchronized failure when that risk materializes. For example, if an AI model incorrectly assesses mortgage risk and is widely used, it could contribute to a housing bubble or sudden credit crunch. This is known as model risk, and it is magnified by AI complexity. Furthermore, automated trading algorithms have in the past caused or exacerbated flash crashes, rapid, deep market falls, which could erode trust in markets. The "flash crash" of May 6, 2010, mentioned earlier, is a case where algorithmic trading interactions led the Dow Jones index to plunge nearly 1,000 points within minutes, only to recover shortly after; it raised alarms about the fragility introduced by automated trading. In the AI era, such events could be even more complex. Another systemic concern is the reliance on a few big technology providers: for instance, if many banks run their AI on the same cloud platform. That platform has an outage, which could disrupt financial services broadly (a concentration risk).

To mitigate these issues, financial regulators are conducting stress tests of AI systems and requiring contingency plans. Financial institutions themselves are developing "kill switches" for rogue algorithms and conducting rigorous validation and back-testing of AI models under varied scenarios. Systemic resilience will need new tools, and perhaps even AI itself to monitor AI, for example, meta-algorithms that watch market health and could suspend trading in extreme volatility.

### 2.4.7 Ethical Use of AI and Customer Trust

Beyond bias, there are broader ethical questions about how AI should be used in relation to customers. Informed consent is one issue: do customers understand when they are interacting with a bot versus a human, or how their data is being used to profile them for financial offers? There's a risk of eroding trust if people feel decisions are made in inscrutable ways or that they've lost control over their financial identity. Another concern is that algorithmic marketing and persuasion AI could identify psychological weaknesses in consumers and market financial products to them in manipulative ways. For example, an AI could find a user who is likely to overspend and then entice them with easy credit offers, leading to debt traps. While companies may not set out to do so, profit incentives could drive the unethical use of AI insights. Therefore, developing a strong ethical culture around AI use is essential. Many leading banks and fintech firms are establishing AI ethics boards or principles, such as committing not to exploit vulnerable customers with AI and ensuring transparency in AI-assisted advice. Regulators might also step in with guidelines on explainability to consumers, e.g., requiring that if an AI sets a price or limit, the consumer gets a basic explanation. Maintaining customer trust in an AI-driven environment is both a challenge and an absolute necessity; without trust, customers will not adopt these new services, and the benefits of Finance 4.0 cannot fully materialize.

### 2.4.8 Financial Exclusion and Digital Divide

Interestingly, the flip side of AI aiding financial inclusion is the risk of a digital divide that excludes those who lack digital access or literacy. As financial services move onto digital platforms and AI-powered apps, people without smartphones or reliable internet, or those not comfortable with technology (such as some elderly individuals), might struggle to access basic services. There's a danger that brick-and-mortar bank branches closing in favor of digital channels could leave sections of society behind. Additionally, if AI models become the gatekeepers for credit or insurance, individuals who lack a sufficient digital footprint (no social media, limited online presence) might be harder to score and thus could be marginalized. Addressing this requires conscious effort: multi-channel approaches (keeping some physical presence, at least through agents or kiosks), digital literacy programs, and designing inclusive user experiences that are accessible to people with disabilities or low tech-savviness. Governments and NGOs also play a role in ensuring marginalized groups are not left behind, for instance, subsidizing internet access or providing community training on digital finance tools. The goal should be that the benefits of Finance 4.0 reach everyone, not just the tech-equipped middle and upper classes. This is both a social responsibility and, ultimately, good business sense, as the more people included, the larger the market and overall economic participation.

In confronting these challenges, a recurring theme is balance: balancing innovation with oversight, automation with human judgment, efficiency with equity. The financial industry is increasingly aware that failing to manage these risks could lead to backlash, reputational damage, or crises that erode the progress made. Conversely, proactively managing them can pave the way for sustainable, trustworthy innovation. As one report summarized, the dual nature of AI in finance is that it is "both an extraordinary opportunity and a mounting risk" [4]. It is incumbent on all stakeholders to install the proper "guardrails," be they technical fixes, ethical guidelines, or regulations, to steer Finance 4.0 in the right direction (Table 2.2).

**TABLE 2.2**
**A Table Listing Challenges versus Potential Mitigation**

| Challenges | Potential Mitigation |
|---|---|
| Bias | Fairness audits |
| Black box | Explainable AI tools |
| Cybersecurity | Encryption/AI security |

## 2.5 THE FUTURE OUTLOOK (LIFE 4.0 SCENARIO)

Looking ahead, we enter what might be called the LIFE 4.0 scenario, where AI and related technologies are deeply interwoven into the fabric of everyday life, including how we manage wealth and financial well-being. In this outlook, we imagine how Finance 4.0 could evolve and what the world of finance might look like over the next few decades if current trends continue to mature. This scenario is speculative but grounded in today's emerging technologies and behaviors. The aim is to sketch an inspiring yet realistic picture of "Wealth in the Age of AI" as part of LIFE 4.0.

In the LIFE 4.0 era, financial services become nearly invisible, yet omnipresent. Much as electricity quietly powers modern life in the background, financial functions will be seamlessly embedded into other activities, enabled by AI.

### 2.5.1 Seamless Embedded Finance

People may no longer need to consciously "do banking"; instead, banking happens automatically as part of routine life. For example, consider an individual's morning routine in 2035. As they commute in an autonomous vehicle, an AI financial assistant (integrated in their AR glasses or smart car interface) provides a quick update: "Your savings are on track for the month. I noticed you have a higher cash balance than usual; would you like me to invest an extra $500 into your diversified index portfolio?" The person can say "yes," and the AI executes the investment instantly. Such ambient finance means your money continuously works for you with minimal effort, guided by AI optimizations. Paying for goods might be entirely hands-free, with biometric authentication or devices in the background handling payments as you walk out of a store (already piloted with "just walk out" shops). Loans might be granted at the point of need. If your refrigerator senses it's failing, it may pre-qualify you for an affordable appliance loan via an integrated finance service and even help order a replacement, all behind the scenes.

### 2.5.2 Personal AI Wealth Advisors

Almost everyone could have access to a sophisticated AI wealth advisor that monitors their financial health in real time, something previously available only to the very wealthy through human advisors. These AI advisors (let's call them "FinBots") would use a deep understanding of your life goals, income patterns, spending habits, and risk preferences to provide ongoing advice. They might encourage good financial habits, for instance, nudging you to save a bit more this week because you splurged last week, or advising you to adjust insurance coverage when it detects a life change (like an AI noticing from your calendar and emails that you had a child, and suggesting life insurance or a college fund). With advances in AI, these FinBots could exhibit a high degree of emotional intelligence and personalized communication, making the advice feel genuinely supportive and empathetic. Importantly, such services would be available to a broad population, not just private banking clients, because the marginal cost of providing an AI's attention is low. This democratization of financial advice could lead to better financial literacy and outcomes across society, people making fewer blind mistakes, investing more wisely, and being more prepared for retirement, guided by their AI co-pilots.

### 2.5.3 AI-Augmented Financial Markets

Financial markets themselves may undergo fundamental change. Continuous, AI-driven trading could improve market liquidity and perhaps stability, as AI market-makers provide round-the-clock trading and instantly adjust to new information. However, markets might move faster than ever, with AI algorithms competing, which raises the importance of oversight. We may see new asset classes: for instance, tokenization might turn virtually any valuable item (a piece of real estate, a work of art, a patent's royalty stream) into a digital token that can be traded on global platforms. AI would assist investors in valuing and selling these tokens, thereby opening investment opportunities that were previously illiquid. The concept of fractional ownership (owning 1/1,000th of a Picasso painting via tokens, for example) could become common, allowing more people to participate in wealth-generating assets. Central banks might use AI for more effective monetary policy. Some experts envision "smart" digital currencies in which interest rates or even the money supply could be adjusted in real time by algorithms to stabilize the economy. While such ideas are experimental, they reflect a future in which AI isn't just reacting to financial conditions but actively managing aspects of the economic system under human policymakers' oversight.

### 2.5.4 Global Financial Inclusion and Equality

If the benefits of Finance 4.0 are spread globally, the LIFE 4.0 scenario could be one of vastly increased financial inclusion. By 2030 or 2040, we could see nearly everyone with access to financial services via a smartphone (or successor device), given the dropping costs of technology. A farmer in a rural area might secure crop insurance or a loan for better seeds through an AI-powered platform that evaluates satellite weather data and soil info to offer fair terms, all without requiring a brick-and-mortar bank or lengthy paperwork. As emerging markets continue to adopt technology (many already lead in mobile money), the gap in financial access could narrow significantly. This would help reduce poverty and enable entrepreneurship worldwide, as more people can safely save, borrow, and invest. However, ensuring equity will require conscious effort: if left purely to market forces, there is a risk that AI could concentrate on wealth (for instance, if only certain people have access to the best AI tools). The optimistic outlook is that, with proper governance, AI can be a great equalizer, providing high-quality financial tools to anyone, regardless of their starting wealth or location. Indeed, the World Economic Forum has argued that digital finance is key to achieving the UN Sustainable Development Goals due to its multiplier effects on development [9].

### 2.5.5 New Ethical and Social Contracts

LIFE 4.0 may prompt society to establish new ethical norms and possibly regulations around wealth and AI. For example, the idea of a data dividend or reward: since AI in finance benefits from everyone's data (to train models, etc.), there could be mechanisms to share the gains, perhaps paying users for their data usage or establishing community funds from the efficiency gains of AI automation. If AI and robots take on a large portion of jobs, some economists propose measures like a Universal Basic Income (UBI) funded by the productivity boost. While UBI is a broader economic policy issue, it becomes intertwined with finance in an AI age where automated systems generate wealth with less human labor input. Additionally, concepts like digital identity and privacy might evolve; individuals might have greater control via personal data vaults, deciding which aspects of their financial life to share with AI services in exchange for value. Financial literacy could also become redefined. Tomorrow's "literacy" might include knowing how to set parameters for your AI agent (like risk tolerance or ethical investing preferences) as much as understanding interest rates.

### 2.5.6 Human-AI Collaboration

Rather than AI simply replacing human roles, the most successful LIFE 4.0 scenario sees collaboration between humans and AI. In the finance context, professionals would work alongside AI to produce superior outcomes. A wealth manager might rely on AI to crunch scenarios and optimize portfolios, but still provide the human touch in understanding a client's unique situation and reassuring them in volatile times. Credit officers could use AI suggestions but exercise discretion in exceptional cases or override the model when human intuition or context indicates something the data doesn't. Regulators, too, might use AI "co-pilots" to scan the financial system for emerging risks, focusing human attention where needed. This partnership model would ideally yield the best of both: AI's speed and analytic power with human values and judgment. It could also provide a check-and-balance system, with humans catching AI's mistakes and AI catching humans' errors.

To illustrate a day in LIFE 4.0: A small business owner might wake up to see that their AI accounting system has closed the books for yesterday and applied for a short-term loan overnight because it anticipates a cash flow gap next week (perhaps a major invoice is delayed). The loan was approved by an AI at a fintech lender, at a fair rate tailored to the business's profile, all while the owner slept. Over breakfast, the owner's finance app shows a simple summary: "All set! We secured a $5,000 credit line to cover the gap next week. Also, I noticed you've been paying high fees on international transfers; I've switched you to a cheaper service, saving $200/month." In LIFE 4.0, these behind-the-scenes optimizations happen continuously. The owner can focus on the core business, knowing their AI assistant manages financial minutiae efficiently and intelligently.

However, the LIFE 4.0 scenario is not automatically utopian. The same technology could be misused, for instance, authoritarian use of AI credit scoring to control citizens' behavior (an adverse social credit system scenario), or large tech monopolies dominating finance, stifling competition. The hope is that awareness of these pitfalls, combined with proactive policy and ethical design, steers us toward a more positive outcome.

In conclusion, the future of wealth in the age of AI is likely to be radically different from today's finance, more integrated into our lives, highly personalized, and largely automated. Finance 4.0 technologies hold the potential to create a world where managing one's financial health is as effortless as having an AI autopilot, yet as empowering as having expert advice at one's fingertips. By learning from earlier stages and addressing current challenges, we can work toward a LIFE 4.0 financial system that is innovative and inclusive, efficient and ethical, automated yet human-centric.

## 2.6 CONCLUSION

The evolution from Finance 1.0 through Finance 4.0 reflects a broader story of human progress: leveraging new tools to create more effective and accessible systems for managing wealth. From the humble foundations of ledgers and face-to-face banking, we have arrived at an era of intelligent finance driven by algorithms and data. Finance 4.0: Wealth in the Age of Artificial Intelligence encapsulates both the tremendous potential and the profound responsibility that comes with this stage of development.

In the chapter, we revisited the historical milestones that paved the way for the digitization of finance in the late 20th century and the connectivity explosion of the early 21st century, highlighting how each phase solved specific problems but also introduced new challenges. Finance 4.0 builds on those layers, using AI and automation to address complexities that human capabilities alone struggle with: analyzing massive data in real time, eliminating routine inefficiencies, and tailoring services to individual needs. The opportunities are exciting and numerous. We stand to gain financial systems that are more efficient (fast, cost-effective, automated), more inclusive (bringing services to the underserved), and more innovative (offering entirely new ways to borrow, invest, and transact) than ever before. Indeed, if implemented thoughtfully, Finance 4.0 innovations could contribute to broader economic empowerment and growth, as evidenced by studies projecting trillions added to GDP and millions of jobs from digital finance inclusion [9].

Yet, as we emphasized, these advances come intertwined with challenges. AI is a powerful amplifier; it can amplify positive outcomes, but also negative ones if misapplied. Ensuring fairness, transparency, and security in an AI-driven financial system is not optional; it is a prerequisite for its sustainable success. Trust is the bedrock of finance, and in the age of AI, maintaining public trust will require that stakeholders rigorously address algorithmic biases, protect customer data, and guard against new systemic vulnerabilities. The ethical considerations are as vital as the technical ones. The chapter underscored that managing ethical AI in finance is an active process: financial institutions must implement responsible AI principles, regulators must update oversight frameworks, and society must remain vigilant about how these tools affect human well-being.

Looking at the horizon, we envision a LIFE 4.0 scenario in which wealth management is transformed into a brilliant, largely automated utility in people's lives, empowering yet unobtrusive. It's a future where AI works as an ally to help individuals secure their financial goals and to help institutions allocate capital in socially beneficial ways (such as toward sustainable development or inclusive growth). Achieving this future is not guaranteed; it depends on the choices we make today. Collaboration will be key among technologists, financiers, regulators, and citizens working together to shape the trajectory of Finance 4.0. There is a clear imperative for education and adaptation: we must educate current and future professionals, as well as the public, to navigate an AI-rich financial landscape, and adapt our policies and institutions accordingly. Encouragingly, we see this happening: universities are adding fintech and AI courses, central banks are experimenting with innovation hubs, and international bodies are discussing AI ethics.

In summary Finance 4.0 is more than just a suite of new technologies; it represents a paradigm shift in how wealth can be created, managed, and shared. It has the capacity to augment human intelligence, not just AI, enabling better decisions at both the individual and collective levels. If we steer it wisely, Finance 4.0 could help realize a vision of financial well-being that is more attainable and equitable across the globe. The journey from 1.0 to 4.0 teaches us that each leap comes with challenges that must be mastered. By learning from the past and vigilantly guiding present innovations, we can embrace Finance 4.0 to build a financial system that truly works "intelligently" for all. In the age of AI, wealth and technology will be increasingly intertwined, and it is our role to ensure that this intertwining leads to prosperity, stability, and opportunity in the story of LIFE 4.0.

## REFERENCES

1. Perk, What Is Finance 4.0, in *Perk Blog*, 2023.
2. Singh, R., Finance 4.0: The Race to Rebalance the World, in *Financial IT*, 2019.
3. Mohsen, A., How AI Credit Scoring Models Can Boost Financial Inclusion, in *World Economic Forum*, 2025.

4. RGP Research Professionals, AI in Financial Services: Striking the Balance between Innovation and Regulation in 2025, in *RGP Research Insights*, 2025.
5. FinqUP, From Barter to AI: The Remarkable Journey of Finance through Time, in *FinqUP Insights*, 2024.
6. Hernández de Cos, P., *Financial Technology: The 150-Year Revolution*, 2019.
7. Ernst & Young Global Limited, *Global FinTech Adoption Index 2019*, 2019.
8. Deloitte, *Finance 2025: Digital Transformation*, in *Deloitte Insights*, 2018.
9. Manyika, J., et al., Digital finance for All: Powering Inclusive Growth in Emerging Economies, in *McKinsey Global Institute*, 2016: p. 1.
10. Kabell, P.V., and L.W. Worm, The Future of Finance: Unpacking the Impact of Finance 4.0, in *Copenhagen Business School*, 2024.

# 3 Entertainment 4.0

**Abstract**

Entertainment is undergoing a profound transformation in the era of artificial intelligence (AI). This chapter introduces Entertainment 4.0 as the latest evolutionary stage in the global entertainment landscape, characterized by AI-augmented creativity and immersive experiences. We trace the historical progression from Entertainment 1.0 (Foundations) through Entertainment 2.0 (Digitization) and Entertainment 3.0 (Connectivity and Streaming), leading to the emergence of Entertainment 4.0. Key sectors, including film, gaming, music, and digital media, are examined to illustrate how AI is reshaping content creation, distribution, and audience engagement worldwide. We discuss opportunities such as democratized content creation, personalized experiences, and new revenue streams, as well as challenges such as the ethical use of AI, intellectual property, and the need to protect human creativity. Through a global perspective and a formal yet accessible tone, the chapter aims to educate and inspire readers about the transformative potential of AI in entertainment. Entertainment 4.0 is presented not merely as a technological upgrade, but as a paradigm shift that could redefine storytelling and audience experiences in a future LIFE 4.0 society. The chapter concludes with an outlook, envisioning an entertainment ecosystem where human creativity and AI co-evolve, and highlights the importance of guiding this evolution with ethical and human-centric principles.

## 3.1 INTRODUCTION

In the 21st century, the entertainment industry stands at the cusp of a new era often termed Entertainment 4.0, echoing the concept of Industry 4.0 (the Fourth Industrial Revolution), which is characterized by pervasive digitalization, automation, and artificial intelligence (AI) [1]. Just as Industry 4.0 revolutionized manufacturing and services through cyber-physical systems and AI, Entertainment 4.0 refers to the transformation of media and entertainment through advanced technologies, particularly AI-driven tools and experiences. This chapter explores Entertainment 4.0 as "the new age of experience in the AI era," examining how AI is reshaping the creation, delivery, and consumption of entertainment on a global scale.

Entertainment has continuously evolved alongside technological and social changes. The journey to Entertainment 4.0 can be viewed in terms of several key phases, each building on the previous (analogous to "version" upgrades of entertainment): Entertainment 1.0 (the foundational era of analog and live media), Entertainment 2.0 (the Digitization Era), Entertainment 3.0 (the Connectivity and Streaming Era), and now Entertainment 4.0 (the AI-augmented Era). Table 3.1 provides an overview of these stages, highlighting the approximate time periods, key technologies, and dominant forms of entertainment in each.


DOI: 10.1201/9781003455219-3

**TABLE 3.1**
**Evolution of Entertainment Eras (1.0 through 4.0)**

| Era and Key Phase | Approx. Period | Key Technologies | Dominant Entertainment Forms |
|---|---|---|---|
| Entertainment 1.0 – foundations | Early mid-20th century (and prior) | Analog media (film cameras, radio transmitters, vinyl records); Electric power; Broadcast networks | Live theater and concerts; Silent and early sound films; Radio shows; Analog television (one-to-many broadcasting) |
| Entertainment 2.0 – digitization era | Late 20th century (1980s–1990s) | Digital media formats (CD, DVD); Personal computing; Video game consoles; Computer-generated imagery (CGI) | Music and video in digital formats (CDs, DVDs); Video games (arcade and console); CGI-enhanced films; Cable and satellite TV |
| Entertainment 3.0 – connectivity and streaming | Turn of 21st century (2000s–2010s) | Internet and broadband; Mobile devices; Streaming platforms; Social media networks | Online streaming of music and video (e.g., Spotify and Netflix); Social media content and user-generated videos (YouTube, TikTok); Online gaming and esports; Global franchise blockbusters and fan communities |
| Entertainment 4.0 – AI-augmented era | 2020s and beyond | Artificial intelligence (machine learning, generative AI); Virtual and augmented reality (VR/AR); 5G networks; Internet of Things (IoT) | AI-generated content (films, music, art); Personalized and interactive storytelling; VR/AR immersive experiences; Virtual influencers and digital avatars; Hyper-personalized content recommendations |

Each of these eras reflects a leap in how entertainment is produced and experienced. In Entertainment 1.0, content was limited to analog channels and live venues, created and delivered by humans without computer assistance. Entertainment 2.0 saw the rise of digital production and media, as well as the emergence of video games and computer-generated effects. Still, distribution remained constrained by physical media and scheduled broadcasts. Entertainment 3.0 brought the internet and global connectivity, shattering distribution barriers and enabling on-demand, streaming, and participatory media on a worldwide scale. By the late 2010s, for example, streaming and social platforms were capturing an increasing share of audience attention in the US, and traditional TV and film fell to only about 50 percent of total video viewership (down from 61 percent in 2019) as digital and social video content gained ground [2]. This connectivity era has dramatically changed consumer behaviors and industry business models.

Now, at the advent of Entertainment 4.0, we are witnessing AI's transformative impact on entertainment, potentially the most significant shift since the rise of internet streaming [2]. AI technologies are being infused into nearly every aspect of media and entertainment, from content creation (e.g., scriptwriting, music composition, and visual effects (VFX)) to content distribution (recommendation algorithms, personalized marketing) and user experience (interactive AI characters, immersive environments). The key question is no longer if AI will play a role in entertainment, but "when, how, and just how much," as industry observers note [3]. Enthusiasm for AI's potential runs high: generative and "agentic" AI tools are viewed as strategic assets that can drive new levels of scale, creativity, and efficiency in content production and delivery [3, 4]. At the same time, this rapid change raises essential considerations about the human side of entertainment creativity, jobs, ethics, and cultural impact. Labor and creative communities are actively debating how to ensure AI augments rather than undermines human creativity [5].

The chapter takes a global perspective to explore the emergence of Entertainment 4.0. We begin by reviewing the earlier eras (1.0 through 3.0) to understand the foundations upon which the AI era builds. We then delve into the defining features of Entertainment 4.0, highlighting AI's role in film, gaming, music, digital content, and audience experiences. Throughout, real-world examples and developments from around the world are used to illustrate the trends: from Hollywood studios using AI for VFX to Asian pop stars created as virtual avatars, and from global streaming platforms' algorithms to indie game developers using AI for more imaginative game worlds. We discuss the opportunities and transformational potential that AI brings, such as the democratization of content creation and hyper-personalized entertainment, as well as the challenges and ethical issues, including intellectual property dilemmas, fears of job displacement, and the need for consent and fairness when using someone's digital likeness. Finally, we look to the future, envisioning how Entertainment 4.0 might evolve in a LIFE 4.0 scenario in which AI is deeply integrated into daily life and culture.

By understanding Entertainment 4.0 in context, readers from all backgrounds (general audiences, students, professionals, and academics) can better appreciate how AI is reshaping the entertainment landscape. The goal is to inform and inspire: to show how Entertainment 4.0 can empower creativity and enrich audience experiences globally, if we navigate the transition thoughtfully and ethically. In the sections that follow, we will chart this evolution era by era, then dive into the heart of Entertainment 4.0 and its implications.

## 3.2 ENTERTAINMENT 1.0–4.0

### 3.2.1 Entertainment 1.0: Foundations

Entertainment 1.0 refers to the foundational era of modern entertainment, predating the digital revolution. Roughly covering the late 19th century through the mid-20th century (and building on millennia of prior human creative activities), this era was defined by the rise of mass entertainment via analog technologies and live performance. Key developments of Entertainment 1.0 included the invention of motion pictures, the expansion of radio and broadcast television, and the establishment of popular music and theatrical industries. These innovations provided the infrastructure and cultural bedrock for everything that followed.

The birth of cinema in the late 1800s and early 1900s was a landmark of Entertainment 1.0. Pioneers such as Thomas Edison and the Lumière brothers introduced the world to moving pictures, turning simple sequences of images into a new storytelling medium. Early films were short, silent novelties, but they captivated audiences with the magic of moving images. By the 1920s, cinema had evolved to include sound ("talkies"), marking a significant leap forward in immersive storytelling. The cinematic experience, from silent black-and-white films to the golden age of Hollywood in the 1930s–1950s, became a cornerstone of global entertainment culture. Movies during this foundational era were recorded on photochemical film and projected in theaters; special effects were achieved with practical methods or early camera tricks, and distribution was limited to physical film reels shipped to cinemas.

In the mid-20th century, another revolutionary medium arrived: broadcast radio and television. Radio had already emerged in the 1910s–1920s as a wildly popular form of home entertainment, delivering music, news, and drama to families gathered around wireless sets. The advent of television in the 1940s–1950s brought the visual element into living rooms, forever altering media consumption. By the 1950s, TV sets became everyday household items in many countries, enabling audiences to watch serialized programs and live broadcasts (from news to sports to variety shows) without leaving home. This was a transformative shift: entertainment was now not only something people sought in public venues (theaters, concert halls) but also something that entered the domestic sphere daily. Technologically, television in this era was analog, initially black-and-white, later color; content was delivered over the airwaves by national networks. The one-to-many broadcast model meant that entire populations often tuned into the same program simultaneously, creating a communal cultural experience.

Several characteristics define Entertainment 1.0 content and experiences. Human creativity and performance were at the core of every scripted, acted, directed, and produced radio show, film, or TV program (with no computer assistance, since computing technology was rudimentary or non-existent for media purposes). The delivery platforms were physical or analog: vinyl records for music, celluloid film for movies, and the electromagnetic spectrum for broadcast. Interactivity was minimal; audiences consumed content in a linear, scheduled manner, with limited opportunity for feedback or participation (apart from applauding at a live show or mailing letters to a TV station). Despite these limitations, the foundations laid in this era were profound: enduring genres (e.g., the Hollywood film, the radio drama, and the televised news broadcast) took shape, and the business models of entertainment (box office sales, advertising-supported broadcast, music publishing royalties, etc.) were established.

Moreover, Entertainment 1.0 was characterized by the gradual globalization of culture through media. Hollywood films were exported worldwide from the 1920s onward, creating international stars and shared reference points. Likewise, broadcast innovations spread: for example, the first televised events (like the 1936 Olympics broadcast in Germany or Queen Elizabeth II's coronation in 1953, shown on the nascent BBC TV service) hinted at how television could create global moments. Yet, in this foundational era, geographical distances and technological limits kept entertainment primarily regional. Each country or region had its own film industries, radio stations, and TV networks producing content tailored to local languages and tastes.

In summary, Entertainment 1.0 set the stage (quite literally) for what was to come. It was an era in which entertainment experiences were tangible and synchronous: films on reels, music on records, performances on stages or live airwaves. The creative industries grew around analog tools, and audiences developed a voracious appetite for the new forms of storytelling these tools enabled. By the second half of the 20th century, however, new technologies emerged that would radically change how entertainment was made and delivered. The stage was set for the Digitization Era, Entertainment 2.0, which would build on these foundations and propel entertainment into the computer age.

### 3.2.2 Entertainment 2.0: Digitization Era

Entertainment 2.0 encompasses the digital revolution in entertainment, roughly from the 1980s through the 1990s. In this era, analog technologies gave way to digital formats and computer-assisted production techniques, fundamentally altering how content was created, stored, and disseminated. The Digitization Era introduced the world to personal computers (PCs), video game consoles, Compact Discs (CDs) and Digital Versatile Discs (DVDs), and early special effects driven by computer graphics. It was a time of transition: traditional forms of entertainment were being reinvented through digital means, setting the groundwork for the connected world soon to follow.

One hallmark of Entertainment 2.0 was the shift from analog media to digital media. In music, for example, vinyl records and cassette tapes from earlier eras were gradually replaced by the CD, introduced in the early 1980s as a digital audio format. The CD offered crystal-clear sound and durability, demonstrating how digital encoding could improve audio fidelity and convenience for consumers. Similarly, home video moved from VHS tape (analog) to DVD by the late 1990s, bringing higher resolution and random access to video content. Digital formats enabled copying content without degradation and paved the way for the massive libraries of content storage we take for granted today. As one retrospective noted, "the advent of digital technology in the late 20th century revolutionized entertainment production and distribution," enabling higher-fidelity audio and video and new methods of content delivery.

In parallel, the PC revolution exploded in the 1980s, bringing new forms of entertainment. Video games emerged from simple arcade beginnings (Pong in the 1970s, arcade cabinets in the 1980s) into a mainstream entertainment medium. During the Entertainment 2.0 era, home consoles (such as the Nintendo Entertainment System in 1985 and the Sony PlayStation in 1994) and PC games became hugely popular, adding interactivity as a key component of entertainment. By the late 1990s, video games featured 3D graphics and rich storytelling that began to rival films in cultural impact. Indeed, the era of interactive entertainment was dawning by the turn of the millennium, and video games were a dominant force, with photorealistic experiences that rival Hollywood blockbusters in production scale and revenue. This period saw the emergence of iconic game franchises and the birth of genres (first-person shooters, real-time strategy, etc.) that would later become pillars of digital culture.

Another transformative aspect of Entertainment 2.0 was the introduction of computer-generated imagery (CGI) and digital production techniques in film and television. Early uses of computers in entertainment were primitive (such as the wireframe graphics in 1973's Westworld or the stained-glass knight in 1985's Young Sherlock Holmes). Still, by the 1990s, digital effects had advanced dramatically. Movies like Jurassic Park (1993) showcased CGI dinosaurs integrated with live-action, and Toy Story (1995) became the first fully computer-animated feature film. These milestones proved that digital tools could usher in a new era of special effects and animation, expanding the scope of imagination on screen. Filmmakers began to rely on digital editing, compositing, and eventually digital cinematography (the late 1990s saw the first major films shot with digital cameras). In music, digital synthesizers and sampling in the 1980s changed the sound of popular music (think of the synthesized pop and electronic music boom). By the 1990s, artists were using computer software to record and edit tracks in a Digital Audio Workstation (DAW), replacing bulky analog studio equipment.

During the Digitization Era, we also witnessed the early seeds of the internet's impact on entertainment, though widespread connectivity had not yet arrived. In the late 1980s and early 1990s, technologies like CD-ROMs and multimedia PCs hinted at future interactive content (such as encyclopedia CDs with audio/video, or simple dial-up bulletin boards sharing software). By 1993, the World Wide Web was born. While Entertainment 2.0 is mostly pre-internet in its full scale, by the mid-1990s, one can see precursors to streaming and digital distribution: for example, MPEG video files, early online services distributing games or music in small quantities, and the rise of file formats like MP3 (invented in 1993) that would later revolutionize music consumption. Thus, Entertainment 2.0 laid down the digital foundations and mindset of innovation that would fully blossom in the next era.

Importantly, consumer behavior and industry economics in Entertainment 2.0 began to shift in response to digitization. With home computers and game consoles, audiences, especially younger generations, spend more time on interactive digital entertainment rather than just passive media. The media industry saw new revenue streams: video game sales skyrocketed; DVD sales in the late 1990s gave Hollywood a new cash cow (the "home video" market became as crucial as theatrical releases). However, challenges also emerged: digital content is easily copyable, and by the late 1990s, the music industry faced the phenomenon of MP3 piracy (e.g., Napster, launched in 1999, allowed users to share digital songs peer-to-peer). This was an omen of disruptions to come, suggesting that the control of distribution was shifting due to digitization.

By the end of the 1990s, the stage was set for another leap. If Entertainment 2.0 digitized our content and tools, Entertainment 3.0 would connect them, leveraging the expanding internet and telecommunications infrastructure. The world was about to become much more interconnected, and entertainment content would flow across new channels at lightning speed. The following section explores the Connectivity and Streaming Era, in which the digital foundations of 2.0 combine with global networks to create an entirely new media ecosystem.

### 3.2.3 Entertainment 3.0: Connectivity and Streaming Era

Entertainment 3.0 represents the Connectivity and Streaming Era, spanning roughly the 2000s through the 2010s. In this period, the proliferation of the internet, mobile devices, and high-speed networks radically transformed the entertainment landscape. If the prior era digitized content, Entertainment 3.0 mobilized and globalized it, enabling on-demand streaming, online communities, and interactive engagement worldwide. The defining features of this era include ubiquitous connectivity, user-generated content, social media, and the rise of global streaming platforms that disrupted traditional media distribution.

A key hallmark of Entertainment 3.0 is the advent of streaming media. In the early 2000s, as broadband internet access became common, new platforms emerged to deliver music, movies, and other content directly to consumers over the internet. Napster (1999) and subsequent peer-to-peer networks demonstrated the demand for digital music distribution (albeit illegally, at first), prompting legitimate services like Apple's iTunes Store (launched in 2003) to start selling downloadable music. Before long, the model shifted from downloads to streaming: YouTube was founded in 2005, pioneering user-uploaded video streaming, and Spotify (2008) introduced a streaming music service that gave users instant access to millions of tracks on demand. By the 2010s, streaming had become mainstream across all media domains: Netflix, initially a DVD-by-mail service, launched streaming video in 2007 and went on to produce its own content, disrupting the television industry; Amazon Prime Video, Hulu, and other services followed. Traditional broadcast and cable TV experienced intense competition from over-the-top (OTT) platforms that deliver content via the internet. The "binge-watching" culture emerged, as entire seasons of TV shows could be released at once for streaming.

This era also witnessed explosive growth in mobile entertainment. The introduction of smartphones (beginning with the iPhone in 2007) and later tablets (iPad in 2010) put powerful computers and broadband connections into the pockets of billions of people. Entertainment was no longer confined to living room TV sets or desktop PCs; now, anyone could watch videos, play games, or listen to music anytime, anywhere. Mobile games became a massive industry (the late 2000s saw phenomena like Angry Birds and Pokémon GO, which by 2016 blended mobile with location-based AR). The convenience of mobile streaming further accelerated the shift away from scheduled programming; audiences could consume content on the go and at their convenience.

Another transformative aspect of Entertainment 3.0 is the rise of social media and user-generated content as entertainment. Platforms such as Facebook (2004), Twitter (2006), Instagram (2010), and TikTok (2016) not only connected people but also became primary channels for entertainment content. The concept of the "content creator" or "influencer" emerged from individuals who could attract large audiences on YouTube or TikTok by producing their own videos, comedy sketches, music, or commentary. This democratization of content creation meant that audiences were no longer just consumers but also producers of entertainment. It blurred the line between professional and amateur content. Viral videos and memes became part of the global pop culture lexicon. For instance, Gangnam Style (a music video by Psy) became the first YouTube video to hit 1 billion views in 2012, illustrating how a creator from South Korea could gain worldwide fame through a social platform. Social media also fostered new forms of interactive entertainment: from live-streaming yourself playing video games on Twitch (founded 2011), to global talent shows on platforms, to fandoms organizing on forums and hashtags. The Entertainment 3.0 era truly "blurred the lines between passive consumption and active participation," as audiences engaged with content by commenting, sharing, remixing, and even co-creating.

Connectivity also led to the emergence of new genres and experiences. Online multiplayer games became hugely popular, creating virtual worlds where millions of players could interact (e.g., World of Warcraft in 2004, or later the rise of battle royale games like Fortnite in 2017). These games evolved into social spaces and even spectator events, giving birth to the eSports industry (professional competitive gaming), which, by the late 2010s, filled stadiums and drew streaming audiences of tens of millions for tournaments. Traditional sports and entertainment were also affected: now, fans could follow and discuss events in real time on Twitter, and sports leagues began streaming games online to reach global audiences.

By the end of the 2010s, the impact of Entertainment 3.0 was evident in industry metrics and consumer habits. Digital distribution had overtaken physical in most areas: for example, global music industry revenues from streaming surpassed those from CDs and downloads combined. In video, cord-cutting (cancelling cable TV in favor of streaming) became common, pressuring traditional TV networks to launch their own streaming apps (e.g., Disney+ launched in 2019, consolidating Disney, Marvel, and Star Wars content for streaming). The global nature of connectivity also meant entertainment had a broader reach: a niche show on Netflix could find fans in dozens of countries; a local cultural phenomenon could become worldwide (the Korean pop band BTS leveraged social media to build a massive international fanbase).

An essential technological underpinning in this era was the gradual improvement of recommendation algorithms and personalization. Services like Netflix and Spotify invested in machine learning algorithms to analyze user preferences and consumption patterns, providing tailored content recommendations. Early forms of AI thus quietly entered entertainment during 3.0 as recommendation engines (e.g., the Netflix recommendation algorithm became famous for driving engagement by suggesting what to watch next). By doing so, platforms increased user satisfaction and time spent, but also raised questions about "filter bubbles" and algorithmic influence on culture. Nonetheless, personalized feeds and suggestions became a standard expectation for users. AI-driven content discovery was shifting the industry from search-based consumption to recommendation-based consumption.

By around 2020, Entertainment 3.0 had fully transformed the media environment. People worldwide were hyper-connected, with entertainment accessible on demand across devices and borders. The stage was now set for Entertainment 4.0, building atop this connected digital ecosystem but adding a new catalyst: advanced AI. In the next section, we turn our focus to Entertainment 4.0, exploring how AI is augmenting creativity and experience across film, gaming, music, and beyond, and why it is seen as the next major leap in the evolution of entertainment.

### 3.2.4 Entertainment 4.0: AI-Augmented Creativity and Experience

Entertainment 4.0 refers to the emerging era in which AI and related technologies augment the creativity, production, and consumption of entertainment. It builds on the digital, globally networked foundation of the previous era, infusing it with intelligent automation and new immersive modalities. In Entertainment 4.0, AI is not just a back-end recommendation tool; it is at the front and center of content creation (e.g., writing, composing, and designing), content delivery (personalizing streams, dubbing content in multiple languages), and even content itself (AI-generated characters or interactive narratives). This section will examine how Entertainment 4.0 manifests across different domains: film and television, gaming, music, digital/social media content, and overall audience experience. We will see that AI is accelerating innovation in each of these fields, often in conjunction with other technologies like virtual/augmented reality (VR/AR) and the Internet of Things (IoT), truly making this the "new age of experience."

#### 3.2.4.1 AI in Film and Television Production

In the film and TV industry, AI is poised to reinvent the production process from script to screen [2]. Traditionally, creating a movie or show is labor-intensive and costly, involving many creative and technical steps. AI-powered tools are now emerging to assist or enhance nearly every stage of this pipeline. For instance, generative AI algorithms can help with screenwriting and story development. While AI is not about to win an Oscar for Best Original Screenplay on its own, writers have begun experimenting with AI systems that generate story ideas or even complete draft scripts, which humans can then refine. AI can analyze existing script libraries and suggest plot structures or character arcs that resonate with audiences (though such use remains experimental and somewhat controversial, especially amid writers' guild concerns). Moving into preproduction, AI tools are streamlining tasks like storyboarding and pre-visualization. Filmmakers can input a script and have AI generate rough storyboards or concept art for scenes, helping them visualize the project before expensive shooting begins [2]. According to a McKinsey analysis, AI-assisted storyboarding and 3D modeling of sets can "front-load" work in preproduction, reducing the length and cost of physical production [2]. Directors who once had to imagine scenes in their heads can now use AI to render different visions quickly, and even A/B test them, for example, generating multiple possible set designs or camera angles and evaluating which is more effective [2].

During production, AI's role has been more limited (given that capturing performances is inherently a human process), but there are notable examples. "Virtual cinematography" can use AI to optimize camera placements or even automate camera movements (drones with AI tracking for dynamic shots, for example). On set, AI-driven tools can monitor continuity (flagging if an actor's appearance or positioning doesn't match between takes) or even generate synthetic extras to fill out a crowd scene without hiring hundreds of background actors. One cutting-edge development in some film projects is the use of AI-based real-time effects; for instance, on the Disney+ series The Mandalorian, a game engine (Unreal Engine) renders digital backgrounds in real time on a volume stage. While that's more game tech than AI, the blending of game AI to adjust lighting or background details on the fly is starting to happen.

It is in post-production where AI has really taken off in filmmaking. "Studios are already deploying gen AI tools in editing suites to handle microtasks that once took hundreds of hours," reports an industry executive [2]. VFX artists now leverage AI for tasks like rotoscoping, object removal, and in-between frame generation, massively speeding up what was traditionally painstaking manual work. AI-based software can automatically erase a boom microphone that dipped into a shot, or replace a green-screen background with minimal human rotoscoping. AI has revolutionized de-aging and digital cosmetics. Instead of animators tweaking frames, neural networks can learn a face and apply a younger appearance or minor facial adjustments (removing wrinkles, etc.) across frames. One studio executive noted, "vanity fixes [in post-production] that used to be incredibly manual are now pretty easy to do with AI," such as smoothing skin or removing blemishes [2]. Perhaps the most headline-grabbing use of AI in recent films is the resurrection or recreation of actors. For example, the Star Wars franchise has used AI-driven deepfake techniques to recreate young Princess Leia and Grand Moff Tarkin in Rogue One (2016), and to de-age Mark Hamill for Luke Skywalker's appearances in The Mandalorian (2020) and The Book of Boba Fett (2022). These were achieved with a combination of CGI and AI facial mapping on stand-in actors, pointing to a future in which an actor's digital likeness can be part of new performances long after they have aged, or even posthumously. This technology is powerful but raises ethical questions that we will explore later.

AI is also transforming audio post-production. Dialogue dubbing and translation are prime examples. Traditionally, to release a film globally, you'd hire voice actors to dub dialogue into various languages, and sometimes faces wouldn't perfectly match the new speech. Now, AI can generate remarkably human-like voices in multiple languages that match an actor's vocal qualities, and even use lip-syncing algorithms to adjust the on-screen lip movements to fit the new language better [5]. Companies like Flawless AI (mentioned earlier) use a combination of deepfake video techniques and linguistic AI to alter mouth movements so that, say, an English-speaking actor appears convincingly to be speaking Japanese or Spanish [5]. This AI-driven localization means audiences around the world can have a more authentic experience, and it significantly speeds up the localization process for studios, which historically was slow and expensive. In fact, at industry events like NAB 2025, experts highlighted real-time AI dubbing as an emerging solution, envisioning a near future in which a live broadcast or streaming event can be instantly available in many languages via AI voices, "making global distribution instantaneous" for live content [3].

Even marketing and distribution for film/TV benefit from AI. AI can automatically edit trailers by analyzing a film and picking exciting snippets (IBM famously did this with the horror movie Morgan in 2016, using Watson to cut a trailer). AI can also predict which scenes or promotional images will appeal most to specific demographics by analyzing data, thus guiding marketing campaigns. On streaming platforms, AI curates which thumbnail to show each viewer for the same movie (Netflix uses different images from the film as the "cover" depending on a user's profile). These subtle AI-driven customizations can significantly enhance engagement by catering to individual tastes.

A crucial point in Entertainment 4.0's approach to film/TV is that AI is seen as an augmenting tool, not a replacement for human creativity. "Great stories and great storytelling will always matter. AI can speed the workflow, but creativity still defines the outcome," as one AI product leader emphasized [2]. Creative professionals and guilds strongly echo this sentiment. In 2023 and 2024, Hollywood's writers and actors unions (WGA and SAG-AFTRA) raised concerns about uncontrolled use of AI. They pushed for regulations ensuring AI doesn't undercut writers' credit/compensation or allow studios to reuse actors' likenesses without consent. Indeed, SAG-AFTRA's position, articulated by its negotiator Duncan Crabtree-Ireland, is that AI should "serve human creativity, not replace it" and that performers must retain control and be compensated when their digital replicas are used [5]. In new contracts, they have secured guardrails, such as requiring explicit informed consent and fair pay for any AI-generated use of an actor's voice or image [5]. This illustrates a broader principle: Entertainment 4.0 is most potent when AI empowers creators, reducing drudgery, expanding what's possible on screen while humans continue to provide the heart and vision.

Concrete examples of AI-augmented production are already surfacing. In 2024, a film titled The Brutalist made news for using AI to subtly alter the lead actors' spoken dialogue accents (the film, about Hungarian architects, used AI to make the actors' English sound more Hungarian) [4]. The director intended to enhance authenticity for creative reasons, and notably, the actors were involved, and their performances remained their own; AI just adjusted the timbre of the voice. This kind of use case shows AI as a "post-production assistant" to achieve a creative vision that otherwise might be impossible without re-casting or months of voice training.

Perhaps one of the most ambitious examples of Entertainment 4.0 in film is coming from outside Hollywood: an Indian startup is producing what may be the world's first AI-generated feature film, Maharaja in Denims, slated for release in 2025 [6]. The filmmakers are using a suite of generative AI tools to create nearly everything from characters and scenes (via image and video generation models) to dialogue (via AI voices), essentially animating the author's novel without a whole live-action shoot [6]. They fine-tuned AI models on Indian cultural data to get authentic-looking results, and only plan minimal live filming (e.g., recording a real dance to incorporate via AI) [6]. While technical challenges like maintaining visual continuity of characters are significant (AI outputs can vary each time, though new techniques allow adding a "digital tag" to keep a character's appearance consistent) [6], this project showcases the frontier of AI in filmmaking. The motivation here is that the story was too expensive/complex to make traditionally. Still, AI might make it possible within a startup budget, a clear example of democratization. Industry experts acknowledge it's a stretch to do a whole movie with today's AI, but the fact that it's being attempted demonstrates how far the tools have come [6].

In summary, AI's role in film and TV under Entertainment 4.0 is multi-faceted: accelerating production workflows, enabling new creative possibilities, and potentially lowering barriers to entry for creators with big ideas but small resources. From Hollywood studios to independent filmmakers in India and beyond, AI is augmenting the way stories are crafted for the screen. As we embrace these tools, the focus remains on using them to serve storytelling, making movies. It shows better or more efficiently produced, rather than turning storytelling over entirely to machines. The result could be a flourishing of creativity, where more voices and visions find expression with the help of AI.

#### 3.2.4.2 AI in Gaming and Interactive Entertainment

The gaming industry has long been a playground for AI, even in earlier eras (game AI controlling non-player characters goes back decades). However, Entertainment 4.0 marks a new chapter in which AI and machine learning are fundamentally expanding what games can do, how they are developed, and the kinds of experiences players can have. Games are increasingly featuring AI-driven content generation, more intelligent and more human-like NPCs (non-player characters), and even AI-assisted game design. This not only improves player experiences but also transforms the production side of gaming.

One major trend is the creation of more intelligent and adaptive NPCs. Traditionally, game NPCs followed predetermined scripts or simple rule-based AI (often resulting in predictable or repetitive behavior). Now, developers are using machine learning and advanced algorithms, so NPCs can learn, adapt, and even exhibit what seem to be "emotions" or social behaviors [7]. For example, games can implement AI techniques such as behavior trees and reinforcement learning to enable enemies or allies to respond dynamically to a player's actions rather than simply repeating pre-coded patterns [7]. A famous early example was the "Nemesis System" in Middle-earth: Shadow of Mordor (2014), where enemy orc characters would remember encounters with the player, dynamically improve or change attributes if they survived, and even generate personalized taunts based on past events [7]. That system was rule-based, but it pointed toward an ideal which machine learning now pursues: NPCs that genuinely "learn" from each interaction. In some cutting-edge prototypes, NPCs are powered by large language models (not unlike ChatGPT) that allow for open-ended dialogue with players. Imagine a role-playing game where every character in a village can carry on a unique conversation or has a simulated life of their own; such possibilities are being actively explored. Early experiments include modders integrating GPT-4 into games like Skyrim so that NPCs can converse fluidly and unscripted. While still experimental, this idea of AI NPCs with believable personalities and memories could bring unprecedented depth to game worlds. A recent analysis noted that AI NPCs "can enable more immersive experiences, deeper storytelling, and stronger player engagement," essentially making game worlds feel alive in a way pre-scripted NPCs never could.

Another transformative use of AI is in procedural content generation. Games have used procedural generation for terrain or levels for years (e.g., the Diablo series random dungeon layouts, or the vast algorithmically generated universe of No Man's Sky). Entertainment 4.0 takes this further by using AI to create entire ecosystems, quests, and even storylines on the fly [7]. Developers set high-level rules or parameters, and the AI fills in the details, creating, say, a forest that grows and changes based on player behavior, or generating side quests that are unique for each playthrough but still coherent within the narrative. The result is games where no two experiences are alike, and the story can "evolve organically with the player" [7]. For example, AI-powered narrative engines could analyze your play style and tailor the game's plot branch or new challenges to suit your interests (a combat-heavy player might find more battles, while a puzzle-solving player encounters more riddles, etc.). Some experimental AI game masters for tabletop-like RPGs have been developed that generate descriptions, outcomes, and even adapt the difficulty on the fly. The combination of procedural generation with ML-based AI can yield much richer content than the old math-driven methods. Indeed, such techniques can be seen as the game analog of an AI improvisational storyteller that collaborates with the player in real time.

AI is also revolutionizing game development workflows. From concept art to level design to quality testing, AI tools are saving developers time and expanding creativity. A 2025 industry survey (the Game Developers Conference State of the Industry) found that over half of game development companies already use generative AI in their workflows [7]. For instance, AI image generators can create textures, character concept art variations, or environment illustrations that artists can refine. This can accelerate iteration, as artists do not start from scratch but instead guide AI outputs. Code-generation AI can help programmers by writing routine code or optimizing performance. Automated game testing is another area: AI bots can play a game tirelessly to find bugs or balance issues, which is especially useful in complex open-world or multiplayer games, where all possibilities are hard to test manually. Major studios use machine learning to simulate how players might approach their game (to identify if one strategy is overpowered, for example). The result is potentially shorter development cycles and more polished releases. It's reported that studios using AI in content creation and testing are managing to "reduce production time and costs while increasing creative output," effectively doing more with less [7]. This is a significant advantage in an industry where making a AAA game can cost hundreds of millions of dollars and several years.

A fascinating frontier is AI-generated game content created in real time with the player. Some researchers and indie developers are experimenting with games that generate dialogue, items, or even entire environments on the fly via AI in response to player input. For example, imagine a text adventure where you can type any action or dialogue and an AI language model dynamically responds, creating new story paths. A few such AI-driven narrative games have appeared (using GPT-3/4 under the hood). In 2023, a game called AI Dungeon did precisely this, offering an endlessly generative text adventure. Now, teams are adding graphics to this concept, aiming for "infinite games" that constantly create new content. While this is early and can sometimes lead to incoherent outcomes (AI can go off the rails), it hints at a future in which game designers set up worlds and rules. Then, the AI does some of the storytelling heavy lifting in response to the player's actions, essentially customizing the game to each player.

The convergence of gaming with other Entertainment 4.0 tech is also noteworthy. VR and AR are natural partners with AI. In VR experiences, AI can populate the world with realistic interactive characters (a VR simulation of, say, a mystery might have AI characters you can interrogate by voice). In AR, AI might recognize real-world objects and incorporate them into gameplay (e.g., a smartphone AR game that uses your actual surroundings dynamically in a puzzle). The metaverse concept, a persistent, shared virtual world if it comes to fruition, will likely rely on AI to generate vast amounts of content and manage the world in real time for millions of users. Gaming companies like Roblox and Epic Games (Fortnite's creator) are already investing in AI to make it easier for users to create their own games or content on their platforms, aiming for a creator-driven metaverse.

From a business perspective, AI is seen as a key growth driver for gaming. The global AI in gaming market is projected to grow massively (from about $3.3 billion in 2024 to over $51 billion by 2033) [7], reflecting how central these technologies are becoming. This includes spending on AI middleware, tools, and AI features in games themselves. Game studios are hiring AI specialists; some even have AI ethics teams due to concerns around AI-generated content (for instance, if an AI inadvertently creates something offensive, the developers need to have guidelines and filters in place).

Players themselves are starting to notice and expect more innovative, richer games. For the gaming community, one exciting manifestation is AI-driven creation of more immersive esports broadcasts. For example, AI can generate live commentary for a match in any language or highlight reels auto-edited from a tournament. It's noted that AI can "simulate opponents, balance matches, and even generate commentary, making esports more immersive and accessible" [7]. This again broadens the audience and global reach of interactive entertainment.

In conclusion, AI in the context of Entertainment 4.0 gaming means games that are more lifelike, dynamic, and tailored to the player, and game development that's more efficient and creative. We're likely to see new genres enabled by AI, perhaps games that are half-game, half-story co-created with an AI dungeon master, or simulation worlds that evolve in unpredictable ways. The overarching theme is that the line between game developer and player might blur. Players could have more influence over how the game world unfolds, essentially serving as co-creators through their gameplay, with the AI as the mediator. As with film, the idea is not to remove the human element (players play games to be challenged and entertained by crafted experiences, not just random chaos), but to enhance the human experience. More intelligent NPCs, infinite game possibilities, and accelerated creativity make interactive entertainment even more engaging and inclusive in the AI era.

#### 3.2.4.3 AI in Music and Digital Content Creation

The music industry and digital content creation at large are also being transformed by Entertainment 4.0, as AI tools become composers, performers, and curators. Historically, technological shifts in music from phonographs to synthesizers to DAWs have expanded artists' creative palettes. AI is perhaps the most radical tool yet, capable of analyzing and generating musical patterns, mimicking instruments or vocal styles, and even personalizing listening experiences in real time. Meanwhile, in the broader sphere of online content (spanning blogs, videos, social media posts, etc.), AI is enabling new forms of creation and curation that make content more accessible and engaging.

**AI in Music Creation**: One of the most publicized aspects of AI in music is generative composition. AI-driven composition tools can produce melodies, harmonies, and even entire songs with minimal human input. For instance, apps like Amper Music and Artificial Intelligence Virtual Artist (AIVA) allow users (even those with no music-theory knowledge) to generate original music by selecting parameters such as genre, mood, or instruments [8]. These tools work by training on vast datasets of existing music and then generating new pieces that statistically resemble those styles. Composers and producers are using AI as a creative assistant, not to replace their work but to spark ideas. For example, an artist facing "writer's block" in composing a melody can have the AI suggest multiple melodies, then pick one as a starting point or weave elements of it into their piece. "Gone are the days of laboriously crafting melodies entirely by hand," as one music technology publication quipped. AI can now help generate musical ideas swiftly [8]. We already see usage in areas such as stock music (for background music in videos, where AI-generated music can suffice and reduce costs) and in experimentation by mainstream artists. Pop star Taryn Southern released an album in 2018 (I AM AI) composed partly with AI tools. Notably, these tools democratize music production: you no longer need years of training to create a passable soundtrack; AI can do heavy lifting, which could lead to an explosion of content from indie creators, game developers, and small businesses who need music.

Beyond composition, AI aids the production process with tools for automated mixing and mastering. Mixing a song (balancing levels of vocals, drums, guitars, etc.) and mastering (finalizing the track's EQ, compression, loudness for distribution) are skilled tasks traditionally done by experienced engineers. Now, services like LANDR and software like iZotope's Ozone use machine learning to analyze audio and apply signal processing to yield a polished sound [8]. These AI systems have been trained on many professionally mixed tracks to learn what settings produce a "professional" sound. The user can upload a raw track, and the AI will output an enhanced version, often in a matter of seconds. While top-tier artists might still prefer human engineers for finesse, AI mastering is enormously popular among independent musicians due to its low cost and decent quality. It's a great equalizer; anyone can achieve near-studio-quality sound without access to expensive studios. This again reflects democratization, a recurring theme of Entertainment 4.0.

AI is also being creatively used in sound design, creating novel sounds and timbres. For instance, tools like Soundful or Google's NSynth allow the generation of new instrument sounds by blending characteristics of others (want something between a flute and a sitar? AI can create that). Soundry.AI and Orb Composer similarly help producers experiment with unique audio textures [8]. This expands the sonic vocabulary available to artists.

Perhaps the most controversial domain is when AI encroaches on human performance, particularly in AI-generated vocals and styles that mimic real artists. Recent advancements in voice cloning mean AI can learn a singer's voice from samples and produce new vocals that sound uncannily like the singer. In 2023, we saw viral examples on social media of "new songs" seemingly by famous artists that were created with AI, such as a track that mimicked the voices of Drake and The Weeknd so well that many listeners were fooled [9]. This raised huge legal and ethical questions: the artists did not sing those vocals or authorize the song, so who owns it? As one expert put it, "AI can be tailored to generate content in the style of a specific person… You can create new Beatles songs" [9]. This led to debates about whether intellectual property law needed to catch up, as current copyright might not protect an artist's style or "creative essence" from being cloned [9]. Some artists, however, are embracing AI voices as a tool. For example, music producer Holly Herndon allowed fans to use an AI model of her voice under certain conditions, almost like licensing a virtual instrument. And on a positive note, AI "digital replicas" of performers can also generate revenue streams if done ethically. For instance, voice actors might license AI versions of their voices for ads or games, provided they are paid and retain control [10]. The music group ABBA recently launched a concert where "ABBAtars" (digital avatars of their younger selves, created with performance capture and AI enhancements) perform old hits; while not exactly AI-generated music, it uses technology to deliver an old performance in a new form, hinting at how legacy artists might use AI to extend their performing life.

In terms of audience experience in music, AI heavily influences what we listen to through personalized recommendations. Streaming platforms like Spotify and Pandora famously use algorithms to analyze your listening history and compare it with others to curate custom playlists (Discover Weekly, Daily Mixes, etc.) [8]. This has fundamentally changed music discovery; we rely less on radio DJs or MTV and more on what the algorithm thinks we'll like. When done well, it feels like it knows your taste intimately, serving as a musical concierge. Some critics argue that it can create filter bubbles (you only hear content you already like). Still, many listeners enjoy the convenience and serendipity of discovering new favorites through AI suggestions. Interestingly, artists now also pay attention to how to "appease" the algorithms, e.g., releasing more singles or ensuring their songs fit into popular playlists, which is a direct impact of AI on creative decisions in music.

**AI in Live Music and Performance**: Entertainment 4.0 also touches the live side of the music industry. AI can assist performers with tools like Neural Mix Pro (which can isolate parts of a song, such as vocals or drums, in real time for DJs) and generative visuals that react to the music during concerts [8]. There have been instances of "hologram" concerts featuring departed artists (such as the holographic Tupac at Coachella 2012 or hologram Whitney Houston tours). Typically, those use pre-existing recordings, but one can imagine AI generating a new performance or improvisation in real time by a virtual avatar of an artist, a truly virtual concert experience. At the Davos 2025 World Economic Forum, an opening concert involved AI-generated visual art accompanying music to highlight climate change, blending AI visuals with human musical performance [11]. These hybrid experiences may become more common.

Beyond music, AI in digital content creation for platforms like YouTube, TikTok, and blogging is streamlining the creator economy. For example, YouTubers use AI tools to automatically edit videos (picking the best takes, cutting dead air), to caption and translate videos (thus reaching global audiences YouTube's auto-caption and translate are AI features. Social media influencers use AI filters for creative effects (enhancing images or transforming video styles). There are AI-powered writing assistants for script writing or creating engaging captions for posts. Even for average users, AI enables "shortcuts" to creativity, e.g., someone can generate a piece of artwork using tools like DALL-E or Midjourney and post it as part of their content without traditional art skills. This raises its own set of debates (e.g., the ethics of AI art trained on human artists' works), but it undeniably lowers barriers to producing polished content.

Another domain is content moderation and management, quietly handled by AI. Platforms deploy AI to identify and filter out hate speech, copyrighted material, etc. While not "entertainment" for the user, it shapes what content remains visible and monetizable. Also, deepfake technology, while often in the news for misinformation, can have entertainment uses, for example, parody content or transforming oneself to look like a celebrity in a comedy video. Some TV shows have experimented with AI to dub actors' lips into different languages (as in the earlier film example) to localize content more effectively for global viewers [10].

AI is even enabling real-time collaboration across distances for content creation. Musicians around the world can jam together using latency-minimizing AI or shared AI-driven music platforms [8]. Similarly, tools like Audiomovers allow producers at different studios to work on the same live audio stream, which proved helpful during COVID-19 lockdowns and is likely to continue, making the industry more global and connected [8].

In summary, AI in music and digital content is marked by both augmentation of creativity and automation of production. Artists can venture into new creative realms with AI-generated ideas or sounds, and everyday creators can produce professional-looking/sounding content without professional training. The global perspective is essential here: someone in a small town with a laptop now has access to virtual orchestras, mixing engineers, and marketing analytics all via AI tools, just like a top studio in Los Angeles might. This could lead to an outpouring of diverse content from regions and communities that have been previously underrepresented in global media, as the gatekeeping effect of expensive production is reduced.

However, these developments also force us to ask: What is the role of the human artist in an age where a machine can make a decent song or painting? Many artists argue that it is the human touch, context, and emotional intent that give art its meaning. A catchy melody by AI is lovely, but knowing it came from a human artist's experiences might make it resonate differently with listeners. As one art historian framed it, AI-created music can lack the "tertiary meaning," the intrinsic content or soul behind it leaving some listeners feeling something is missing Nonetheless, others like musician Nile Rodgers have a more optimistic view: "Any tool that allows an artist to create is an amazing thing," he says, seeing AI as just another tool in the long history of music technology [10]. Just as synthesizers or drum machines were initially controversial but have since become standard tools, AI might become part of the modern creator's palette.

#### 3.2.4.4 Personalized and Immersive Audience Experiences

A key promise of Entertainment 4.0 is that audience experiences will become more personalized, interactive, and immersive than ever before. AI enables not only better content but also the ability to tailor experiences to everyone in real time and to create new forms of engagement that blur the line between audience and participant.

On the personalization front, we've already discussed recommendation algorithms in music and video. This trend extends further with AI-driven dynamic content. For example, consider interactive films or TV episodes (such as Netflix's Black Mirror: Bandersnatch, released in 2018, in which viewers made choices to determine the storyline). That was a manually crafted branching narrative; in the future, AI could allow truly dynamic plots. A streaming service might adjust the pacing or even the tone of a show based on a viewer's preferences. For instance, if it detects (from your past behavior) that you love action, an AI-edited version of a show might slightly trim slower dialogue scenes and emphasize action sequences. While this raises questions about artistic integrity (would directors allow that?), it's technically feasible. Perhaps more viable is personalized storytelling in games or interactive fiction, as mentioned earlier, where the narrative evolves differently for each player. We might also see personalized education-entertainment content (edutainment apps that adapt the narrative to keep a child engaged while learning, for example, using AI to detect when a child is bored and then introducing a new character or challenge).

AI is significantly enhancing immersive experiences in tandem with VR/AR. In VR, beyond games, we have virtual experiences such as concerts and theater performances. AI can help here by controlling crowds of virtual audience members or by allowing virtual performers to respond to audience reactions. For instance, a VR concert could have an AI avatar pop up next to you to explain the backstory of a song if it senses you're curious (maybe based on biometric feedback, which is another element of wearables that monitor heart rate, gaze, etc., feeding that to AI to gauge your engagement or emotion). This is speculative but within reach as various technologies converge.

Another development is virtual influencers and digital celebrities, which are computer-generated characters that amass real social media followings. Examples include Lil Miquela (a virtual Instagram influencer) and the Vocaloid pop star Hatsune Miku in Japan, who performs as a 3D hologram for thousands of fans. Initially, these were entirely controlled by humans (animators, music producers), but increasingly, AI is being used to drive their content. In the Entertainment 4.0 era, a virtual persona could have an AI "brain" that interacts with fans in real time. Imagine a virtual pop star holding an AI-powered conversation with fans on Twitter or even appearing as a hologram in your living room via AR, to give you a personalized shout-out. This kind of participatory, customized fan experience could significantly strengthen engagement. In South Korea, there are already K-pop groups composed of virtual members; in China, AI-driven news anchors present TV news. While not "entertainment" per se, the line between news and entertainment is thin in some cases, and the tech is similar.

**Audience empowerment is another angle**: Entertainment 4.0 potentially makes audiences co-creators. In some video games and streaming communities, fans are feeding content into the experience (think of Twitch Plays Pokémon, where viewers collectively controlled a game via chat voting). With AI moderation and management, such mass participation events can be smoother and more complex than before. There's talk of "responsive narratives" in streaming content, say a TV show that films multiple options for a story arc and lets fans vote on which direction to take (some reality shows have done this). AI could help by gauging audience sentiment from social media in real time and altering certain elements accordingly. In live sports broadcasts, AI can overlay personalized stats or visuals (e.g., if you watch a game in AR glasses, you might see an AI overlay highlighting your favorite player more often).

On the practical side, AI enhances the audience's experience by improving accessibility. Automated subtitles and translations (as earlier discussed) allow content to reach broader audiences across languages and for those who are deaf or hard of hearing. AI image recognition can describe scenes for visually impaired audiences (some apps do real-time scene description). These features ensure entertainment in the AI era is more inclusive globally.

Finally, an interesting convergence is that Entertainment 4.0's personalization might lead to entirely personalized content, not just recommending a film that exists, but creating a movie for you. This remains a futuristic idea, but consider how generative AI could produce a short story or short film on demand in response to a user's request (e.g., "Make me a 10-minute mystery movie with a humorous tone and surprise ending"). Already, AI can generate images and animations; combining that with story generation could yield custom cartoons or comics, for example. In interactive chat-based fiction, users already get personalized stories. As generative video matures (still early tech), one can imagine personalized episodes starring virtual actors that cater to an audience of one. While that may not replace big blockbuster productions with high human craftsmanship, it could become a niche form of entertainment, like how fanfiction (often one fan writing for another's specific interest) thrives alongside professionally authored novels.

In essence, Entertainment 4.0 is shifting the paradigm from one-size-fits-all content to experiences that adapt to each user and making the consumption of entertainment a more active, participatory affair. Audiences can shape their own journeys through content, and content, in turn, can respond intelligently to audience input. The world of entertainment becomes a two-way street, a conversation between creators (human or AI) and consumers, rather than a monologue from studios to passive viewers.

Across all these domains, film, gaming, music, and digital content, the infusion of AI under Entertainment 4.0 is raising entertainment to a new level of interactivity, creativity, and personalization. AI is a catalyst for innovation, enabling things that creators of the past could only dream about (or perhaps wrote about in science fiction). As we've seen, these changes bring tremendous opportunities, but they are not without pitfalls. In the next section, we will consider the broader opportunities and transformational potential of Entertainment 4.0 for society and the industry, followed by a discussion of the challenges, risks, and ethical considerations that must be addressed to ensure this new era develops in a positive direction.

## 3.3 OPPORTUNITIES AND TRANSFORMATIONAL POTENTIAL

The advent of Entertainment 4.0 offers a host of exciting opportunities and transformational potentials for the entertainment industry and society at large. By integrating AI into creative processes and distribution, we have the chance to enhance creativity, accessibility, and global cultural exchange fundamentally. Here, we outline some key opportunities presented by Entertainment 4.0, building on examples from the previous section and considering their broader impact.

### 3.3.1 Democratization of Content Creation

Perhaps the most profound change is that the tools of high-quality content production are becoming available to virtually anyone. AI is lowering the barrier to entry in fields like music production, video editing, game development, and animation. A young storyteller with a laptop and an idea can now leverage AI-driven software to create a short film complete with VFX or a full-orchestra music track, without the backing of a major studio or label. This democratization means we can expect a diversification of voices and stories in entertainment. Historically, access to production facilities and funding has acted as a gatekeeper, often limiting media creation to specific geographies or industries; Entertainment 4.0 breaks down these barriers. For example, we saw how an Indian startup is making an AI-generated feature film that would've been prohibitively expensive otherwise [6]. Similarly, independent game developers are using AI assets to build rich games without big studios. As AI product experts note, "there are a lot of gatekeepers right now to getting anything greenlit or distributed. The more AI is high-quality and responsible, the more great stories get told" [2]. These hint at a future where latent creative talent around the world can find expression, leading to a richer global tapestry of entertainment content.

### 3.3.2 Augmented Human Creativity and New Art Forms

AI is not just automating tasks; it's expanding the very canvas of what creators can do. We can consider AI a "creative collaborator" that can inspire humans to take new directions. Artists have described using generative AI as akin to jamming with a partner or having an infinite idea generator at hand. This can lead to new art forms that blend human and AI creativity. For instance, visual artists are creating works in which an AI generates part of the art and a human refines it, raising questions like "who is the artist?" while also yielding fascinating hybrids. In music, some composers now produce "AI-human duets." These new forms challenge our definition of authorship and creativity, potentially leading to what some have called "co-creativity" as an art form, in which the process and interplay between human and machine become the art. Moreover, AI can venture into creative domains humans might not otherwise explore exploring extremely complex mathematical musical patterns or visualizing higher-dimensional concepts in film, essentially extending the frontier of art. As media artist Refik Anadol, known for his AI-driven installations, said: "AI became a collaborator that enhances imagination at a whole new level," treating algorithms as new colors on his palette [10].

### 3.3.3 Personalized and Enhanced Audience Engagement

From the audience perspective, Entertainment 4.0 means content that can be more closely tailored to individual preferences, leading to greater satisfaction and engagement. Personalization can help consumers navigate the sea of content and find what truly resonates with them, increasing the overall value and enjoyment they derive from media. For example, Netflix's and Spotify's recommendation engines have been credited with keeping users engaged and introducing them to content they love but might never have found on their own [8]. In an Entertainment 4.0 scenario, personalization could go even further, perhaps dynamically adjusting a story's perspective to focus on a character the viewer likes more, or a streaming news app that learns one's interests and curates a "news show" with more of the sports or entertainment segments you enjoy and fewer of the topics you skip. Such experiences could give audiences an unprecedented sense of agency and connection with the content. Additionally, features such as interactive narratives and VR experiences (enhanced by AI, as described) can turn passive viewers into active participants, deepening immersion. A fan in Brazil can not only watch a K-pop band's virtual concert but also have a brief AI-mediated "meet-and-greet" conversation with their avatar after the show. These kinds of engagements build loyalty and emotional investment.

### 3.3.4 Globalization and Cross-Cultural Creativity

Entertainment 4.0 could accelerate cross-cultural exchange and understanding. AI translation and localization tools are eliminating language as a barrier. We can now consume content from anywhere in the world with automatic, high-quality subtitles or even AI dubbing [3]. This creates an opportunity for truly global entertainment phenomena that aren't bottlenecked by the need to produce multiple language versions manually. A hit TV series in Turkey, for example, could quickly gain international fans if AI dubbing makes it available in 20 languages shortly after release. Furthermore, AI can act as a cultural bridge in creative production: for instance, an AI trained on both Bollywood musical styles and Western pop might help a composer fuse element into a new cross-genre hit appealing to both audiences. Collaborative projects might emerge where artists from different countries let an AI system merge their styles, a new form of cultural collaboration facilitated by technology. Additionally, with AI's help in scaling production, small regional content producers can start catering to diaspora communities worldwide (e.g., a Nigerian filmmaker can, with minimal cost, subtitle and promote her film to Nollywood fans in the UK and the US via streaming platforms).

### 3.3.5 Efficiency and New Business Models

The integration of AI can bring high cost and time efficiencies to entertainment production. This has an economic impact: lower production costs can mean viable projects that previously wouldn't recoup investment. It also might shift budgets to other areas, for example, if 50 percent of VFX work becomes automated with AI, yielding "80–90% efficiency gains in VFX and 3D asset creation" [2], studios can redirect resources to develop more content or improve quality elsewhere (or simply reduce overhead). The adage has been that savings often go back into making the movie/series better (e.g., savings in VFX might be used to film on location for authenticity), ultimately benefiting audiences. On the business side, we see new models emerging, such as virtual idols (who can "work" 24/7 and never age, offering consistent ROI over decades) or dynamic advertising in content (AI can insert region-specific product placements or ads into streamed content in real time). Additionally, AI opens the possibility of mass-customization of content as a product: for example, a company could sell personalized children's storybooks with the kid's name and likeness. AI can generate an infinite number of story variations that include the child, effectively turning what was a one-to-many product into a one-to-one product. This might become a revenue stream for content owners (imagine Disney offering an AI service where your child can star alongside Mickey Mouse in a custom animated short, a premium personalization product).

### 3.3.6 Augmented Reality and Smart Environments

As AI and IoT integrate, physical environments like theme parks, cinemas, or museums can become more innovative and more responsive entertainment spaces. Theme parks are experimenting with AI to personalize rides or character interactions (e.g., an animatronic character recognizing a returning guest and greeting them by name through facial recognition and a cloud database, magical for the guest, though it raises privacy considerations). Interactive cinema has been proposed, in which an audience could collectively influence a movie's direction via an app, with an AI moderator subtly adjusting the film. Even at home, AI and smart devices can create ambient entertainment, such as interactive stories delivered through smart speakers that adapt to household mood lighting and sounds.

### 3.3.7 Education and Skill Development

Though not a direct benefit of the entertainment industry, there is an overlap with education (edutainment). Entertainment 4.0 tools can be harnessed to make learning more engaging, for example, AI-driven historical simulations for history class, or interactive science fiction narratives that teach coding or problem-solving. This cross-pollination can create more opportunities for content creators as well (educational content as an entertainment-adjacent market).

### 3.3.8 Enhanced Creativity in Marketing and Audience Insights

Entertainment 4.0 also transforms how content is marketed and how creators understand their audience. AI analytics can process massive amounts of social media and viewing data to glean insights about what audiences want. This might lead to audience-driven development: for example, platforms have begun using AI to analyze viewer reactions (pauses, replays, drop-off points) to identify which scenes or characters are most beloved or where a story might be losing interest. Creators can use this feedback (carefully, so as not to pander solely) to refine ongoing series or plan sequels that resonate better. Moreover, AI can enable more creative marketing strategies, such as personalized trailers (cut differently depending on whether the viewer is, say, an action fan or a romance fan), or even AI-driven AR promotional games (like a movie scavenger hunt in AR leading up to a film release). These kinds of innovative engagements can deepen the connection between content and audience.

All these opportunities paint a picture of a vibrantly transforming entertainment ecosystem. If harnessed well, Entertainment 4.0 could usher in a kind of creative renaissance: more content, more diverse voices, new genres, and more engaging experiences that bring people joy, insight, and connection.

As SAG-AFTRA's leader noted, AI technologies are opening new revenue streams and creative possibilities within the entertainment sector, showing that innovation and ethical collaboration can go hand in hand [5]. He gave examples of new platforms that allow performers to use their digital replicas in controlled, consensual ways to reach audiences online, providing new revenue streams while keeping performers in control [11]. This illustrates an ideal scenario where everyone, creators, talent, audiences, and tech companies can benefit from the transformation.

However, realizing these opportunities requires careful navigation of the challenges that come with them. For Entertainment 4.0 to truly deliver on its promise, issues of rights, ethics, and equity must be addressed. The following section will confront those challenges, examining the risks of AI in entertainment and the responsibility we have to manage them. The tremendous potential of this era can only be fulfilled if we actively mitigate its downsides.

## 3.4 CHALLENGES, RISKS, AND ETHICAL CONSIDERATIONS

While Entertainment 4.0 offers many exciting possibilities, it also brings forth a range of challenges, risks, and ethical dilemmas that cannot be ignored as AI becomes intertwined with creative processes and distribution. Issues surrounding content authenticity, intellectual property, job displacement, bias, and the overall societal impact of algorithm-driven entertainment demand careful consideration. In this section, we discuss the significant concerns and pitfalls associated with AI in the entertainment industry and explore how to address them.

### 3.4.1 Intellectual Property and Ownership

One of the thorniest issues is determining who owns content created with AI. Traditional copyright law was not designed for scenarios in which an algorithm with minimal human input can generate a song or artwork. If an AI writes a piece of music, does the person who prompted the AI own it, or the company that made the AI, or is it not copyrightable at all? Likewise, if an AI is trained on existing artworks or songs (often without explicit permission from the original creators), using those pieces as "learning" materials, are the outputs a form of derivative work or something new? These questions are the subject of active legal and academic debate. The World Economic Forum has warned that copyright laws need to evolve to handle AI's capabilities [9]. For example, when AI-generated "deepfake" songs imitating Drake and The Weeknd went viral, Universal Music Group intervened to have them taken down, citing copyright and trademark violations. Still, the legal basis was shaky because the song wasn't a copy of an existing recording (so not traditional piracy) [9]. Lawmakers are considering whether new rights should be created, such as a "right to one's creative likeness or style," which might protect artists from unauthorized AI mimicry [9]. Until legal frameworks catch up, there's a risk of frequent disputes and uncertainty. Artists fear their styles can be "cloned" and monetized by others (e.g., an AI "in the style of Taylor Swift" writing songs). If not addressed, this risk could discourage creatives from sharing their work, resulting in a cultural loss. On the flip side, overly restrictive laws might hamper legitimate creative uses of AI. Striking a balance is crucial. As Professor Arun Sundararajan pointed out, Once an AI can replicate someone's creative process, it challenges the notion that you own your style that you spent years honing, suggesting we may need to extend IP protection to the very process or style, not just finished works [9]. Policymakers will need to clarify how copyright applies to AI-generated works and, if necessary, create new categories of rights.

### 3.4.2 Consent and Digital Likeness Rights

Related to IP is the ethical use of digital replicas of real individuals, whether actors, musicians, or others. Deepfakes (AI-generated synthetic media where someone's likeness is convincingly swapped or generated) present a dual-use dilemma. On the one hand, they enable cool effects in movies (e.g., bringing back Paul Walker in Fast & Furious after his death to finish scenes, with family consent). On the other hand, they can be used maliciously, for example, by creating non-consensual explicit content featuring an actor's face, or by making it appear that a celebrity said or did something they never did (which can constitute defamation or fraud). The entertainment industry must grapple with when and how it's acceptable to use someone's likeness. Actors' unions have made this a key issue, as we saw with SAG-AFTRA's insistence on requiring "explicit, informed consent and fair compensation" for any use of a performer's digital replica [5]. No performer should wake up to find their exact voice or face being used in a new film or advertisement without their permission, effectively competing against their own digital double [5]. Laws are starting to emerge: California and New York passed laws requiring consent for certain digital recreations of performers (including after death, protecting one's image posthumously) [5]. At the federal level, the proposed NO FAKES Act seeks to outlaw creating or distributing an AI-generated replica of someone without consent for commercial use or to cause harm [5]. These are positive steps to ensure exploitation is prevented. However, enforcement may be difficult, especially online, where creators can be anonymous or based overseas. Another nuance is artistic tribute versus exploitation: is it okay to make an AI Elvis Presley sing a new song as a homage? Probably yes if rights holders approve, but an AI Elvis releasing "new" albums for profit could undercut the legacy and revenue of the estate. We'll likely see a robust market for licensed digital likeness usage (as in, estates or individuals renting out their digital self for projects under strict terms). Ethically, transparency is key; viewers should be informed when something is AI-generated or when someone on screen is not honest. Already, after a recent Star Wars series used an AI voice for Darth Vader (mimicking James Earl Jones, who is aging), it raised eyebrows, despite being licensed, because audiences realized that even iconic performances might be simulated going forward. In sum, for trust to remain, creators must handle digital replicas with respect and clarity, always centering the consent of those being replicated.

### 3.4.3 Job Displacement and Labor Impact

Automation by AI inevitably raises concerns about the displacement of human workers. In entertainment, this can affect a wide range of roles: background artists, VFX artists, video editors, sound engineers, and even writers and actors to some degree. If a studio can use AI to generate a convincing crowd of extras, they might hire fewer background actors. If AI can cut a trailer, that's work a trailer editor might lose. The Writers' Guild of America (WGA) went on strike in 2023, partly over concerns that studios might use AI to draft scripts and then have a human do minor rewrites, reducing opportunities and pay for writers. Likewise, voice actors worry that producers might use an AI-generated voice for minor characters or for dubbing rather than hiring extra talent. The key fear is replacement: that AI will do the jobs humans used to do, resulting in fewer creative industry jobs and a race to the bottom in wages for remaining jobs (since the threat of AI replacement could be used as leverage against labor). During negotiations, the creative guilds have demanded assurances that AI "should be used to augment roles, not replace them," echoing the stance that technology must remain a tool under human direction [2]. In final agreements, for example, the WGA got a tentative agreement that AI will not be considered a "writer," and that scripts written by AI will not be used to undermine their pay standards. However, the fast pace of AI development means this will be an ongoing negotiation. It's also worth noting that new jobs will emerge for engineers, AI ethicists on creative teams, etc., but whether they offset losses is uncertain. Reskilling is a vital response: industry professionals may need to learn to work alongside AI (e.g., a concept artist learning to use AI for ideation and to focus on refining AI outputs). Another angle is that, historically, new technology tends to create new jobs rather than destroy existing ones. Still, the transition can be painful for those whose skills become less in demand. The challenge is ensuring humans remain at the heart of creativity, both philosophically and economically. If entertainment companies treat AI as a cost-saving tool only, they might short-term cut jobs but long-term impoverish the creative pool and the uniqueness of content (which could become homogenized). Balancing efficiency with human creativity is not easy; it requires enlightened management and possibly new labor regulations. Some have suggested ideas like a "percentage of AI content revenue goes to a fund for displaced creatives" or mandatory credits for AI usage (so that human contributions are distinguished). The discussion is active and must include the voices of creators, not just tech companies, to find fair outcomes.

### 3.4.4 Bias and Fair Representation

AI systems are trained on existing data, and thus, they can inherit biases present in that data. In entertainment, this could mean several things. Music AI trained predominantly on Western music might underserve non-Western musical traditions in its outputs. A casting AI (imagine a system picking "ideal actors" for a script based on analysis of past movies' success) could inadvertently reinforce racial or gender stereotypes (e.g., it might systematically suggest fewer women or people of color for specific roles if historical data were biased). Recommender algorithms can also create feedback loops that reinforce biases. For example, if a platform's algorithm biases toward pushing specific genres or creators over others due to initial popularity, it may further marginalize niche or minority voices, leading to less diversity in what gets widely seen. There's also a risk of cultural homogenization if AI content generation favors majority preferences gleaned from global data, potentially drowning out local or alternative content. Another example: deepfake and face-swap technologies have struggled with accuracy across ethnicities (some early systems performed poorly on non-white faces due to skewed training data), which could lead to lower-quality or problematic outcomes in entertainment if not carefully addressed. The entertainment industry needs to ensure AI tools are audited for fairness and inclusivity. This means diverse training datasets and involving diverse voices in AI development. It might also mean setting guidelines: e.g., if an AI script generator is used, make sure it's not perpetuating cliches or excluding specific perspectives (maybe by curating the training material or having human-sensitivity readers review AI outputs). Similarly, when algorithms decide what content is recommended or produced, they should be oversighted to maintain a plurality of voices. Ethically, there's an imperative to use AI to reduce bias, not amplify it. Some encouraging signs: companies like Netflix have used algorithms to identify representation gaps in their content (though ironically, that's an AI highlighting human bias, which is good). As AI-generated characters become possible, one potential upside is that they might allow underrepresented groups to create relatable characters without gatekeeper approval (e.g., an independent creator could generate an animated series with a diverse cast using AI voices if they can't afford voice actors in multiple languages). But overall, vigilance is needed to ensure Entertainment 4.0 doesn't inadvertently reinforce stereotypes or the dominance of already dominant cultures.

### 3.4.5 Authenticity and Misinformation

The line between reality and fiction blurs with sophisticated AI mimicry. This has implications not only for news and politics (deepfake concerns) but also for entertainment. Will audiences trust what they see and hear on screen? Some foresee a potential audience distrust if AI gets too seamless. For example, if people know an actor on screen might not actually be physically present or say those lines (it could be AI), does it undermine the emotional connection? There's a psychological aspect: authentic human performance has an ineffable quality, and some fear that AI replicas might fall into the "uncanny valley" or feel emotionally hollow even if they look perfect. A related issue is misinformation in entertainment formats, e.g., deepfake celebrity videos that spread false information (such as a fake video of an actor "confessing" something, which might be tabloid fodder). The industry might have to actively label or watermark AI-generated content to maintain trust. Already, companies and organizations are developing technical solutions for content authentication (such as cryptographic signing of original footage to detect deepfakes). Entertainment companies might adopt such measures to protect their assets from misuse and to assure audiences of what is real. Additionally, there's a cultural question: as AI-generated content floods the internet (memes, videos, etc.), human-made art might need to draw distinctions that some artists now proudly label "hand-drawn" or "human-made" to stand out as a mark of authenticity. The concept of authenticity in art may shift from how it was made to why and by whom it was made. If an AI writes a pop song, can it convey genuine human experience? Many would argue no, it can only imitate patterns. So, we might see a greater premium on authentic storytelling and true human stories as a counter to algorithmic content. The challenge for creators is to use AI to amplify their authenticity, not replace it with soulless replication.

### 3.4.6 Creative and Cultural Loss

Another risk voiced by some creators is potential stagnation or loss of human creative skills. If AI handles many technical tasks (like mastering music or generating concept art), those skills might atrophy among new artists. Also, if producers lean heavily on AI's formula (e.g., if data says "Marvel-style superhero movies get clicks, so make more of those"), we could get even more formulaic content. It's possible that a drive for optimization (via AI predicting what will be popular) could narrow the creative field to what's safest. This is not a new phenomenon; Hollywood has chased trends forever, but AI might supercharge trend-chasing. The risk is a kind of creative homogenization, where everything starts to feel the same because AI (learning from the past hits) regurgitates similar patterns. Humans can break patterns and innovate intentionally; an AI would need humans to push it toward the novel (or be programmed to inject randomness). Ensuring that human vision leads and AI supports rather than AI dictating creative choices will be essential to avoid this. We should encourage the use of AI to explore and expand creativity (as in the opportunities mentioned) rather than as an oracle of "what the audience wants" that execs slavishly follow. Additionally, over-reliance on AI could mean fewer apprenticeships and learning-by-doing opportunities in specific crafts (e.g., junior editors might not get to cut trailers if AI does it). That could create a gap in the talent pipeline over time. The industry might need to consciously preserve pathways for new creators to develop their skills, possibly incorporating AI literacy as a core skill, while ensuring they learn fundamentals without always pressing the "AI button."

### 3.4.7 Data Privacy and Surveillance

Personalization and interactive tech often rely on collecting user data. Suppose Entertainment 4.0 experiences are gathering detailed data (your viewing habits, biometric reactions, etc.), which raises privacy concerns. Netflix's algorithm knows a lot about your preferences; VR systems might know your physical reactions. With AI, there's a temptation to gather as much data as possible to refine algorithms. The industry must respect privacy and secure the data to avoid breaches or misuse. Regulations such as the GDPR in Europe and California's CCPA place constraints on data use that entertainment companies must navigate. They'll need transparent policies and may give users more control (such as opting out of certain personalized features if they don't want that data collected). Also, algorithmic transparency is an emerging ethical principle: companies might be pressured to explain why a particular piece of content was recommended or how an AI decided something, especially if it leads to controversy (e.g., if an AI moderation wrongly censors a piece of art because it misinterpreted it as violating guidelines, creators will want an explanation).

### 3.4.8 Ethical Use of AI and Liability

If an AI system generates something harmful, say, a comedy script that inadvertently includes hate speech, or an image that offends, who is responsible? The human who deployed the AI? The AI developers? Accountability frameworks will be needed. Already, in other fields, we see debates about AI liability (e.g., driverless car accidents). In the media, if an AI plagiarizes (unintentionally recreates an existing work too closely), studios will face lawsuits. So, they must put in place checks (like Google's policies for its AI, which forbid output that is too close to the training data, requiring a certain level of transformation). Also, deepfakes again: if a show used a deepfake and someone was harmed by it (reputationally or otherwise), liability needs to be sorted out. The legal system is behind, so media companies that adopt AI need strong internal ethics review boards and practices to proactively catch these issues.

Considering these challenges, it's clear that while technology evolves rapidly, governance and norms are still catching up. Both industry self-regulation and public policy will play roles. Encouragingly, stakeholders across entertainment are engaging with these issues: creators are unionizing and negotiating guidelines for AI use, large companies are forming ethics panels, and governments are holding hearings on AI (the EU is drafting the AI Act, which might ban specific uses, such as deepfakes, without disclosure).

Duncan Crabtree-Ireland of SAG-AFTRA, in advocating for performer rights, said, We don't want to stop innovation; we want to be part of guiding it, emphasizing collective action to set standards [11]. This captures the ethos needed to include creative professionals in shaping how AI is used, rather than letting tech or business interests unilaterally decide. By standing together, as he put it, industry professionals and audiences can push for 'responsible innovation' that amplifies human artistry rather than undermines it [11].

Ultimately, addressing these challenges is about maintaining the trust and values that underpin entertainment: respect for creators, respect for audiences, and respect for truth and artistic integrity. If we manage that, then Entertainment 4.0 can be a boon. If we fail, we risk devaluing art and damaging the very fabric of creative culture.

In the final section, we will look ahead to the future, envisioning a LIFE 4.0 scenario and considering how these opportunities and challenges might play out in the broader context of human life and society.

## 3.5 THE FUTURE OUTLOOK (LIFE 4.0 SCENARIO)

As we peer into the future, we can envision a LIFE 4.0 scenario. In this world, AI and advanced technologies are deeply integrated into all aspects of daily life, including how we entertain ourselves, learn, work, and socialize. In such a world, Entertainment 4.0 is not a distinct concept but rather an integral part of LIFE 4.0, reflecting the blending of digital and physical realities and the central role of AI in shaping human experiences. What might this future look like, and how will entertainment continue to evolve? Here we sketch an outlook based on current trends, mindful that any prediction is speculative but valuable as a thought exercise to guide present decisions.

In LIFE 4.0, entertainment becomes ubiquitous and hyper-personalized. With the proliferation of IoT and innovative environments, entertainment is no longer confined to screens or venues; it seamlessly blends with our surroundings. Imagine homes equipped with intelligent ambient systems: your living room can transform into a holodeck-like space on demand. If you feel like an adventure, an AI might project a dynamic, interactive 3D scene around you (using holographic displays or AR glasses), turning your space into, say, a medieval castle for a role-playing game you can physically move through. Perhaps your smart speakers and lights coordinate with this experience, providing 360-degree sound and lighting effects. This is an extension of current VR/AR trends, projected into a whole environmental context. Immersive storytelling could reach new heights. You might step into your favorite movie scene and interact with AI-driven characters, effectively becoming a participant in the story. This could be the evolution of theme parks or video games available right at home.

The boundary between creator and audience may blur further. In LIFE 4.0, it's conceivable that everyone becomes a creator to some degree, and AI acts as a universal creative partner. Social media already makes consumers into producers (every TikTok user is, in a sense, an entertainer to their followers). Future platforms might take this to the next level: for example, a streaming service could allow viewers to tweak plot lines or even co-create alternate episodes of a show that others can view. Entertainment might resemble a collaborative game or an open world, where fans and official creators build content together. Audiences could effectively co-create canon, a bit like open-source storytelling. AI moderators ensure narrative consistency and quality, serving almost like dungeon masters in a massive, never-ending interactive story universe.

The concept of virtual companions and characters might become commonplace in LIFE 4.0. Already, we see primitive versions (virtual YouTubers, AI chatbots like Replika that people befriend). In the future, you could have a stable of AI-driven characters, some may be personal only to you (an AI friend who knows your life and preferences, engaging you in conversation or playing games with you), and others might be globally recognized IP. For instance, someone could "license" an AI avatar of Harry Potter to be their interactive tutor in magic for fun, not in official canon, but as a personalized experience. People may develop genuine emotional bonds with these AI personas (raising new psychological and ethical questions, of course). Entertainment here crosses into companionship, a LIFE 4.0 phenomenon where the lines between entertainment, social interaction, and even mental health blur. Is your AI companion a toy, a friend, or something in between? Society will have to adapt to relationships that are part real, part fiction.

One likely development is the convergence of entertainment, education, and work. Gamification and AI could turn traditionally non-entertainment tasks into engaging experiences. A LIFE 4.0 scenario might have us learning new skills through narrative-driven simulations that are as entertaining as blockbuster games. Companies might train employees using VR/AR scenarios that feel like cooperative games. In such a world, the phrase "all the world's a stage" takes on a new meaning; much of what we do could have an entertainment component. This raises a caution that maybe needs balancing (not everything should be a dopamine-inducing game, or we risk fatigue or loss of depth), but judiciously applied, it could improve quality of life and engagement.

From a global perspective, LIFE 4.0 could be quite democratizing for entertainment consumption. Virtually anyone with an internet connection (or whatever network technology is available) can access the richest media experiences, perhaps via cloud rendering and AI that adapts to bandwidth. Language barriers might be eliminated in real time. You could join a virtual concert by a band from another country, and your AR glasses might seamlessly translate the lyrics into subtitles in your language (or even adjust the audio to your language using the singer's AI-cloned voice), matching lip sync, a scenario building on the Flawless AI example [5]. This points to a truly global culture, where entertainment content flows freely and is enjoyed simultaneously by multilingual audiences, perhaps knitting people together. However, cultural identities might need preservation efforts. LIFE 4.0 might see a push in many countries to produce local content with local AI (trained on local culture) to ensure distinct cultural voices remain strong amidst a global flood. Japan's concept of Society 5.0 (human-centered, integrating cyberspace and physical space) envisions technology solving social problems [12]. Entertainment can contribute here by spreading empathy, knowledge, and cross-cultural appreciation through globally shared AI-mediated experiences.

Entertainment 4.0, as part of LIFE 4.0, also plays a role in mental and emotional well-being. Future entertainment might be prescribed as therapy, for example, personalized AR narratives that help reduce anxiety or AI-driven music tailored to your mood in the moment to help with stress (we see early steps in wellness apps). As AI better understands individuals, it could curate not just what we want but perhaps what we need for healthy mental stimulation, maybe encouraging us to try more uplifting content when we're down, or giving us challenging art that broadens our perspective. This is a positive spin, but it does require that AI platforms be designed with ethics in mind, not purely to maximize engagement (which can lead to mindless bingeing or echo chambers). If done thoughtfully, entertainment could become a more constructive force in daily life, enriching us rather than merely distracting.

Another future component is artificial general intelligence (AGI). If AI becomes vastly more capable and autonomous, it could transform entertainment in unpredictable ways. Could an AGI become a master storyteller or director, creating films or games of incredible depth on its own? Possibly, though some argue creativity has an inherently human aspect that AGI might replicate technically but not experientially. Nevertheless, if something like that happens, one can imagine a world where new "blockbusters" or masterpieces come not from human minds but from machine creativity. That raises philosophical questions about art. Is art valuable without human intention behind it? Some might welcome it as just greater content; others might see it as hollow. The value may shift to experiencing and interpreting art rather than lionizing the artist, since the artist could be non-human. Alternatively, AGIs might collaborate with networks of people (like a real-time audience brain-machine interface or something far out like that) to create communal art experiences.

We should consider socio-economic impacts too. If entertainment is highly AI-driven and abundant, how does economics work? Maybe content becomes extremely cheap or subscription-based for unlimited AI content. Human-created live experiences (like a real concert or live theater) might gain even more cachet as "authentic experiences" in a world full of virtual ones. There could be a bifurcation: readily available AI media everywhere, and niche artisanal human experiences that people pay a premium for (analogous to how digital photography made photos ubiquitous, but people still pay photographers or buy film photography for exceptional artistic value). In LIFE 4.0, we might thus see a coexistence of AI mass entertainment and human-crafted boutique entertainment, each valued for different reasons.

Ethically, a LIFE 4.0 future should strive for what some call human-centric AI, aligning with ideas like Society 5.0 in Japan [12]. That means always asking: How does this technology benefit humans? In entertainment, that translates to: Does it inspire us? Connect us? Heal us? Educate us? Give us joy? If those remain the metrics of success (rather than just profit or click-throughs), then technology will be guided in a positive direction. Perhaps in the future, recommendation AIs will optimize for long-term satisfaction or cultural value instead of immediate watch time. Already, there are calls in the tech ethics community for algorithms to consider well-being metrics.

The outlook also must factor in regulatory evolution. By the time LIFE 4.0 is in full swing, governments will likely have instituted laws on deepfakes, AI transparency, and data usage. This might set some boundaries on entertainment uses (for instance, maybe a label requirement: content must disclose if any main elements are AI-generated, like a "nutrition label" for digital content, as one product leader suggested) [2]. Hopefully, regulation finds the right balance to protect rights without stifling innovation.

In a LIFE 4.0 world, collaboration between human creativity and AI will hopefully be normalized and celebrated. The narrative might shift from AI versus Human to AI + Human, a synthesis in which something greater emerges. Jaron Lanier, a tech pioneer, once said that the most beautiful music will be made by humans and machines working together, complementing each other's strengths. We can foresee creative ensembles that include AIs as "band members" or writing partners, acknowledged and embraced. Perhaps future award shows will consist of categories such as "Best Human-AI Collaborative Film." We might laugh now, but it could happen.

Finally, LIFE 4.0 might bring changes in the philosophy of entertainment. With AI blurring the line between creation, some age-old questions resurface in new forms: What is art? What does it mean to be moved by something synthetic? But also new ones: Can an AI have a creative intent, or is that always human? Society's relationship with entertainment could become even more intimate (as entertainment entitles with our lives via AR/VR companions, etc.), which might elevate its importance but also its potential to influence (for good or ill). Perhaps media literacy education will expand to teach people how to critically engage with AI-generated content and how to recognize their own bubble, as part of being a LIFE 4.0 citizen.

In conclusion, the future of Entertainment 4.0 within a LIFE 4.0 scenario holds immense promise: a world where creativity knows no bounds, where storytelling becomes a shared living experience, and where technology genuinely enriches human life. But it also challenges us to uphold human values, creativity, empathy, and fairness in the face of powerful, transformative tools. If we manage to do so, the entertainment of the future could be more inclusive, immersive, and inspiring than ever before, truly earning the moniker of a new age of experience.

## 3.6 CONCLUSION

Entertainment 4.0 represents a pivotal chapter in the ongoing story of human creativity and technological advancement. As we have journeyed from the analog foundations of Entertainment 1.0 through the digital and connected revolutions of 2.0 and 3.0, we find ourselves now at the dawn of an era defined by AI-augmented creativity and experience. The chapter has explored how AI is reshaping the global entertainment landscape in film, gaming, music, digital media, and audience engagement, and what opportunities and challenges arise as a result.

In reviewing the historical progression, we saw that each era of entertainment was built on major technological shifts: the invention of cinema and broadcast (1.0), the adoption of digital media and computing (2.0), and the expansion of the internet and streaming (3.0). Entertainment 4.0 builds upon all these layers, adding to the potent catalyst of AI. AI's contributions span the entire entertainment value chain: it is enhancing content creation (through generative tools for scripts, visuals, music, and more), optimizing distribution (via personalized recommendations and localization), and transforming consumption (with interactive, personalized, and immersive experiences). The result is an entertainment ecosystem that is more prolific and accessible than ever, but also more complex.

From a global perspective, Entertainment 4.0 has a unifying potential. We now have the means to bridge language and cultural gaps in real time, bringing content from any corner of the world to a global audience with AI-assisted translation and customization [5]. AI-driven platforms have already connected creators and fans across continents, creating truly international fandoms and collaborations. This new age can be one of global cultural enrichment, where creativity is shared and celebrated beyond boundaries. Yet it also reminds us that we must use technology to amplify, not erase, the wide variety of human expression.

The opportunities presented by Entertainment 4.0 are indeed transformative. Creativity is being democratized: more people can create high-quality art and entertainment with AI as their ally, leading to a blossoming of content and stories that would otherwise never be told [2]. Audiences stand to benefit from deeper engagement, whether through interactive narratives that put them in the story or personalized content feeds that seem almost telepathic to their tastes [8]. Sectors such as education, tourism, and live events are finding novel ways to blend entertainment and utility via AI, making experiences more informative and enjoyable. Economically, new markets and job roles are emerging around AI content creation, from "virtual idol managers" to "narrative designers" for branching AI-driven stories. As with past technological leaps, Entertainment 4.0 has the potential to spur growth and innovation across creative industries.

At the same time, we have examined the challenges and ethical imperatives that come with this territory. The integration of AI in entertainment surfaces hard questions about authorship, authenticity, and agency. We risk devaluing human artistry if AI is used carelessly to replace rather than enhance creative jobs [2]. Protecting performers' and creators' rights, their voices, likenesses, and unique styles is critical, as underscored by recent industry actions to require consent and compensation for digital replicas [5]. The specter of deepfakes and misinformation looms, demanding vigilance in how we create and consume media in the AI era. Ensuring AI algorithms are fair and inclusive is another non-negotiable task; the stories and recommendations AI provides must not entrench biases but rather help break them, pushing us toward greater understanding and representation. Ultimately, human oversight, transparency, and a strong ethical framework are the safeguards we need as we navigate this new age. The entertainment industry's stakeholders, from tech developers and business leaders to artists and regulators, must continue to engage in dialogue and set standards so that technology serves creative human endings, not the other way around [5].

Looking toward the future, the LIFE 4.0 scenario suggests that entertainment will be even more seamlessly woven into our lives as a constant, adaptive presence that can educate, heal, and connect as much as it amuses. This vision is inspiring: imagine a world where anyone can summon any experience, explore any story, and tap into endless creativity at will. But positively reaching that vision will require that we remember what entertainment is fundamentally about: human experience. Whether an epic film, a catchy song, or a gripping game, entertainment moves us because it speaks to human truths, our emotions, dreams, fears, and aspirations. AI can help illuminate those truths in a new light, but it is our responsibility to infuse them with meaning and values.

In closing, Entertainment 4.0: The New Age of Experience in the AI Era is both a celebration and a call to action. It celebrates the remarkable innovations enabling a new golden age of creativity and experience. It also calls on creators, consumers, and policymakers to actively shape this evolution with wisdom and care. If we succeed, we will usher in an era of entertainment that not only dazzles and delights but also enriches the human spirit, fostering empathy, knowledge, and unity across the globe. The story of Entertainment 4.0 is still being written, and we, collectively, are its authors and audience. Let us write this next chapter so that future generations regard it as a time when technology and artistry came together to elevate life to new heights, truly exemplifying what LIFE 4.0 can be.

## REFERENCES

1. Jatkiewicz, P., An Attempt to Define the Concept of Entertainment 4.0 by Analogy to Other Concepts, e.g., Industry 4.0, Education 4.0, etc. *Economic Affairs (New Delhi)*, 2023. 68: p. 123–134.
2. Brodherson, M., A. Wrubel, and J. Vickers, How AI Could Reinvent Film and TV Production, in *McKinsey & Company*, 2025.
3. Krakowski, O., Three AI Trends Reshaping the Future of Media & Entertainment, in *Panasonic Connect*, 2025.
4. Schwab, K., 1.14 The Fourth Industrial Revolution: What It Means and How to Respond, in *World Economic Forum*, 2016.
5. Crabtree-Ireland, D., Labour Leaders: Artificial Intelligence Must Serve Human Creativity, Not Replace It, in *World Economic Forum*, 2025.
6. Gent, E., Coming to Theaters: An AI-Generated Bollywood Movie, *IEEE Spectrum*, 2024.
7. University, C.T., AI in Video Game Development: From Smarter NPCs to Procedural Worlds, in *Capitology Blog*, 2025.
8. Dahl, J., 7 Ways AI Is Revolutionizing the Music Industry, in *Point Blank Music School – Blog*, 2024.

9. Broom, D., Who Owns the Song You Wrote with AI? An Expert Explains, in *World Economic Forum*, 2023.
10. Pomeroy, R.H., AI vs Art: Will AI Rip the Soul Out of Music, Movies, and Art, or Help Express Our Humanity? in *World Economic Forum – Radio Davos*, 2024.
11. SAG-AFTRA, *TV/Theatrical Contract Highlights*, 2023.
12. Nakanishi, H., Modern Society Has Reached Its Limits. Society 5.0 Will Liberate Us, in *World Economic Forum*, 2019.

# 4 Education 4.0

**Abstract**

Education is undergoing a profound transformation in the wake of rapid technological advances. This chapter examines Education 4.0, an emerging paradigm of learning aligned with the Fourth Industrial Revolution (4IR) and characterized by artificial intelligence (AI)-augmented and personalized education. We trace the evolution from traditional Education 1.0 (teacher-centric, industrial-era schooling) through Education 2.0 (early digitization) and Education 3.0 (online and connected learning) toward the current Education 4.0 landscape. In doing so, we highlight how AI and related innovative technologies are reshaping classrooms, pedagogy, assessment methods, and lifelong learning on a global scale. Key opportunities such as personalized learning, intelligent tutoring, and data-driven educational decision-making are discussed alongside the challenges and ethical considerations, including equity gaps, data privacy, and the changing role of teachers. An outlook (termed LIFE 4.0) envisions an education ecosystem fully integrated with AI, fostering continuous learning and human-AI collaboration. By exploring these themes with a formal yet accessible approach, the chapter aims to educate and inspire readers about the transformative potential of AI in education, while grounding the discussion in global perspectives and current research. Ultimately, Education 4.0 is presented as a catalyst for more inclusive, relevant, and lifelong education if stakeholders navigate its opportunities and risks responsibly.

## 4.1 INTRODUCTION

The convergence of rapid technological innovation and evolving societal needs is driving a transformation in education systems worldwide. The term Education 4.0 has emerged to describe a vision for the future of education that aligns with the Fourth Industrial Revolution (4IR), leveraging innovative technology, artificial intelligence (AI), and robotics to prepare learners for a world in which cyber-physical systems permeate all industries [1]. Just as Industry 4.0 denotes a new era of intelligent automation and data exchange in manufacturing, Education 4.0 signifies a shift toward intelligent, personalized, and technology-enhanced learning experiences. This paradigm envisions classrooms where AI-powered tutoring systems, data analytics, and immersive digital tools enhance human teaching, making learning more adaptive to individual needs and more responsive to the demands of the future workforce [1, 2]. Crucially, Education 4.0 is not about replacing teachers with machines, but about empowering educators and students with advanced tools, fostering a more engaging, customized, and lifelong learning process [2, 3].

To contextualize Education 4.0, it is helpful to understand the historical evolution of modern education. Educational practices have evolved in tandem with technological and social changes. In broad terms, we can distinguish four stages:

 DOI: 10.1201/9781003455219-4

- **Education 1.0** – a foundational, pre-digital model grounded in teacher-centric delivery and rote learning.
- **Education 2.0** – the early digitization era, introducing basic technology into classrooms (e.g., television, radio, and computers) but largely retaining traditional pedagogies.
- **Education 3.0** – an era of connectivity and online learning, marked by the rise of the internet, multimedia resources, and more learner-centered approaches.
- **Education 4.0** – the current emerging phase of AI-augmented learning and personalized education, which builds upon global connectivity and data-driven innovations to profoundly change the learning experience.

It must be noted that these stages often overlap, and progress has been uneven globally. Many countries' educational systems today still primarily reflect characteristics of Education 2.0, with only incremental adoption of interactive or personalized learning, while relatively few countries are pushing forward with proper Education 3.0 or 4.0 reforms [2]. This global disparity, sometimes termed a "digital divide" in education, means that discussions of Education 4.0 inevitably involve questions of equity and access, which we will address in later sections. Nonetheless, the trend is clear: around the world, educators, policymakers, and institutions are increasingly looking to harness advanced technology (especially AI) to improve educational outcomes and relevance [2, 4].

In the sections that follow, we explore each stage of this evolution in detail. We begin by describing Education 1.0, the traditional model that underpinned education for much of the 19th and 20th centuries. We then examine subsequent phases, Education 2.0 and Education 3.0, to understand how digitization and the internet catalyzed changes in teaching and learning. The core of the chapter focuses on Education 4.0, unpacking how AI and related technologies enable new forms of knowledge (e.g., intelligent tutoring, predictive analytics, and virtual reality (VR) classrooms) and empower a shift toward personalization and skills-centric education [1, 2]. We then discuss the manifold opportunities presented by Education 4.0, its transformational potential to improve efficiency, equity, and student engagement, as well as the challenges, risks, and ethical considerations that accompany AI-driven education. A forward-looking section envisions a LIFE 4.0 scenario in which lifelong learning and human-AI symbiosis in education become the norm. Finally, we conclude with reflections on how stakeholders can navigate this transition to ensure that the promise of Education 4.0 is realized globally and ethically.

By combining academic analysis with practical examples and global perspectives, this chapter provides a comprehensive overview of Classroom Transformation in the Age of AI. We draw on international frameworks and research from UNESCO guidelines on AI in education to World Economic Forum (WEF) initiatives to ground the discussion in current thinking. Through this exploration, readers will gain insight not only into what changes are happening in classrooms due to AI but also into why these changes matter for the future of learning and society.

## 4.2 EDUCATION 1.0–4.0

### 4.2.1 Education 1.0: Foundations

Education 1.0 refers to the traditional model of education that dominated formal schooling from its early mass implementation up through the pre-digital era. In an Education 1.0 environment, learning is typically teacher-centered and authoritarian, meaning the teacher is the primary (often sole) source of knowledge and authority in the classroom [2]. The role of the student is passive mainly: students are expected to sit, listen, and absorb information delivered by the teacher, who lectures or dictates as the unquestioned expert. This model is often likened to a one-way transmission of knowledge, much as Web 1.0 was a one-way broadcast of information on the early internet, Education 1.0 is a one-way imparting of knowledge from teacher to student [2, 5].

Classroom dynamics in Education 1.0 are rooted in rote learning and memorization. The curriculum is usually standardized, with a focus on core literacies and factual knowledge. Students are evaluated primarily through examinations and written tests that emphasize recall of information and mastery of well-defined procedures. Critical thinking, creativity, or individual learning preferences receive little emphasis in this paradigm. The implicit goal of Education 1.0 was to efficiently impart a set body of knowledge to many students, a model that fit the industrial-age need for basic, educated workers and citizens with homogeneous skills.

A defining feature of Education 1.0 is the absence of technology in the learning process (aside from fundamental tools like chalkboards and textbooks). In fact, technology is often forbidden or discouraged in traditional classrooms of this type [5]. The emphasis is on face-to-face instruction and printed materials, reflecting the era before digital devices were available. Even overhead projectors or educational television, which became common later, were not present in the earliest form of this model. The learning experience was confined to the physical classroom and library, with minimal multimedia or interactive content.

From a global perspective, Education 1.0 has been the foundation of schooling in virtually every country at some point. Its principles trace back to ancient tutorial models and were formalized in the systems of mass education that spread worldwide in the 19th and 20th centuries. The strengths of this model include its clarity of structure and relative simplicity of implementation. One teacher can handle many students in a controlled setting, but its limitations have become more evident over time. By the late 20th century, educators recognized that a purely teacher-driven, exam-focused approach often failed to develop higher-order skills and could disengage students. This recognition, combined with societal and technological shifts, paved the way for the next phase of evolution, Education 2.0.

Before moving to Education 2.0, it's worth noting that elements of Education 1.0 persist in many places. In some regions or schools around the world, resource constraints and cultural expectations mean that teacher-centric methods and rote learning remain common. However, the momentum of change has firmly set in. The following sections illustrate how educational practice began to adapt slowly at first to new technologies and pedagogical ideas, marking the transition from the foundational Education 1.0 model toward more modern paradigms.

### 4.2.2 Education 2.0: Digitization Era

Education 2.0 represents the initial integration of digital technology and multimedia into education, roughly corresponding to the late 20th century, as schools began to incorporate electronic resources. In this stage, traditional teaching methods are augmented but not fundamentally transformed by technology. Classrooms might have had televisions and VCRs for educational broadcasts, language labs with audio tapes, or early computers used for specific subjects. The use of such tools enabled a more engaging and interactive learning experience than in Education 1.0, as visual and audio resources complemented lectures [2]. For example, a history lesson could be supplemented with documentary footage, or a science class with an educational video, providing students with a richer context than textbooks alone.

Despite these additions, Education 2.0 still maintained much of the traditional structure. The teacher maintained control over the flow of information and classroom activities; technology was typically used to support the teacher's lesson plan rather than to empower student-driven learning. In many cases, schools in the Education 2.0 era continued to emphasize lectures and examinations, with memorization of content being the main path to success [5]. The idea of "student-centered learning" might have been introduced in educational discourse, but in practice, it was often "more of a concept on paper than a reality" in this phase [2]. In other words, while educators talked about focusing on learning outcomes and active learning, the implementation lagged behind the rhetoric. Class schedules, curricula, and assessment methods remained essentially unchanged. Students still followed a fixed syllabus and proved their learning through standardized tests and grades.

A hallmark of Education 2.0 was the emergence of collaboration and communication tools, albeit in nascent forms [5]. In some progressive schools, group projects and classroom discussions became more common, acknowledging that learning can be a social, interactive process. However, these efforts were still bound by the teacher's oversight and the physical constraints of the classroom. In the late 1990s, email or basic online forums might have been used to facilitate communication, but widespread interactive online learning was not yet the norm.

Globally, the Digitization Era of Education 2.0 unfolded unevenly. Wealthier schools and countries in the 1980s and 1990s began investing in computer labs, educational TV programming, and, eventually, internet connectivity. At the same time, under-resourced regions often continue to rely on chalkboards and print materials. Nonetheless, the paradigm of incorporating technology in some form spread worldwide, influenced by international programs and the falling cost of digital devices. Education 2.0 can thus be seen as a transitional phase; it preserved the familiar comfort of the teacher-led classroom but started to infuse it with the possibilities offered by new technology.

In summary, Education 2.0 expanded the toolkit available to educators. It allowed students to experience learning through multiple media (visual, auditory, interactive software), which could improve engagement. However, it did not radically change the roles of teachers and students. Teachers still largely "delivered" education, and students received it, albeit with flashier presentations or occasional computer-based tasks. The limitations of this approach became more apparent as the internet and personal computing grew. By the turn of the millennium, the stage was set for a more significant change in the education model, one that would more fundamentally alter how, where, and whom students learn, leading to what we call Education 3.0.

### 4.2.3 Education 3.0: Connectivity and Online Learning

Education 3.0 emerged in the early 21st century alongside the explosion of the internet, social media, and mobile computing, bringing about an era of unprecedented connectivity. In contrast to the modest tech enhancements of Education 2.0, this phase fundamentally reshaped access to information and the learning environment. Classrooms were no longer constrained to four walls or limited to resources provided by a single teacher or textbook. With the advent of widespread internet access, students could tap into a vast network of knowledge: online libraries, educational websites, forums, and later massive open online courses (MOOCs) and video tutorials. This shift paralleled the Web 2.0 revolution in society at large, in which users became content creators and collaborators online. Similarly, Education 3.0 is often described as a stage where learning has become more participatory, learner-driven, and networked [1].

A key feature of Education 3.0 is the student-centered approach. In this paradigm, the role of the teacher shifts from a lecturer to a facilitator or mentor [2, 5]. Teachers guide students in how to learn and how to evaluate information critically, rather than simply delivering facts. Classrooms adopting Education 3.0 methods often employ blended learning models that combine face-to-face instruction with online components and embrace strategies such as the flipped classroom (where students watch lectures or review content at home, and use class time for discussion, projects, or problem-solving) [2, 5]. The use of collaborative projects increased, leveraging online tools to enable students to work together, sometimes even across different locations. In this stage, it became increasingly common for lesson plans to be called "Learning Plans" rather than lecture notes, reflecting a focus on the learning activities of students rather than the teaching script of instructors [5].

Technology during Education 3.0 is ubiquitous; "technology is everywhere" in the learning process [5]. Students might use laptops or tablets in class, access cloud-based assignments from home, and interact via educational platforms (like learning management systems or social learning networks). Information Technology (IT) is widely used, enabling not just access to content but also new modes of learning such as interactive simulations, educational games, and virtual labs [2]. The relationship between students and information fundamentally changed: students in Education 3.0 often know how to find information quickly (e.g., via Google) and are not entirely dependent on teachers for answers [2]. This democratization of information sources means educators must adapt, acknowledging that part of their role is now to help students develop skills to navigate and critically assess abundant information, rather than being the single source of truth.

Assessment in Education 3.0 also began to diversify. While traditional exams remained in use (especially for high-stakes testing), there was a growing emphasis on continuous assessment, project outcomes, and portfolio work. Feedback loops could be shortened: for instance, online quizzes could provide instant feedback, and teachers could use data from e-learning platforms to see where students struggle and adjust their teaching in near-real time. This era also saw the beginning of learning analytics: the fundamental data analysis of student performance to inform instruction.

From a global standpoint, Education 3.0 was accelerated by international movements to increase internet connectivity in schools and the creation of open educational resources (OER). Some countries led the way by implementing one-to-one device programs (providing each student with a laptop or tablet) and updating curricula to include digital literacy and collaborative skills. Notably, Education 3.0 expanded educational access, for example, through online courses; a student in a remote area could, in principle, learn from top instructors around the world. This set the stage for greater educational equality, though it also depended on addressing infrastructure gaps.

In summary, Education 3.0 transformed the educational landscape by making learning more interactive, decentralized, and student-empowered. The classroom began to open, knowledge could flow from student to student, and even from student to teacher, as much as from teacher to student. The skills emphasized (sometimes called 21st-century skills) included critical thinking, communication, collaboration, and creativity, attributes necessary for a connected world. However, Education 3.0 is not the endpoint. With the rise of AI, big data, and advanced automation in the 2010s and 2020s, education is moving into an even more dynamic phase. The following section delves into Education 4.0, where AI and personalization become central, promising to revolutionize further how learning occurs.

### 4.2.4 Education 4.0: AI-Augmented Learning and Personalized Education

Education 4.0 represents the frontier of educational innovation, characterized by the deep integration of AI and other advanced technologies into learning environments. In this paradigm, the focus is on AI-augmented learning where AI tools work in tandem with educators to deliver personalized, adaptive, and efficient education, and on cultivating the skills and mindsets needed for life and work in the 4IR era. Education 4.0 builds upon the connected infrastructure of Education 3.0 but adds a layer of intelligence and personalization that was previously not possible at scale.

A defining feature of Education 4.0 is personalized learning. AI enables the creation of adaptive learning systems that can tailor educational content and pacing to individual students' needs and abilities [3]. For instance, AI-driven tutoring platforms can analyze a student's performance in real time and adjust the difficulty or style of questions accordingly, much like a personal tutor would. Research has long shown that one-on-one tutoring is highly effective (often cited as a two-standard-deviation improvement in learning outcomes, known as Bloom's 2 Sigma problem). Still, it's economically impractical to provide every student with a personal human tutor. AI offers a potential solution: intelligent tutoring systems and educational software can provide customized practice and feedback to each learner, essentially scaling the benefits of one-on-one instruction to entire classrooms or online cohorts [3]. This AI-augmented personalization means that students who grasp material quickly can move ahead to more challenging topics. At the same time, those who struggle receive additional support and alternative explanations, all within the same class. Such systems not only improve academic performance but also cater to diverse learning styles and needs, for example, by offering multi-sensory content for different preferences or by being responsive to neurodiverse learners who may need a unique approach [3].

In Education 4.0, the role of the teacher evolves further. AI does not replace teachers; instead, it augments and supports them. Routine tasks like grading objective assignments, taking attendance, or scheduling can be largely automated, freeing teachers' time for more meaningful interactions with students [3]. AI-driven analytics dashboards might highlight which students need extra help on which topics, enabling a teacher to focus remedial efforts more efficiently. In this sense, the teacher in Education 4.0 becomes more of a designer of learning experiences and a coach, orchestrating a rich environment in which human and AI resources are combined to maximize learning. The teacher can concentrate on the human elements of education mentorship, motivation, and socio-emotional support, which machines cannot provide, while trusting AI to handle specific analytical or repetitive tasks [3]. This partnership aims to enhance human-led pedagogy, not diminish it, a point emphasized by proponents who insist AI should "augment, not replace" the teacher's role [3].

Curriculum and content in Education 4.0 also transform. Given that information is now easily accessible (and even trivially retrievable via AI assistants), there is a greater emphasis on teaching students how to think, learn, and adapt. The WEF's Education 4.0 framework underscores skills like problem-solving, collaboration, adaptability, and digital literacy as critical outcomes of modern education [2]. These are often termed 21st-century skills, and they align with what employers and communities will need in an AI-rich world. For example, problem-solving in Education 4.0 might involve students working on interdisciplinary projects, using AI tools as is necessary to analyze data or simulate scenarios, thereby also learning about the AI itself. Collaboration could extend beyond one classroom; students might collaborate with peers in other countries via telepresence or joint online projects, reflecting a globalized context. Adaptability is both a skill to be learned and a necessity in this fast-changing environment; Education 4.0 encourages continuous learning and comfort with change, preparing students to update their skills throughout life as technology evolves [2].

The learning environment in Education 4.0 is often described by the mantra "learning anytime, anywhere." This points to the breakdown of traditional time and space constraints: with cloud platforms, mobile devices, and AI assistants, learning can extend beyond school hours and physical classrooms. A student might learn coding after school with an AI tutor, or a professional might take an online course with AI-driven feedback on weekends. Lifelong learning becomes a practical reality, supported by technology. Formal education providers (schools, universities) are thus increasingly seen as part of a broader ecosystem of learning that also includes online providers, employer training programs, and informal learning via digital media.

Another aspect of Education 4.0 is the use of immersive and interactive technologies such as VR, augmented reality (AR), simulations, and educational games to enhance experiential learning. AI plays a role here by either powering these experiences (for instance, AI characters in simulations) or by personalizing them to the user. These tools enable students to, say, perform virtual science experiments, visit historical sites via VR, or practice skills in a risk-free simulated environment. The increased use of VR and simulations is noted as a characteristic of Education 4.0 environments, where technology is "free and/or easily accessible" to learners [5].

Crucially, Education 4.0 also encompasses teaching about AI and technology, not just with it. Recognizing that AI will be a pervasive force in society, educators and policymakers argue that students should graduate with a basic understanding of AI and data science, as well as the ethical implications surrounding these tools [3]. This means including topics like algorithmic thinking, how machine learning works at a conceptual level, and issues of bias and privacy in algorithms as part of the curriculum. Some countries have already begun introducing AI literacy programs in K-12 education, and UNESCO has developed AI competency frameworks for students and teachers to ensure the workforce of tomorrow is AI-savvy [4]. In Education 4.0, teaching about AI is considered equally crucial to teaching with AI [3].

From a global perspective, Education 4.0 is gaining traction through various international initiatives. The WEF's Education 4.0 Alliance and reports (such as Shaping the Future of Learning: The Role of AI in Education 4.0) are documenting case studies and guiding principles for AI in education [3]. UNESCO has called for a human-centered approach to AI in education, prioritizing equity and inclusion and empowering teachers rather than marginalizing them [3]. Around the world, there are examples of Education 4.0 in action:

- In Asia, countries like Singapore and China are heavily investing in smart classrooms, AI-driven learning apps, and nationwide AI curricula for students. China, for instance, has experimented with AI tutors for math and language learning at large scales, while Singapore's Smart Nation initiative includes personalized learning systems in schools.
- In Europe, nations such as Estonia and Finland have introduced coding and AI basics in early education, and pilot programs use AI to support teachers in grading and in identifying students' learning gaps.
- In parts of the developing world, Education 4.0 offers a chance to "leapfrog" traditional models, for example, in Africa, some remote schools use solar-powered tablets with adaptive learning software to reach learners with no access to quality textbooks or teachers. However, these efforts also face hurdles of infrastructure and cost.

To summarize, Education 4.0 is an aspirational model that seeks to transform education to be more adaptive, skill-oriented, and relevant to the contemporary world. It leverages AI and digital tools to create learning experiences that are radically different from the one-size-fits-all factory model of Education 1.0. Students in an Education 4.0 setting ideally take charge of their learning trajectories (with guidance), gain not only knowledge but the skills to continue learning, and are comfortable collaborating with both humans and AI. Meanwhile, teachers assume new professional identities as facilitators, data-informed decision-makers, and content curators in a tech-rich environment. The next sections of the chapter will delve into the opportunities that Education 4.0 brings, the tangible benefits and improvements it offers, and the challenges and ethical dilemmas we must navigate to implement it responsibly.

Before proceeding, Table 4.1 provides a concise comparison of the four education paradigms (1.0 through 4.0) discussed so far, highlighting how various aspects of education have evolved. This summary underscores the increasing student-centricity and technology integration as we move from 1.0 to 4.0, providing a reference point for understanding the profound changes that Education 4.0 entails.

**TABLE 4.1**
**Comparison of Education 1.0, 2.0, 3.0, and 4.0 Paradigms (characteristics and trends)**

| Aspect | Education 1.0 (Traditional) | Education 2.0 (Digitization Era) | Education 3.0 (Connected Learning) | Education 4.0 (AI and 4IR era) |
|---|---|---|---|---|
| Dominant era | Pre-digital, Industrial Age (19th–20th century) | Early digital integration (late 20th century) | Internet era, Web 2.0 (2000s–2010s) | Intelligent tech era (2020s onward) |
| Teacher's role | Knowledge transmitter; authoritative lecturer | Lecturer using some tech aids; incremental change | Facilitator, guide (blended and flipped learning) | Mentor, coach, co-learner (augmented by AI tools) |
| Student's role | Passive recipient of information | More engagement, but still largely passive | Active learner, researcher, collaborator | Central participant, self-directed co-creator |
| Pedagogy | Teacher-centered; rote memorization; one-size-fits-all | Some multimedia and group work; still exam-focused | Student-centered; project and problem-based; flipped classrooms | Personalized and adaptive learning; project-based and experiential; learning anytime/anywhere |
| Technology | None or minimal (chalkboard, books only) | Basic ed-tech (TV, radio, overhead projector) | Ubiquitous ICT: computers, internet, e-learning platforms | Ubiquitous smart tech: AI tutors, data analytics, VR/AR, IoT in classrooms |
| Assessment | Standardized tests; focus on recall and exams | Predominantly exams (content knowledge) | Mix of exams and continuous assessment; online quizzes, feedback loops | Continuous adaptive assessment; real-time analytics; portfolios and competency-based evaluation |
| Key emphasis | Basic literacy, discipline, and factual knowledge | Integrating multimedia resources; computer literacy | 21st-century skills: critical thinking, digital literacy, collaboration | 21st-century skills + AI literacy; creativity, innovation, adaptability; lifelong learning orientation |

## 4.3 OPPORTUNITIES AND TRANSFORMATIONAL POTENTIAL

Education 4.0, driven by AI and advanced technologies, holds immense transformational potential for improving educational outcomes and experiences. In this section, we explore the key opportunities that AI-augmented learning environments offer. These opportunities range from enhancing teaching efficiency and personalized support to expanding access and fostering new skills. Collectively, they illustrate why many educators and policymakers are optimistic about the promise of Education 4.0 to address long-standing challenges in education and to create more engaging, effective learning for all.

### 4.3.1 Personalized and Adaptive Learning

The most celebrated advantage of AI in education is its ability to facilitate personalized learning paths for students. As discussed earlier, AI-driven systems can adapt in real time to a learner's level, providing tailored exercises and feedback. This means students can learn at their own pace, accelerating when concepts come easily and slowing down for more difficult material, which is a stark contrast to the one-speed-fits-all approach of traditional classrooms [3]. Adaptive learning platforms can also adjust their instructional mode (e.g., switching between text, video, or interactive simulations) to match how a student learns best, addressing diverse learning styles. The benefit of this personalization is improved learning outcomes and higher engagement. Students are neither bored by material that's too easy nor overwhelmed by material that's too hard. By customizing content, AI makes learning more student-centric, embodying the Education 4.0 ethos that each learner is on a unique journey. Early studies and pilot programs have shown that such personalized approaches can notably increase students' mastery of subjects and retention of knowledge, as compared to traditional methods [6].

### 4.3.2 Enhanced Teaching and Administrative Efficiency

AI offers tools to reduce the administrative and repetitive burdens on teachers, thereby allowing them to focus more on high-value teaching activities. For example, AI can automatically grade multiple-choice tests, fill in assessment rubrics for certain types of assignments, and even evaluate written essays to a preliminary degree (providing scoring suggestions or flagging errors). It can also assist with tasks such as attendance tracking, timetable scheduling, and generating progress reports. By automating these routine duties, AI gives teachers more time for lesson planning, one-on-one mentoring, and professional development [3]. This augmentation of the teacher's role not only potentially improves teacher job satisfaction (as educators can spend more time on the creative and interpersonal aspects of their work) but also improves the quality of instruction, as teachers can dedicate more attention to understanding students' needs and innovating their teaching strategies. In places facing a teacher shortage, AI tools can help fill gaps by handling larger class loads for administrative tasks, thus partially mitigating the effects of too few staff [3]. It is important to stress that these efficiencies are meant to enhance human teaching, not replace it;

the ideal scenario is a human-AI symbiosis in the classroom where mundane tasks are offloaded to machines and teachers concentrate on personal interactions with students.

### 4.3.3 Data-Driven Insights and Improved Assessment

AI and analytics enable educators to harness data to gain insights into student learning like never. Learning analytics can track a student's interactions with digital learning materials in fine detail, such as which questions they got right or wrong, how long they spent on a task, which resources they accessed, etc. By analyzing this data, AI systems can identify patterns that might be invisible to teachers in a traditional setting. For instance, AI might notice that a student performs better on algebra word problems when they involve real-world contexts, suggesting a way to improve that student's engagement. It might detect early warning signs that a student is falling behind (such as repeated failed attempts on a concept), allowing for timely intervention. Overall, AI refines assessment and feedback by moving beyond the one-time exam score to a continuous, formative assessment model [3]. Teachers can receive dashboards that highlight class and individual performance in real time, enabling a more responsive teaching approach. This real-time feedback loop helps personalize support: for example, if an AI tool flags that a group of students didn't grasp a particular subtopic, the teacher can revisit it the next day in a different way. Furthermore, AI opens possibilities for evaluating non-traditional skills, such as collaboration and creativity, by analyzing, for example, group chat logs or project development processes, thus broadening what can be assessed beyond standardized tests.

### 4.3.4 Accessibility and Inclusion

A central promise of Education 4.0 is making learning more accessible to diverse populations. AI tools can be designed to accommodate special educational needs and overcome barriers. For example, speech recognition and natural language processing enable voice-activated assistance and real-time transcription, invaluable for students who are visually impaired (who can listen to written text) or hearing impaired (who can read live captions of lectures). AI-based translation can break down language barriers; students can access content in their native language or have an AI tutor that converses with them in that language, which is especially useful in multilingual societies or for refugee children adjusting to a new language [6]. Additionally, AI can support students with learning disabilities: an AI reading tutor might help dyslexic students by providing personalized phonics exercises, or an AI coach could help autistic students rehearse social interactions in a safe, controlled simulation. The concept of "learning anytime, anywhere" facilitated by technology also means education can reach students in remote or rural areas via online platforms. An AI tutor doesn't require a local, trained teacher to be present, which can be transformative for underserved regions. Collectively, these innovations can make education more inclusive, addressing the needs of learners who might be left behind in a one-size-fits-all system.

### 4.3.5 Engagement through Innovative Pedagogy (VR, Gamification, etc.)

AI, in combination with other emerging technologies, allows for more engaging and experiential learning approaches. For instance, incorporating gamification and AI can create highly motivating learning environments. An AI can act as a game master or adaptive engine in an educational game, adjusting challenges to keep students in a state of "flow," not too easy to be boring, not too hard to cause frustration. VR and AR experiences can bring abstract or distant concepts to life: students can take a virtual field trip to ancient Rome in history class or explore the inside of a cell in biology. AI enhances these experiences by, for example, controlling virtual characters that interact with students or by analyzing student interactions in VR to tailor the scenario in real time [6]. Moreover, the interactivity offered by these tools encourages active learning; students who learn by doing (even in a virtual environment) often understand and retain concepts better than those who learn passively through passive listening. By making learning more fun and interactive, these approaches can increase student motivation and reduce dropout rates.

### 4.3.6 Global Collaboration and Cultural Exchange

The connected nature of Education 4.0 means classrooms can transcend geographic boundaries. AI translation and communication platforms can connect classrooms across the world, enabling global projects and cultural exchange programs that operate in virtual spaces. For example, students in different countries might jointly conduct a science project, using an AI platform that helps bridge language differences and manage their collaboration. This opportunity nurtures global citizenship skills, assisting students to become more culturally aware and improve their ability to communicate with diverse teams, which are crucial competencies in the globalized 4IR workplace [1]. Additionally, global connectivity means the best educational content can be shared widely: a talented teacher in one part of the world can deliver a live or recorded lesson to students elsewhere, effectively democratizing access to quality instruction. AI can personalize these shared resources (e.g., by adding subtitles, adjusting content difficulty for different age groups, or providing supplemental explanations based on a learner's background knowledge).

### 4.3.7 Lifelong Learning and Reskilling

Beyond formal K-12 or tertiary education, AI is transforming lifelong learning, which is increasingly important in a world where industries evolve quickly, and workers need to update skills continually. AI-driven learning platforms for professionals can recommend personalized learning pathways for career advancement or reskilling, for instance, identifying skills a user lacks for a desired role and then providing targeted modules to learn those skills. These platforms often use AI to adapt to the busy schedules of adult learners, breaking learning into micro-lessons that can be consumed on the go. The idea of a "LIFE 4.0" scenario, which we discuss later, is one in which learning is seamlessly integrated into daily life and work, often through AI assistants that prompt you to learn something new or help solve a problem. The opportunity here is to create a culture where education doesn't stop at graduation but is an ongoing, self-directed process supported by intelligent tools.

In highlighting these opportunities, it's worth noting concrete examples:

- **Intelligent Tutoring Systems (ITS)**: Platforms like Carnegie Learning's AI for math or Duolingo's AI for language learning illustrate how personalized tutoring can reach millions of learners, adapting in real time to each one's progress.
- **AI in Educational Content Creation**: AI is also being used to generate practice questions, summarize textbook chapters, or even create interactive lab scenarios. This can significantly reduce the cost and time to develop high-quality educational materials, making it easier to update curricula and share resources globally.
- **Improving Equity through AI**: Initiatives exist where AI is used to map underserved regions' educational needs (using data on learning outcomes, school resources, etc.) and guide policy or interventions, essentially using AI at a systemic level to allocate support where it's needed most.

Overall, the opportunities afforded by Education 4.0 suggest a reimagining of what learning can be: more efficient, more engaging, and more equitable. By personalizing learning and automating tedious tasks, AI can help each student reach their potential, and each teacher can focus on what humans do best: inspire, mentor, and innovate. Providing data-driven insights enables continuous improvement of educational strategies. And by extending the reach of education beyond traditional limits, it has the potential to bring quality learning to communities that have historically been left behind.

However, realizing this transformational potential is not automatic or guaranteed. With new technology comes new challenges and risks. The following section addresses the challenges, risks, and ethical considerations that accompany the integration of AI into education, acknowledging that, to truly benefit from Education 4.0, these issues must be thoughtfully managed.

## 4.4 CHALLENGES, RISKS, AND ETHICAL CONSIDERATIONS

While Education 4.0 offers exciting opportunities, it also raises significant challenges and ethical concerns that need to be addressed to ensure that the transformation of education via AI is beneficial and fair. As schools and universities worldwide experiment with AI tools, stakeholders have identified numerous risks from privacy and bias issues to the potential exacerbation of inequalities that must be mitigated.

This section discusses the key challenges and ethical considerations in adopting AI and advanced technologies in education.

### 4.4.1 Digital Divide and Inequity

A foremost concern is that AI in education could widen the digital divide, the gap between those who have access to technology and those who do not. Implementing AI-driven learning requires infrastructure: reliable electricity, internet connectivity, and modern devices, as well as maintaining software and hardware. Many regions of the world, and even underprivileged communities within developed countries, lack these prerequisites. If Education 4.0 is only available to well-funded schools or wealthier students, it risks amplifying educational inequality. For instance, an urban school with high-speed internet can employ adaptive learning software and give students take-home tablets. In contrast, a rural school with no connection cannot leave its students at a relative disadvantage. UNESCO has pointed out that at least one-third of the world is not digitally connected, meaning a significant portion of students could be left behind if education becomes too tech-centric [4]. Ensuring equitable access to AI tools is therefore a critical challenge. This might involve public investments in infrastructure, subsidizing devices for low-income students, and developing offline or low-bandwidth versions of AI educational software for communities with limited connectivity. Without deliberate action, Education 4.0 could inadvertently create an AI-driven learning elite while others remain in analog Education 2.0 or 3.0 environments, exacerbating the global achievement gap [6].

### 4.4.2 Privacy and Data Security

AI systems in education often rely on collecting and analyzing detailed data on students' performance and behaviors, as well as biometric data (such as attention tracking via cameras or emotion-detection software). This raises serious privacy concerns. Students (especially minors) are a vulnerable population, and their personal data, including academic records, learning difficulties, or behavioral patterns, must be protected. There are risks that sensitive information could be misused if it falls into the wrong hands or that commercial entities could exploit student data for profit (e.g., by profiling students for advertising). Additionally, schools might not always have robust cybersecurity measures, making them targets for data breaches. A related issue is data ownership: Who owns the data generated by an AI tutor, the student, the school, or the software provider? Ethical guidelines suggest students and their guardians should have control over educational data, but enforcing this can be challenging. Ensuring compliance with regulations such as GDPR in Europe or FERPA in the US is necessary when implementing AI solutions [7]. Moreover, even when data is used internally for learning analytics, there needs to be transparency about what is being collected and how it influences decision-making (e.g., if an algorithm labels a student as "at-risk," how is that information used?). Protecting student privacy and security is thus paramount, requiring encrypted data storage, strict access controls, clear consent protocols, and perhaps even algorithmic audits to ensure that AI systems handle data appropriately.

### 4.4.3 Algorithmic Bias and Fairness

AI systems learn from data, and if that data reflects human biases or systemic inequalities, the AI can inadvertently perpetuate or even amplify those biases. In education, this could manifest in several ways. For example, if a predictive algorithm is used for college admissions or identifying gifted students, it might be biased against certain demographic groups if not carefully designed, because historical data may reflect past discrimination (e.g., underrepresentation of minorities in advanced courses). Similarly, language processing AI might not understand dialects or non-native accents well, disadvantaging certain students in, say, an automated oral language assessment. Bias in AI could also affect disciplinary actions. Imagine an AI surveillance system in a classroom that more frequently flags certain students as misbehaving due to biased facial recognition or contextual training data. Ensuring fairness means that AI tools should be rigorously tested for disparate impacts and trained on diverse datasets. Educators and developers should use principles of responsible AI: for instance, including fairness as a goal in model development and possibly employing explainable is so that decisions (such as a particular personalized learning recommendation) can be understood and questioned. UNESCO and other international bodies have emphasized that inclusivity and equity must be built into AI in education by design [3]. If not addressed, algorithmic bias could undermine trust in educational AI and harm the very students the technology is meant to help, particularly those from marginalized communities.

### 4.4.4 Overreliance and Diminished Human Interaction

Another challenge is striking the right balance between technology and human elements in education. There are concerns that overreliance on AI might erode critical thinking and creativity in students [6]. If students come to rely heavily on AI for answers or direction, they may not develop the grit to solve problems independently or think outside the solutions provided. For example, if a writing assistant AI always suggests sentence improvements, students might not learn to refine their writing through their own thought process. Moreover, learning is not purely cognitive; it's also social and emotional. Human interaction, encouragement from a teacher, camaraderie among classmates, and the development of social skills are vital components of education. If too much learning happens in isolation with a screen or an AI tutor, students might miss out on developing teamwork, communication skills, and empathy. There's also a risk of reduced engagement if AI tools are poorly designed; not every student will respond to a digital tutor as positively as to a passionate human teacher. Thus, an ethical consideration is to ensure AI complements rather than replaces human connection in classrooms [3, 6]. Schools need to actively preserve time for peer collaboration, discussions, and teacher-student mentorship, even as they integrate AI tools. Some educational theorists argue for a "balanced diet" of learning experiences: a mix of AI-guided practice and rich human-mediated activities. In essence, maintaining the human touch in Education 4.0 is crucial for holistic development.

### 4.4.5 Teacher Readiness and Changing Roles

The advent of AI in classrooms is also disruptive to teachers, and not all educators are prepared or willing to adapt to these changes. Many teachers fear that AI could render their role redundant or force them to play second fiddle to machines. There's an urgent need for teacher training and professional development focused on AI literacy and integration. Teachers should be trained not just in using new tools, but in understanding their underlying principles and limitations (what AI can and cannot do). According to UNESCO's new competency frameworks, teachers need knowledge of AI ethics, data management, and the critical evaluation of AI educational products [6]. Without such support, there's a risk that AI tools will be misused or underused in classrooms. Additionally, the transition can be stressful: mastering a new adaptive learning platform or interpreting analytics reports might feel burdensome on top of existing duties. Resisting change is a natural response when teachers perceive that technology is imposed or as diminishing their professionalism. Ethically, it's important to involve teachers in the design and implementation of AI in education as co-creators and decision-makers so that solutions address real classroom needs and have buy-in from those on the front lines [3]. Ignoring teacher perspectives can lead to implementation failures or even active pushback (e.g., educators lobbying against specific ed-tech policies). Ultimately, teacher agency should be respected; the ethical approach is to empower teachers with AI rather than subjugate them to it.

### 4.4.6 Academic Integrity and Misuse of AI by Students

With AI tools such as advanced language models (e.g., GPT-based chatbots) becoming increasingly accessible, new forms of academic dishonesty are emerging. Students might use AI to write essays, solve assignments, or cheat on exams. This poses a challenge to academic integrity. If a student submits AI-generated work, it undermines the learning process and fairness in evaluation. It can also make it harder for teachers to assess a student's accurate understanding. Educational institutions are scrambling to update their policies and detection methods (such as AI detectors) to handle this scenario [7]. There is a fine ethical line: on the one hand, using AI to aid learning (such as brainstorming ideas or getting feedback on drafts) can be a legitimate educational practice; on the other hand, using it to do the work without learning is problematic. Schools need to establish clear guidelines on the ethical use of AI for schoolwork and ensure students understand them. A related issue is plagiarism and intellectual property: AI can generate content that may remix existing works without proper citation, leading to inadvertent plagiarism. Also, how do we credit AI's contribution in student work? These are new questions that educators and academic bodies are now grappling with. The risk of misuse is a barrier to adoption as well; some schools reacted to AI's emergence by outright banning AI tools on campus. However, outright bans may be unenforceable and deprive students of learning about a technology that will be part of their world. A more balanced approach is needed: integrating AI literacy, ethics, and integrity training into the curriculum so students learn to use AI responsibly, as we teach responsible internet use.

### 4.4.7 Reliability and Quality of AI Outputs

Current AI, especially cutting-edge generative AI, is not infallible. AI tutors or answer engines can sometimes give incorrect information or problematic explanations, e.g., a math solution with a subtle error or an explanation of a historical event that's factually off. There is a phenomenon known as AI "hallucination," in which an AI might produce a confident-sounding answer that is entirely fabricated. In an educational context, this is dangerous if unchecked, because students might trust the AI's output. Ensuring content accuracy and reliability of AI-driven instruction is a challenge. It may require having humans in the loop (e.g., teachers reviewing AI-generated materials) or designing AI with transparency so that it provides sources or reasoning steps that can be verified [6]. Moreover, the dynamic nature of AI means that as models update, their behavior can change, posing is consistency issue. If a student gets help from an AI on Monday and sees a different approach from the AI on Friday for the same problem, it could be unclear which approach is correct. These uncertainties mean educators must be cautious and ideally treat AI as a supportive tool rather than an unquestionable authority. Part of the solution is also educating students in critical thinking and in cross-verifying information from AI (turning this into a learning experience in itself).

### 4.4.8 Ethical Use of Student Data and AI Decision-Making

Another ethical layer is how decisions made by AI affect students' futures. For instance, if an AI system is used to stream students into different ability groups or recommend them for specific programs, it must do so fairly and transparently. Students and parents should have the right to know why an algorithm made a recommendation and can appeal or correct it if it's wrong. Transparency in AI decision-making is a core ethical principle often summarized as the need for AI to be explainable (XAI). A black-box model that labels a student as "low potential" could stigmatize or limit opportunities for that student without due justification, which is ethically unacceptable. Schools should be careful in how they use predictive analytics. For example, predicting dropout risk can help intervene to support the student, but it should never be used to pre-emptively exclude a student from opportunities. Maintaining a stance that AI predictions are just one factor among many and that human educators have the final say in critical decisions can help mitigate this.

### 4.4.9 Environmental and Economic Considerations

An often-overlooked challenge is the environmental impact of AI. Training and running large AI models consume significant energy, contributing to carbon emissions. If educational institutions widely deploy AI, the climate impact is a sustainability concern [6]. Ethically, this calls for using energy-efficient technologies and offsetting emissions. Economically, there's the challenge of cost: advanced AI systems and the IT infrastructure to support them can be costly. Schools with tight budgets might struggle to afford the latest AI tools, raising equity issues again. Long-term, open-source, and affordable AI solutions for education need to be encouraged to avoid a situation in which only wealthy schools can fully implement Education 4.0.

Considering these challenges and ethical issues, what can be done? Several global and local initiatives are already in motion:

- **Policy and Guidelines**: UNESCO released guidance for policymakers on AI in education, highlighting principles like inclusion, safety, and human-centric values [8]. Similarly, many countries are drafting national AI-in-education strategies that include ethics charters.
- **Principles and Frameworks**: Organizations have proposed principles for responsible AI use in schools. For instance, a WEF article outlines seven principles for practical AI guidance in education, including purpose alignment with educational goals, ensuring privacy, promoting AI literacy, maintaining a balance between benefits and risks, and upholding academic integrity [7].
- **Professional Development**: Investment in teacher training on AI, as mentioned, is growing. Conferences, workshops, and online courses for teachers about AI are more common, reflecting recognition that teacher buy-in and understanding are crucial.
- **Community and Stakeholder Engagement**: Ethically integrating AI also means talking to students and parents. Some schools have begun to involve student voice in selecting or reviewing AI tools and to inform parents clearly about what data is collected and how it benefits their child's learning.

In conclusion, the journey to Education 4.0 must be navigated with care and responsibility. The transformative power of AI in learning comes with potential pitfalls. By acknowledging these challenges and actively working to address them through thoughtful policy, inclusive design, transparency, and continuous dialogue among educators, technologists, students, and parents, we can strive to implement AI in education in a way that is ethical, equitable, and aligned with the core mission of education. The final part of the chapter will look ahead to the outlook of education in the context of LIFE 4.0, imagining how these pieces might come together in the world of tomorrow.

## 4.5 THE FUTURE OUTLOOK (LIFE 4.0 SCENARIO)

Looking forward, we envision a LIFE 4.0 scenario, a future in which the integration of AI and advanced technologies into all aspects of life, including education, is ubiquitous and seamless. In this future, education is not confined to formal institutions or early life stages; instead, learning becomes a continuous, lifelong endeavor supported by intelligent systems and a culture that values adaptability and growth. This section paints a picture of how classrooms, workplaces, and communities might evolve in the coming decades under the influence of

Education 4.0 principles, and what new developments we might anticipate as LIFE 4.0 unfolds.

### 4.5.1 Education as a Lifelong, Lifewide Experience

In the LIFE 4.0 future, the concept of a "classroom" expands beyond physical or even virtual schoolrooms. Learning modalities will be embedded in everyday life, often referred to as ubiquitous learning. Imagine AI-driven learning companions that stay with individuals throughout their lives, much like a highly personalized mentor accessible at any time. These AI mentors could know an individual's learning history, career goals, and even personal interests, continuously suggesting learning opportunities (articles, courses, experiences) to help them grow. Need to acquire new job skills? Your AI assistant curates a micro-learning module and perhaps even arranges a quick VR practicum. Curious about a topic after watching a movie? The AI can answer questions or connect you with an online expert community. Micro-credentials and stackable certifications might replace or augment traditional degrees, as people assemble portfolios of skills and knowledge over a lifetime. This aligns with the idea that, in an AI-transformed economy, workers may need to retrain multiple times; education systems will support mid-career transitions and upskilling through flexible, just-in-time learning opportunities, many of which are powered by AI personalization.

### 4.5.2 Hybrid Human-AI Collaboration in Learning

As AI becomes more advanced, we can foresee more sophisticated forms of human-AI collaboration in learning tasks. For example, Large Language Models (LLMs), such as future iterations of GPT or other AI collaborators, might become creative partners in projects, helping students brainstorm, iterate, and even venture into areas beyond human expertise. A LIFE 4.0 classroom could have AI-driven laboratories where students and AI work side by side to conduct experiments or design solutions to real-world problems. The AI might generate hypotheses or run simulations at super-human speed, while the students interpret results, ask new questions, and apply human judgment and ethics. This kind of partnership can accelerate innovation and deeply personalize the learning experience (each student effectively has an always-available "teaching assistant"). There is ongoing research into how LLMs and AI companions can function as knowledgeable peers or tutors, and by the time LIFE 4.0 is in full swing, these might be common. The key is that learners will need to master the art of working effectively with AI, knowing when to rely on AI, when to question it, and how to leverage its strengths while providing human oversight.

### 4.5.3 Global Learning Networks and Cultural Fusion

The future may bring about truly global classrooms where geographical distance is irrelevant. Building on the connectivity of Education 3.0 and the translation and personalization capabilities of AI, global learning networks could become the norm. A high school project team might include students from five different countries, meeting in a VR space to collaborate on an environmental science study, guided by an AI facilitator that ensures everyone contributes despite language differences. Such interactions from a young age can foster unprecedented levels of cross-cultural understanding and global citizenship. Education 4.0 thus contributes to a LIFE 4.0 society that is more globally cohesive, as shared learning experiences knit people together across borders. We might even see the emergence of "world curricula" with common foundational content (such as climate change science or universal human rights concepts) delivered globally, supplemented by local context. AI could help localize and adapt these curricula for each context while ensuring a shared global baseline of knowledge, striking a balance between international and regional education.

### 4.5.4 Customized Learning Environments and Spatial Transformation

Classrooms and learning spaces themselves will likely transform. The traditional rows of desks might become obsolete. Instead, flexible spaces equipped with technology that can change configuration for various learning activities will dominate. Smart classrooms will automatically adjust lighting, sound, and resources based on the activity, e.g., dimming lights and projecting a simulation when students are doing a virtual lab, or rearranging furniture via robotics for a collaborative workshop. AI will be the invisible orchestrator of these environments. Furthermore, AR could turn any location into a learning space: a student on a field trip to a forest might use AR glasses to identify plant species with AI's help, turning the whole forest into a "classroom." In cities, public spaces might carry educational AR overlays, such as historical facts, when you point your device at a monument, effectively blending education into daily life for citizens of all ages.

### 4.5.5 Emphasis on Meta-Learning and Human Skills

With AI handling more routine aspects of knowledge delivery, human educators and curricula will likely emphasize meta-learning, learning how to learn, as well as uniquely human skills. Critical thinking, creativity, ethical reasoning, and emotional intelligence will be front and center. For instance, schools might have dedicated classes on "AI Ethics and Society" to ensure that, from a young age, students reflect on the implications of AI (thus preparing them to navigate an AI-rich life). The arts and physical education might also gain renewed importance as we recognize that holistic human development requires more than cognitive skills. The performing arts may evolve with AI (think of AI-assisted music composition being taught alongside traditional instruments), and sports might integrate AI for personalized coaching. Ultimately, LIFE 4.0 education is about cultivating versatile, emotionally resilient, and morally grounded individuals who can thrive in synergy with intelligent machines.

### 4.5.6 Continuous Feedback and Improvement of Education Systems

At a systemic level, data from Education 4.0 implementations can inform the continuous improvement of education policy and practice. Governments and institutions could use AI analytics on a large scale (with privacy safeguards) to understand which methods work best, which areas or demographics need more support, and how educational innovations propagate. Evidence-based policy could reach new heights of sophistication, as real-time data allows quick course corrections and targeted interventions. For example, if a particular AI tool is found to help students with dyslexia across thousands of classrooms significantly, policymakers could quickly allocate resources to deploy that tool to all schools in need. Conversely, if data shows a widening gap due to tech, measures can be taken swiftly to address it. The education system itself becomes more "intelligent" in learning and adapting, akin to a living organism.

### 4.5.7 Ethical and Equitable Lifelong Learning Society

In the ideal LIFE 4.0 scenario, society recognizes education as a cornerstone not just for economic productivity but for personal fulfillment and democratic participation in an AI-enhanced world. Lifelong learning could be supported through policies like learning credits or sabbaticals for adults to re-educate, and through community learning hubs where technology is accessible to all (ensuring that even those not in formal education or high-paying jobs can benefit from Education 4.0 resources). Ethical considerations remain central: there might be regulatory bodies that ensure AI in education vendors adhere to standards, as we have the FDA. Public awareness of AI's role in life will be higher; ideally, every citizen will have basic literacy in how AI works, enabling informed consent and critical engagement with the AI systems they encounter.

Of course, the LIFE 4.0 vision is aspirational. There are uncertainties and varying opinions on how far and how fast these changes will occur. Some futurists imagine a scenario with Artificial General Intelligence (AGI) that could revolutionize education even further, for example, an AGI that can truly understand and teach any subject matter, essentially democratizing the best quality education everywhere. Others are more cautious, focusing on augmenting human teachers rather than supplanting them and emphasizing the unpredictable, creative aspects of human learning that machines may never fully replicate.

One can anticipate new roles and professions emerging as well: perhaps "learning experience designers" who specialize in creating AI-enhanced curricula, "education data analysts" who continually refine algorithms to ensure fairness and effectiveness, or "digital ethics coaches" in schools who ensure students develop healthy relationships with technology.

In summary, the future of education under a LIFE 4.0 scenario is one of profound transformation and continuity in values. The core values of education to empower individuals, to transmit culture and knowledge, to foster critical and compassionate citizenship, remain unchanged. Still, how we achieve them will evolve dramatically. Education 4.0 technologies, especially AI, will be key enablers in making learning more personalized, immediate, and far-reaching than ever before. If guided with wisdom and inclusivity, this future holds the promise of an education system that truly meets learners where they are, helps each person develop to their fullest potential, and equips humanity to navigate the complexities of an AI-rich world.

As we prepare for this future, it will be essential to keep a vigilant eye on the ethical compass and the human purpose of education. The next generation of students will live in a world where AI is everywhere. Our task today is to ensure that by the time the future arrives, we have laid the groundwork for AI to be a positive force in their learning and their lives.

## 4.6 CONCLUSION

The advent of Education 4.0 marks a pivotal chapter in the story of learning, one where the synergy of human ingenuity and AI is poised to redefine how we teach, learn, and grow. In the chapter, we journeyed from the roots of traditional Education 1.0 with its chalk-and-talk classrooms and one-size-fits-all ethos through the incremental technological additions of Education 2.0, into the interactive and connected world of Education 3.0, and finally arrived at the cusp of the intelligent, personalized learning revolution that is Education 4.0. Each stage in this evolution has layered new capabilities onto the old. With Education 4.0, we see an unprecedented fusion of pedagogy and technology that holds both great promise and profound responsibility.

AI in the classroom is not a distant science fiction scenario; it is already here, augmenting language learning apps, powering tutoring bots, grading essays in pilot programs, and helping teachers to identify students in need of help. The global perspective taken in our discussion shows that countries and communities worldwide are grappling with what this means. On the one hand, AI offers developing nations a chance to leapfrog educational hurdles by providing quality resources where human experts are scarce, and offers developed nations tools to personalize learning and make education more relevant to the 21st-century economy. On the other hand, it raises urgent questions about equity. Will Education 4.0 benefit all, or only a privileged few? We have highlighted the need to bridge the digital divide and implement policies to ensure that rural and underserved populations are not left behind in this transformation [6].

Throughout the chapter, a recurring theme has been the balancing of technology with the human touch, innovation with ethics, and efficiency with equity. Education 4.0 is not about dehumanizing education or handing everything over to algorithms. As stressed, the vision is AI with teachers, not instead of teachers [3]. The ideal classroom of the future is one where AI empowers teachers, freeing them from drudgery to focus on creative and emotional labor, and allows students, freeing them from the limitations of pace and resources to focus on deeper learning and skill mastery. Achieving this ideal requires careful implementation: robust data privacy safeguards, bias checks in AI systems, strong digital literacy curricula, and ongoing support for educators' professional development. It also requires continuous dialogue among stakeholders to align AI tools with educational values and goals [7].

We also discussed how assessment and curriculum are evolving. The traditional examination-heavy approach is giving way to more nuanced methods thanks to AI analytics and adaptive assessment [3]. Meanwhile, curricula are expanding to include not just STEM skills but also digital ethics, AI literacy, and the soft skills of teamwork and adaptability, which are so crucial in an AI-rich society [2]. In the LIFE 4.0 scenario we envisioned, learning becomes lifelong and omnipresent, and

Education 4.0 is a foundation for that reality, creating learners who know how to learn continuously and work alongside intelligent machines.

The chapter's goal to educate and inspire is rooted in the belief that while technology will shape the future of education, it is human agency and vision that will ultimately determine whether those changes are for the better. Around the world, inspiring examples are emerging: a rural teacher using AI translation to include immigrant students in class, a government deploying an AI tutor program to improve literacy rates in far-flung villages, a university using AI to predict and prevent student dropouts, and international collaborations in which students collectively tackle global challenges with AI support. These stories give a glimpse of a future where Education 4.0 truly humanizes learning by making it more inclusive, engaging, and attuned to each learner's needs.

In conclusion, Education 4.0: Classroom Transformation in the Age of AI is not merely an educational reform agenda; it is part of a larger societal shift toward what we have termed LIFE 4.0. The classroom is a microcosm of society. Changes in how we educate the next generation will echo in how we work, how we communicate, and how we solve problems together. By embracing the positive potential of AI to enrich learning and by vigilantly addressing its challenges, we can ensure that this transformation leads to a more enlightened and equitable global society. Education has always been the key to unlocking human potential. In the age of AI, it is heartening to realize that education itself is being reimagined and unlocked, becoming smarter, more adaptive, and hopefully more empowering than ever before.

The journey is just beginning, and its success will depend on all of us educators, learners, technologists, and leaders working collaboratively to shape an Education 4.0 that truly serves humanity. In doing so, we carry forward the timeless mission of education into a new era: to prepare each generation to thrive in their world, with wisdom, compassion, and the capacity to build a better future.

## REFERENCES

1. Symonds, Q., Everything You Need to Know about Education 4.0, in *QS Insights*, 2019.
2. Talentra, From Education 1.0 to Education 4.0, in *Talentra Blog*, 2023.
3. Milberg, T., The Future of Learning: How AI Is Revolutionizing Education 4.0, in *World Economic Forum*, 2024.
4. UNESCO, AI *and Education: Guidance for Policy-Makers.* UNESCO Publishing, 2023.
5. Mondaca, D., Differences between Education 1.0 to 4.0, in *Educational Games*, April 2020.
6. *UNRIC, What AI Thinks about Its Role in Global Education*, 2025.
7. Partovi, H., and P. Yongpradit, AI and Education: Kids Need AI Guidance in School. But Who Guides the Schools? in *World Economic Forum*, 2024.
8. Elhussein, G., E. Hasselaar, and O. Lutsyshyn, Shaping the Future of Learning: The Role of AI in Education 4.0, in *World Economic Forum*, 2024.

# 5 Ethics 4.0

**Abstract**

Ethical systems have continually evolved through history from ancient moral codes and religious laws to Enlightenment principles of universal rights, and onward to new dilemmas of the digital age. Today, we stand at the cusp of an "Ethics 4.0," an era defined by artificial intelligence (AI), automation, and algorithmic decision-making. This chapter surveys the trajectory of ethical thought through Ethics 1.0 (classical and religious frameworks), Ethics 2.0 (Enlightenment humanism and human rights), and Ethics 3.0 (digital-era ethics), setting the stage for Ethics 4.0, the novel challenges and opportunities posed by AI and autonomous systems. We explore how time-honored values are being tested and reshaped by AI technologies, from ensuring algorithmic fairness and accountability to safeguarding human autonomy in an age of intelligent machines. Global perspectives are emphasized, recognizing that different cultures and societies approach AI ethics with diverse values and policy priorities. Throughout, we highlight the transformational potential of AI to advance human well-being, such as improving healthcare, education, and sustainability, while also addressing the profound risks (bias, privacy erosion, loss of human oversight, and even existential threats). By examining both opportunities and challenges, this chapter aims to provoke thoughtful discussion on how our moral compasses can be realigned for the AI epoch. In the concluding outlook, a "LIFE 4.0" scenario is envisioned, where humanity navigates a future of pervasive AI by leaning on enduring ethical principles reinterpreted for a new age. Ethics 4.0 calls for an adaptive, inclusive, and forward-looking ethical framework, one that keeps humanity at the center of technological progress and ensures that intelligent systems serve the global common good.

## 5.1 INTRODUCTION

The rapid emergence of artificial intelligence (AI) and autonomous technologies is fundamentally redefining how we think about ethics in the modern world. We are entering what might be called Ethics 4.0, analogous to the "Industry 4.0" technological revolution, which demands recalibrating moral frameworks to keep pace with AI's transformative impact [1]. Just as previous eras saw ethics shaped by new social realities from the codes of ancient societies to Enlightenment ideals, to the digital information boom, the current AI epoch raises unprecedented ethical questions that challenge traditional compasses. This chapter explores the evolution of ethical systems through four primary stages (Ethics 1.0–4.0), culminating in an in-depth look at the ethical landscape of the AI era. The goal is to educate readers about how age-old moral principles are being tested by contemporary technology and to provoke thoughtful discussion on guiding values that can steer us through the promises and perils of AI-driven life.

Ethics does not develop in a vacuum; it responds to transformations in society, knowledge, and power. With each great leap in human capability, the advent of cities

DOI: 10.1201/9781003455219-5 

and organized religion, the rise of reason and individual rights, the global connectivity of the Internet, our understanding of right and wrong, of duties and rights, has been forced to adapt. Today's leap, characterized by machine learning algorithms, big data, and autonomous decision-making, is no different. AI systems now perform tasks ranging from trivial to life-altering: recommending news and products, driving vehicles, diagnosing patients, approving loans, and even influencing democratic processes. These innovations carry enormous opportunities to improve human life, but also risks of misuse or unintended harm. As one Harvard study noted, while AI's growing utility is undeniable, concerns abound that these complex and opaque systems "may do more societal harm than economic good" if left unchecked [2]. We must ask: How can society reap AI's benefits without sacrificing core values like fairness, accountability, and human dignity?

Crucially, the ethical challenges of AI are global in scope. Cultures and nations bring different philosophical and legal perspectives to questions such as data privacy, algorithmic bias, or robot autonomy. A truly comprehensive approach to Ethics 4.0 must therefore include global perspectives and engage a diverse chorus of voices. International efforts, from the UN's AI governance initiatives to cross-cultural dialogues on digital rights, indicate a growing recognition that no single country or philosophy has all the answers [1]. Just as the effects of AI transcend borders, so too must our ethical deliberations. Throughout this chapter, examples from around the world will illustrate how various societies are grappling with AI ethics, whether through policy (like Europe's regulatory push for "Trustworthy AI") or through cultural lenses (such as Asian and African viewpoints on community, harmony, and technology).

In the sections that follow, we journey from the past to the future: Ethics 1.0 revisits classical and religious moral frameworks that provided humanity's first ethical compasses. Ethics 2.0 discusses the Enlightenment shift toward reason and universal human rights. Ethics 3.0 examines the late 20th-century emergence of digital ethics amid computing and information technology. Building on these foundations, Ethics 4.0 delves into the novel terrain of A issues of autonomy, algorithmic decision-making, and accountability when machines act on their own. We then explore the transformative opportunities AI offers for advancing human well-being, as well as the challenges and risks it poses to societal and moral order. Finally, The Future Outlook (LIFE 4.0) speculates on a scenario in which AI is deeply woven into the fabric of life, posing questions about the future of work, the nature of consciousness and personhood, and the ethical framework that will guide a world of humans and intelligent machines coexisting.

By examining where our ethical compasses have come from and how they are being tested today, we can better chart a course for the future of AI. Ultimately, Ethics 4.0 calls on us to carry forward the wisdom of past ethical systems and principles of justice, compassion, autonomy, and rights, and to reinterpret them for an AI-driven world. In doing so, we aim to ensure that technology serves humanity's highest values, rather than undermining them. The stakes are enormous: the character of the AI epoch will be defined by the ethical choices we make today.

## 5.2 ETHICS 1.0–4.0

### 5.2.1 Ethics 1.0: Classical and Religious Moral Frameworks

Ethics 1.0 represents the foundational moral framework that guided human societies in the classical age and through religious teachings. Long before the advent of modern science or global connectivity, ancient civilizations grappled with questions of right and wrong, virtue and vice, often grounding their answers in a mix of philosophical reasoning and spiritual belief. These early ethical compasses were diverse across cultures; yet, we can trace common threads and enduring themes that still influence moral thought today.

In the cradles of civilization, religion and morality were deeply intertwined. The earliest recorded codes of ethics often took the form of divine commandments or cosmic order. For example, in ancient Mesopotamia, the Instruction of Amenemope and similar texts urged compassion for the less fortunate ("leave some grain for the poor gleaners") and promised divine favor as a reward [3]. Many religions articulated some version of the Golden Rule as a fundamental guide: essentially, treat others as you would wish to be treated. Variations of this principle emerged independently across cultures: a negative form appears in Confucian teachings ("do not do to others what you do not want done to yourself"), in ancient Greek thought, and in the sayings of Rabbi Hillel, among others [3]. This near-universal rule encapsulated a basic empathy and reciprocity that could bind communities together, suggesting an early convergence of ethical insight across humanity's spiritual traditions.

Classical philosophy, particularly in ancient Greece, marked a shift toward rational inquiry about ethics. Figures like Socrates, Plato, and Aristotle moved moral discussion from myth and edict to reasoned argument. Socrates famously questioned his fellow Athenians in search of definitions of virtues like justice and courage, laying the groundwork for ethics as a field of philosophical study [3]. Plato envisioned a transcendent Form of the Good as an ultimate moral truth and, in The Republic, argued that a just society reflects a higher order of virtue [3]. Aristotle, in his Nicomachean Ethics, took a more practical approach: he defined ethics as the pursuit of eudaimonia, often translated as human flourishing or happiness achieved through virtuous living. Aristotle's virtue ethics emphasized character traits (virtues such as courage, temperance, and generosity) and the idea of the "Golden Mean," finding moral virtue in a balance between extremes. These classical Greek frameworks were human-centered and reasoned, but they did not entirely reject the divine. Aristotle, for instance, believed in a natural order and purpose (telos) inherent in the world, which a virtuous person aligns with.

Religious moral frameworks around the world provided authoritative codes of conduct believed to be ordained by a higher power. Abrahamic religions (Judaism, Christianity, Islam) grounded ethics in divine law: the Ten Commandments in the Hebrew Bible, for example, set out fundamental prohibitions (against murder, theft, falsehood, etc.) and duties (honoring parents, the Sabbath) that have guided believers for millennia. Similarly, the Quran provides moral guidance and legal principles for Muslims, and the teachings of Jesus in the New Testament (such as the blessings on the meek, merciful, and peacemakers in the Sermon on the Mount) have profoundly shaped Christian ethical understanding. These frameworks often rely on the concept of a covenant or binding moral order, the idea that humans are bound to God (or gods) by promises and commands [4]. The etymology of "religion" (Latin religare, "to bind") reflects how closely morality was tied to religious duty [4]. In many of these traditions, morality was inseparable from spirituality: to live a good life meant to live in accordance with God's will or the sacred order of the universe.

Beyond the West, other rich ethical systems were developing. In ancient China, Confucianism offered a distinctly social and family-centric moral vision. Confucius (Kongzi), around the 5th century BCE, taught that a harmonious society rests on ren (humaneness or benevolence) and li (ritual propriety). Notably, Confucius did invoke a version of the Golden Rule ("What you do not wish for yourself, do not do to others") as a guiding principle [3]. However, he also stressed the importance of context; different relationships carry different duties (filial piety toward parents, loyalty to rulers, etc.), so ethical behavior means fulfilling one's role appropriately [3]. Meanwhile, in ancient India, traditions like Hinduism, Buddhism, and Jainism approached ethics through concepts such as dharma (moral duty or cosmic law) and karma (the moral consequences of actions). Buddhism, for instance, articulated ethical precepts (such as nonviolence, truthfulness, and the avoidance of intoxicants) and encouraged compassion for all living beings, grounded in an understanding of suffering and the interdependence of life. The Buddha's Noble Eightfold Path included right action, right speech, and right livelihood, essentially an ethical code aiming at the cessation of suffering. Jainism took the principle of ahimsa (non-harm) to its strictest interpretation, promoting extreme care to avoid violence to any creature.

Despite their differences, these classical and religious frameworks share common ethical themes that continue to resonate: value for life, honesty, justice, compassion, and the importance of intention. They often posit an objective moral order, whether rooted in divine decree, nature, or reason, to which humans are accountable. For example, Socrates famously posed the question (in Plato's Euthyphro) of whether something is good because God wills it, or God wills it because it is good, illustrating a debate on whether morality is external to the divine or defined by it [3]. In either case, early thinkers assumed moral truth was discoverable and universal in some sense.

Another enduring concept from Ethics 1.0 is Natural Law, particularly developed in the classical and medieval West. This idea, later articulated by Christian theologians such as St. Thomas Aquinas, holds that morality is woven into the fabric of reality and human nature, accessible to reason and aligned with God's plan. Aquinas synthesized Aristotelian philosophy with Christian doctrine, arguing that what is "natural" (contributing to human flourishing) is morally right [3]. For example, preserving life is natural (so murder is wrong), seeking knowledge is natural (so education is a right), etc. [3]. Such views show the blending of classical reason with religious belief, illustrating how Ethics 1.0 was not monolithic but dynamic, as seen when medieval scholars tried to reconcile ancient philosophy with religious ethics.

In summary, Ethics 1.0 provided humanity's first moral compass. Whether through the commandments of gods or the arguments of philosophers, pre-modern societies established foundational concepts of virtue, duty, justice, and care that still ground ethical thought. These classical and religious frameworks were often community-oriented, stressing obligations to family, tribe, or faith community and character-oriented, encouraging the cultivation of virtue and piety. They assumed an ordered cosmos where morality ultimately made sense (under divine providence or rational natural law). Of course, these systems also had limitations: they could be hierarchical, exclusionary (with in-groups versus out-groups), and slow to adapt to social change. Nevertheless, they form the bedrock on which later ethical "upgrades" would build. As we shall see, Enlightenment thinkers both drew upon and reacted against these traditional moral visions, ushering in Ethics 2.0, a more individualistic and universal conception of ethics in the age of reason.

### 5.2.2 Ethics 2.0: Enlightenment and Human Rights

Ethics 2.0 marks a profound shift in moral philosophy and social thought, emerging from the Enlightenment era (17th–18th centuries) and maturing through modern movements for human rights and democracy. In this stage, ethics increasingly moved away from external authorities like religion or tradition and toward human reason, individual autonomy, and universal principles. The Enlightenment project spread across Europe (and later the world), championing ideas such as the inherent rights of individuals, the value of liberty and equality, and the use of reason and science to reform society. This was a time when thinkers attempted to recode ethics in secular, rational terms, and its legacy includes cornerstone documents like the Declaration of the Rights of Man (1789) and the Universal Declaration of Human Rights (1948).

One of the pillars of Enlightenment ethical thought was Immanuel Kant's deontological philosophy. Kant (1724–1804) argued that morality is grounded in rational duty and the intrinsic dignity of persons. He famously formulated the Categorical Imperative, a supreme moral law based on reason: one formulation instructs to act only according to that maxim whereby you can, at the same time, will that it should become a universal law [3]. In other words, one should do only what one could consistently encourage everyone to do, a test of universality and consistency in moral rules. Another formulation says we must treat humanity, whether in oneself or others, always as an end and never merely as a means. This was a clear call to respect the autonomy and worth of everyone. Kant's approach was secular and rationalist. He believed that all rational beings could discern moral law through reason alone, without reference to divine command or consequences. An action, for Kant, is truly ethical only if done from the motive of duty, not out of emotion or self-interest [3]. This emphasis on intention and principle was a stark departure from earlier frameworks that might stress obedience or virtue in a more habitual sense. Kant thus laid the groundwork for modern notions of human rights and dignity, asserting that every person, by virtue of rational agency, has inviolable moral worth.

Alongside deontology, the Enlightenment also gave rise to utilitarianism, another influential ethical theory that spread in the 18th and 19th centuries. Thinkers such as Jeremy Bentham and John Stuart Mill argued that the morality of actions should be judged by their consequences, specifically by the principle of utility, or the greatest happiness. Bentham succinctly framed it as seeking "the greatest happiness of the greatest number," turning ethical decision-making into something like a calculation of pleasure versus pain. Mill refined utilitarianism by emphasizing qualitative differences in pleasures (intellectual and moral pleasures being higher than mere physical ones). By the 19th century, utilitarianism had become a powerful tool for social reformers. It was invoked in debates about political policy, economics, and law to maximize overall well-being. Importantly, utilitarianism and Kantian ethics, despite their differences (outcome-focused versus duty-focused), both supported Enlightenment-era "liberal" values. Each, in its way, underpinned arguments against the feudal hierarchy and tyranny: Kantian ethics underscored individual rights and equality (since every rational being is an end in itself), and utilitarianism offered a common currency of welfare that applied to all individuals (each person's happiness counts equally in calculation). Together, these theories provided moral justification for expanding suffrage, education, and humane legislation. Indeed, it has been noted that both Kantianism and Utilitarianism buttressed the liberal political developments of the Enlightenment, which reconceived the individual as worthy of rights and capable of self-governance [3].

The Enlightenment period also witnessed the first articulations of what we now call human rights. Thinkers like John Locke had earlier argued for natural rights to "life, liberty, and property," influencing revolutionary movements in America and France. In 1776, the American Declaration of Independence asserted self-evident truths about equal creation and endowed rights (to life, liberty, and the pursuit of happiness). The French Declaration of the Rights of Man and of the Citizen (1789) proclaimed liberty, property, security, and resistance to oppression as "natural and imprescriptible" rights. These documents were revolutionary in grounding political order not in divine right or tradition, but in the inherent entitlements of individuals as members of the human family. The concept of universal human rights was further solidified after the horrors of World War II. In response to the atrocities of the Holocaust and the war, the world came together to affirm a baseline of human dignity that must never be violated. The United Nations' Universal Declaration of Human Rights (UDHR) of 1948 is a milestone of Ethics 2.0, declaring that "all human beings are born free and equal in dignity and rights" and enumerating rights to life, education, work, free expression, and more. This was, in effect, a global moral consensus that specific values like the sanctity of life, freedom from torture and slavery, and equality before the law are not culturally relative but universal. The UDHR asserted "inalienable" moral rights of every person to basic goods and freedoms [3]. It represented the culmination of Enlightenment ethical ideals on a global stage, especially the principle that everyone has inherent worth that society and government must respect.

It's worth noting that Ethics 2.0 was not purely a product of Western thought; it had echoes and parallels worldwide. For instance, concepts of human equality and anti-imperial critique emerged in various colonies during their struggles for independence (inspired, in part, by Enlightenment rhetoric). Leaders like Mohandas Gandhi, though rooted in different religious and cultural contexts, advocated universal principles of nonviolence and human dignity that resonated with global human rights ideals. Nonetheless, Enlightenment ethics did entail a secularization of morality, a move to base ethics on reason or shared human values rather than on religious doctrine. This sometimes put Enlightenment thinkers at odds with traditional authorities (Kant, for instance, faced pushback for his critical philosophy and was a devout

Christian in his personal life, but argued ethics should stand on reason alone). Over time, secular ethics expanded into fields like social justice, abolitionism, and feminism. The 18th and 19th centuries saw ethical arguments made against slavery (on grounds that it violates natural rights and human equality) and for women's rights (asserting the equal moral and rational nature of women). These were practical extensions of the core Enlightenment idea: every human being has a moral status that demands respect.

Another crucial development in this era was the notion of moral progress and the perfectibility of society. Enlightenment optimism suggested that through education, science, and sound governance, humanity could overcome superstition, intolerance, and cruelty. This belief fueled reforms in criminal justice (Beccaria's fight against torture and cruel punishments), in economics (critiques of poverty and early socialism), and in expanding the circle of moral concern. The idea of "progress" was itself a new lens: medieval and ancient worldviews often saw history as cyclical or declining from a golden age, whereas Enlightenment thinkers posited that humanity could advance morally as well as technically. This mindset carries over into modern narratives about abolishing war, ending hunger, or achieving global human development.

By the mid-20th century, Ethics 2.0 principles had deeply influenced international norms and laws. The UN Charter (1945) and subsequent human rights treaties (such as the International Covenants on Civil and Political Rights and on Economic, Social and Cultural Rights in the 1960s) created an international moral/legal framework primarily based on Enlightenment-humanist values. Philosophers continued to refine these ideas: for instance, John Rawls's A Theory of Justice (1971) sought to reconcile liberty and equality through a thought experiment (the veil of ignorance), in the spirit of rational universalism. Rawls essentially updated the social contract tradition for a modern age, arguing that just principles are those that everyone would agree to in a fair initial situation, leading to affirmations of equal fundamental liberties and socioeconomic equality (unless inequality benefits the least well-off). His theory echoes Kant's respect for persons and the utilitarian consideration of welfare, blending them into a contemporary view of justice.

In summary, Ethics 2.0 established the primacy of individual rights, universal moral laws, and human equality as the basis for ethics. This was the era that enshrined freedom of conscience, expression, and belief, that fostered the end of institutionalized slavery and movement toward democratic governance, and that led to the concept of crimes against humanity (implying universal moral standards that no government can transgress). It moved moral authority from heaven to earth into parliaments, courts, and the court of public reason. Of course, the lofty ideals were (and are) often betrayed or unrealized in practice, and critiques have been made (from Nietzsche's dismissal of universal morality as herd mentality, to postcolonial critiques that Enlightenment "universalism" sometimes masked Western imperial biases). Yet the core notion that every person carries inherent dignity and rights is arguably the most potent moral legacy of Ethics 2.0. As we turn to Ethics 3.0 and 4.0, we will see that these Enlightenment values remain crucial touchstones; for instance, debates about digital privacy or AI bias often invoke rights and equality. The challenge is to apply those 18th–20th-century ideals to the novel problems of the 21st-century digital and AI revolution.

### 5.2.3 Ethics 3.0: Digital Ethics and Information Age Concerns

With the late 20th century and the rise of computing and the Internet, humanity entered the Information Age, bringing a host of new ethical questions that constitute Ethics 3.0. This period, roughly from the 1970s onward, saw the advent of personal computers, global telecommunications, and eventually the World Wide Web and social media. Technology began to mediate every aspect of daily life communication, commerce, education, and healthcare, creating unprecedented ethical challenges around information privacy, intellectual property, digital rights, and the power of algorithms. Scholars and professionals responded by developing fields like computer ethics, cyber ethics, and information ethics to grapple with these issues. Ethics 3.0 is characterized by extensions of earlier moral principles to the digital realm, as well as recognition of entirely new dilemmas that have no analogue in pre-digital times.

One of the earliest explicit recognitions of digital-era moral issues came with the rise of computer ethics in the 1970s and 1980s. As computing moved from a niche enterprise into business, government, and home use, thinkers like Norbert Wiener (as early as the 1950s) and later James Moor (with his seminal 1985 paper "What Is Computer Ethics?") pointed out that computers introduced policy vacuum situations where existing ethical and legal frameworks didn't yet address the capabilities of technology. By the mid-1970s, new applied ethics fields were burgeoning: bioethics addressed medicine and biology, while computer ethics addressed issues such as software piracy, data privacy, and the social impacts of automation [3]. Professional associations (such as the Association for Computing Machinery) began drafting codes of ethics for computing professionals, highlighting responsibilities in designing and using information systems. Notably, a 1970s insight was that technology is not value-neutral; how we build and use digital systems can embed biases and value choices. Thus, the seeds of later debates on algorithmic fairness were planted in these early computer ethics discussions, even if the technologies then were far simpler than today's AI.

As the digital revolution accelerated, several core ethical concerns of the information age became apparent:

#### 5.2.3.1 Privacy

The capacity of computers to store, process, and share vast amounts of personal data raised acute concerns about privacy and surveillance. In the 1990s, when the Internet became widespread, people realized that digital footprints (web browsing, emails, search queries) could be tracked and mined. Legal frameworks such as data protection laws (e.g., Europe's GDPR, enacted in 2018) are essentially ethical responses to this concern. Privacy, once protected largely by practical obscurity (information being hard to collect), became something that required explicit safeguarding in an online world. Scholars like Alan Westin defined information privacy as the right to control one's personal data. The Cambridge Analytica scandal in 2018, in which a political consulting firm harvested data from 87 million Facebook users without consent to influence elections, exemplifies the magnitude of privacy violations in the digital age. Public outcry over such incidents underscores how strongly societies value privacy as an ethical issue. Indeed, the Cambridge

Analytica case fueled "ongoing concerns about the negative consequences of digital technologies," demonstrating how the misuse of personal data can undermine democratic processes [5, 6].

#### 5.2.3.2 Data Security and Cybersecurity

Alongside privacy, ensuring the security of digital information became a moral imperative. Massive data breaches that expose millions of users' information (from credit card numbers to health records) not only harm individuals but also erode trust in institutions. There is an ethical onus on organizations to protect the data they collect; negligence in cybersecurity can be seen as a moral failing. Cybersecurity also touches national security, and civil liberties debates rage over government surveillance versus citizens' rights, exemplified by the Snowden revelations of 2013, which exposed extensive government data collection programs.

#### 5.2.3.3 Intellectual Property and Access

The digital age made copying and distributing information trivially easy, forcing a rethinking of intellectual property (IP) rights. Ethical questions arose about software piracy, music and video sharing (remember Napster), and later, the balance between protecting creators and providing public access to knowledge. The concept of the Digital Divide, the gap between those with access to digital technologies and those without, became a social-ethical issue, highlighting potential inequalities in information access. If knowledge is power, then unequal access to the internet and digital tools can exacerbate social inequalities, raising issues of justice. Many views bridging the digital divide (e.g., bringing internet to developing regions and ensuring tech accessibility for people with disabilities) are seen as an ethical obligation in pursuit of global fairness.

#### 5.2.3.4 Freedom of Expression versus Harmful Content

The internet connected the world, enabling unparalleled freedom of speech but also the rapid spread of harmful content like hate speech, extremist propaganda, and disinformation. The ethical tightrope involves protecting free expression (a core human right from Ethics 2.0) while mitigating real harms that speech can inflict in a networked public sphere. The rise of social media in the 2000s intensified this debate: companies like Facebook and Twitter have had to develop policies on content moderation, grappling with whether and how to remove content that is violent, false, or incites hatred. The "fake news" phenomenon and election interference via social media (e.g., Russian bots and trolls influencing the 2016 US election) show that misinformation online isn't just a nuisance; it can undermine democratic discourse and public trust in knowledge. As one UNESCO analysis notes regarding deepfakes (AI-generated fake videos), we are facing a "crisis of knowing itself," not just isolated falsehoods, but a broader erosion of society's ability to agree on truth. Deepfakes blur the line between reality and fiction so convincingly that seeing and hearing are "no longer believing," posing a fundamentally new epistemological and ethical challenge. The moral response may require bolstering media literacy, developing verification technologies, and perhaps legal frameworks for AI-generated content, all of which are active areas of discussion now [1].

#### 5.2.3.5 Algorithmic Bias and Discrimination

As algorithms began to guide high-stakes decisions (who gets a loan, what news people see, where police patrol, etc.), evidence mounted that they could reflect and even amplify human biases. Ethics 3.0 began asking: If a hiring algorithm is biased against women or minorities due to skewed training data, who is accountable, and how can it be fixed? For instance, investigative work in the 2010s showed that machine learning models used in US court systems for recidivism prediction were biased against Black defendants. Scholars like Cathy O'Neil (in Weapons of Math Destruction, 2016) highlighted numerous such cases where algorithms, sold as "objective," in fact were a "veneer of objectivity" that concealed biases. Michael Sandel, a political philosopher, pointed out that many algorithms "replicate and embed the biases that already exist in our society," giving them a scientific aura that can make discrimination harder to spot. For example, an AI that learns from historical hiring data might inadvertently learn sexist or racist patterns if past hiring was biased. The ethical challenge is ensuring algorithmic fairness, meaning AI systems should be designed and tested to avoid unlawful or unjust discrimination. This is part technical (developing bias-detection and mitigation methods) and part moral (agreeing on what "fair" outcomes look like in context, which can be complex and context-dependent) [2].

#### 5.2.3.6 Accountability and Transparency

With decision-making increasingly delegated to complex software, accountability became a pressing issue. If an algorithm denies someone a credit card or parole, on what basis did it decide, and who is responsible if the decision is wrong or unfair? The concept of algorithmic accountability emerged, referring to the idea that those who design and deploy algorithms must be answerable for their outcomes [2]. Ideally, algorithms should be designed to eliminate bias and evaluate only relevant factors, but in reality, they often act as "black boxes"; their inner workings are opaque even to their creators [2]. This opacity challenges individuals' ability to contest decisions that affect them (a due process concern) and societies' ability to trust AI systems. In 2016, the European Union addressed this by introducing a "right to explanation" in the GDPR for algorithmic decisions, reflecting an ethical stance that people deserve transparency and recourse when automated systems make essential judgments about them. There is also movement in courts: for example, a Wisconsin Supreme Court case ruled that if judges use risk assessment algorithms in sentencing, defendants must be informed about the algorithm's accuracy and limitations. This points to a growing norm: AI systems should be auditable, transparent, and subject to oversight, aligning with principles explicitly stated in guidelines, such as UNESCO's recommendation on AI ethics (which calls for AI to be traceable and explainable) [7]. Algorithmic accountability is about both assigning responsibility when AI causes harm and ensuring due diligence in AI design to prevent harm in the first place.

#### 5.2.3.7 Cyber-social Issues and Digital Rights

The internet also gave rise to new forms of social interaction and identity, raising questions like online harassment (ethically, how to balance free speech with protecting individuals from cyberbullying and abuse?), anonymity (when is it ethical to remain anonymous online, and how to handle anonymous wrongdoing?), and digital selfhood (do we have a "right to be forgotten" online, to delete our digital traces after a time?). The European Court of Justice recognized such a right in 2014, requiring search engines to de-index certain personal information upon request, framing it as a matter of privacy and reputation in the digital age. Additionally, concepts of digital rights now extend human rights into cyberspace. For instance, some argue that access to the internet itself is a human right, given its essential role is freedom of expression and access to information.

During the Ethics 3.0 era, global variations in digital ethics approaches also became evident. Different countries prioritized different values: the EU has tended to emphasize privacy and rigorous data protection (e.g., GDPR), seeing privacy as foundational to personal dignity and autonomy. The US, by contrast, historically leaned more on free-market and free-speech principles, often letting tech companies self-regulate and only belatedly considering federal privacy legislation, an approach that Sandel and others criticized as leaving potential "redlining" and discrimination by AI systems unchecked without proactive rules [2]. In China, the approach has been to harness digital technology for both economic innovation and social control, leading to ethical debates over the government's extensive surveillance systems and social credit scoring, which are officially justified by values of social stability and public security but elsewhere are critiqued as Orwellian. These differences underscore that even in the digital realm, cultural values and political systems shape what ethical issues are prioritized. A 2025 study of government AI ethics policies found broad consensus on some core principles (such as promoting responsible technology and protecting digital rights), but significant divergence in specific topics and priorities across countries [5]. For example, while most agree on principles such as transparency and fairness, some governments emphasize digital sovereignty and economic development, while others emphasize individual rights and civil liberties [5]. This calls for international dialogue to find common ground, something we'll see intensifying as we move into Ethics 4.0.

In sum, Ethics 3.0 addressed the moral terrain of computers and information networks. It forced society to reconsider how classical rights (privacy, free expression, property) apply in a world of instant communication and data ubiquity. It also introduced new ethical categories: data ethics, cyber warfare ethics (think of questions around hacking in conflict), virtual reality ethics (can virtual actions be immoral if no one "real" is harmed?), and more. The development of digital ethics has been reactive in many cases; regulations and norms often follow crises or scandals. For instance, only after major privacy breaches and public scandals did robust data protection laws gain momentum [5]. Nonetheless, by the 2020s, digital ethics had matured into a recognized field intersecting law, philosophy, and technology design. Governments worldwide were issuing AI ethics guidelines and creating digital councils. Key themes like transparency, accountability, fairness, and human-centric design became common in these guidelines [7].

Importantly, Ethics 3.0 set the stage for the challenges of Ethics 4.0. As AI systems grew more advanced, moving from rule-based algorithms and statistical models into self-learning neural networks, the ethical issues of the digital age did not disappear; they intensified. Bias in algorithms became increasingly challenging to detect in black-box neural networks; privacy threats loomed larger as AI's ability to infer sensitive information from seemingly innocuous data grew; misinformation took on new dimensions with AI-generated fake content. Thus, the transition to Ethics 4.0 can be seen as a continuity of Ethics 3.0's issues at a greater scale and with higher stakes, plus the addition of qualitatively new dilemmas around autonomy and agency of AI. We now turn to Ethics 4.0, where we will examine how the rise of intelligent, autonomous systems is challenging us to further evolve our ethical frameworks.

### 5.2.4 Ethics 4.0: Artificial Intelligence, Autonomy, and Algorithmic Accountability

Ethics 4.0 is the emerging ethical paradigm for an era in which AI is ubiquitous and increasingly autonomous. If Ethics 3.0 dealt with computers as tools managing information, Ethics 4.0 grapples with AI as agents that can make decisions, learn, and act in the world with a degree of independence. This raises profound questions: How do we ensure AI systems align with human values and do not cause harm? Who is responsible when an autonomous machine causes damage or injury? How can we preserve human agency and oversight when algorithms execute many decisions faster and more complexly than any human could follow? And even as AI grows more sophisticated, might we owe ethical duties to machines (e.g., if they were to become sentient)?

One central focus of AI-era ethics is algorithmic accountability and transparency. As discussed, AI algorithms, especially those using machine learning, often operate as "black boxes" whose internal logic is complex to interpret. Yet these algorithms increasingly influence consequential outcomes: hiring, college admissions, loan approvals, medical diagnoses, legal sentencing, and more. The ethical stance widely advocated is that AI systems should be transparent, explainable, and subject to human oversight [7]. The European Commission's High-Level Expert Group on AI, for example, issued Ethics Guidelines for Trustworthy AI (2019), which outline seven key requirements for AI, derived from four foundational ethical principles: respect for human autonomy, prevention of harm, fairness, and explicability [8]. Concretely, this means AI should include human agency and oversight (people should ultimately be in control, able to intervene or understand AI decisions), be technically robust and safe, ensure privacy and data governance, be transparent (explainable to those affected), be non-discriminatory and fair, and be socially and environmentally beneficial while being accountable. These principles echo long-standing values but are tailored to AI's context; for instance, "explicability" reflects the new need to shine light into the black box, so to speak. Achieving this is technically challenging; researchers in AI are developing explainable AI (XAI) techniques, and some jurisdictions consider requiring algorithmic impact assessments or audits (analogous to financial audits) to verify compliance with ethical norms.

Human oversight remains a cardinal principle in AI ethics. No matter how autonomous or intelligent an AI system is, many argue there should always be a human ultimately accountable for its actions, sometimes phrased as keeping "human-in-the-loop" or at least "human-on-the-loop." UNESCO's 2021 Recommendation on AI Ethics explicitly states the importance of human oversight such that AI systems do not displace ultimate human responsibility and accountability [7]. For example, in healthcare AI, while an algorithm might scan X-rays and flag possible tumors, the final diagnosis and treatment plan should ideally be confirmed by a human doctor, both for safety and to ensure empathy and contextual judgment. In judicial settings, when risk assessment algorithms suggest sentences or parole decisions, a human judge should review them rather than blindly defer to the machine. The efficiency or performance gains that AI might offer sometimes challenge this principle: if an AI consistently makes more accurate decisions than humans (say, in driving or diagnostics). Should we still insist on human control at the expense of some effectiveness? Ethicists essentially say yes, at least in matters involving fundamental rights or life-and-death stakes, because autonomy and moral responsibility are at issue. Machines do not (yet) bear moral responsibility; humans do. Thus, retaining meaningful human agency in AI decision-making processes is crucial to avoid what some call the "responsibility gap," in which something goes wrong, and humans shrug and say, "the computer made the decision."

Bias and fairness in AI have also become high-profile ethics 4.0 issues. Investigations into commercial AI systems have revealed problematic biases from facial recognition systems that perform poorly on darker-skinned or female faces (leading to wrongful identifications), to natural language models that exhibit gender or racial biases learned from the internet text they were trained on. For instance, studies found that specific AI hiring tools discriminated against women by learning from past hiring data that favored men in tech roles. The ethical imperative is to ensure AI does not perpetuate or worsen societal inequalities. This has led to calls for diverse and representative training data, bias testing, and inclusion of fairness criteria in algorithm design. Some researchers propose formal fairness definitions (such as parity in specific error rates across groups), though implementing them in practice can be tricky, as different fairness criteria can conflict. Still, the consensus is that AI ethics must prioritize fairness and non-discrimination, aligning with the broader human rights framework. In fact, global bodies are weighing in: the UNESCO AI ethics recommendation emphasizes Fairness and Non-Discrimination as one of its ten principles, urging that AI "promote social justice, fairness, and non-discrimination" and benefit all, regardless of race, gender, or other characteristics [7].

Another key concern is autonomous decision-making in critical domains. Take autonomous vehicles as a prominent example: self-driving cars must make split-second decisions that carry moral weight (the classic "trolley problem" has been invoked to imagine scenarios in which a car must choose between hitting two different obstacles, each with potential human casualties). Though some argue that the trolley problem is an oversimplification for AVs [9], it sparked debate about how to encode ethical decision rules into machines. Manufacturers and regulators must consider questions such as: Should an AV prioritize the safety of its passengers or pedestrians in an unavoidable crash? How should it value the lives of many versus one? One might hope to avoid these situations altogether through better design (some analysts suggest the trolley problem is a misguided framing for AVs, since real accidents might not allow for such conscious "choice" by the AI) [9]. Nonetheless, society expects that AVs will be at least as safe, and preferably safer, than human drivers, and that they will be transparent in how they make critical decisions. Germany, for instance, published ethical AV guidelines stating that an AV should not discriminate based on personal features (age, gender, etc.) in an accident scenario, essentially forbidding, say, the explicit valuing of one person's life over another's based on such traits. They also said that property and animal damage should be preferred to human injury, if possible. These efforts show an attempt to pre-program ethical priorities into autonomous systems. More broadly, as autonomy spreads (delivery drones, automated trading agents, AI medical triage), similar issues arise regarding predefining ethical boundaries and fail-safes.

The concept of "meaningful human control" has also emerged, especially in the context of autonomous weapons (lethal autonomous weapon systems, or LAWS). Many ethicists and policymakers insist that life-or-death decisions in warfare must not be fully delegated to algorithms; a human should make the final kill decision. This is both a moral stance (taking a human life is such a grave act that it demands human deliberation) and a safety concern (preventing unanticipated escalation or mistakes). There is ongoing debate at the United Nations about banning or regulating killer robots, precisely because fully autonomous weapons raise questions of accountability (if an AI drone mistakenly bombs civilians, who is accountable? The commander who deployed it? The programmers? The manufacturer? The AI itself cannot be punished or held liable). This accountability gap is a strong argument for keeping humans in control of lethal force decisions. It also illustrates a general rule for Ethics 4.0: the greater the stakes and the uncertainty, the stronger the case for human oversight.

Transparency in AI goes hand in hand with accountability. One interesting frontier is AI explainability: developing methods that enable AI to explain its reasoning in human-understandable terms. This is challenging with complex deep learning models. Some progress involves creating simplified surrogate models or highlighting which input features most influenced a decision. For instance, a medical AI might highlight sections of an X-ray it considers indicative of a tumor to explain its diagnosis. Explainability is ethically significant because it is tied to trust; people are more willing to accept and use AI if they understand why it behaves as it does. It's also tied to rights; as noted, the EU's regulations hint at a "right to explanation," aligning with a citizen's right to contest decisions. Moreover, transparency is a partial antidote to biases: if we can inspect an algorithm's process, we may catch undesirable patterns or decision rules (like if a loan AI was effectively redlining by considering zip code as a proxy for race in lending decisions – an explainable model could reveal that and allow a correction).

However, complete transparency can conflict with other principles. Sometimes exposing an AI's inner workings might violate privacy (if it reveals information about training data) or enable gaming the system (if malicious actors learn how to fool it). So, there is an ethical balancing act: appropriate transparency enough to ensure accountability and trust, but not so much as to cause other harms. UNESCO's principle on Transparency and Explainability acknowledges such tensions. It suggests that the level of transparency should be appropriate to the context, given that there may be trade-offs with safety or privacy [7].

Another Ethics 4.0 theme is the impact of AI on autonomy and human well-being. By independence here, we mean human autonomy, our ability to make choices without undue manipulation. AI's prowess in nudging behavior (through targeted ads, recommendation algorithms that shape what information we see) puts human autonomy at risk in subtler ways. For example, personalized content algorithms on social media create echo chambers and can gradually radicalize individuals by feeding them more extreme content to maximize engagement. Ethically, this raises concerns about manipulation and the erosion of the informed, free decision-making that underpins a democratic society. If AI systems optimize purely for engagement or profit without ethical constraints, they might exploit human psychological biases (such as showing incendiary content to capitalize on outrage, which captures attention) at the cost of social cohesion and individual critical thinking. Some ethicists argue for algorithmic ethics in design that platforms should have a duty to consider effects on users' autonomy and society, not just their bottom line. This could mean adjusting algorithms to promote diverse content exposure (to counter echo chambers) or explicitly demoting proven falsehoods even if they are "clicky." It's a debate between pure laissez-faire tech and a more human-centric approach where technology is built to support human values, not undermine them.

The global governance of AI is another facet of Ethics 4.0. Because AI technology and its impacts cross borders, there is recognition that international cooperation and norms are needed. UNESCO's Recommendation on the Ethics of AI (2021) is one example of a global standard-setting effort that calls on all member states to implement core values such as human rights, fairness, sustainability, and peace in their approaches to AI. It also goes beyond principles to propose concrete policy areas (such as data governance, education, culture, and the environment) in which ethical action is needed. The Recommendation emphasizes that human rights and dignity are the cornerstone, and principles such as transparency, fairness, human oversight, accountability, privacy, and sustainability must guide AI development. The fact that 193 countries agreed to this document shows a global convergence on high-level ethical intent, even if implementation will vary. Other global initiatives include the OECD's AI Principles (2019), the Global Partnership on AI (GPAI), which fosters international collaboration on AI ethics, and work toward a possible international treaty on AI (though none exists yet specifically for AI, unlike, for example, treaties on landmines or climate change). A challenge is reconciling different political philosophies: for instance, China's emphasis on collective security and social harmony can conflict with Western liberal emphasis on individual privacy and freedom. Yet on issues like preventing catastrophic risks posed by super intelligent AI, there may be common ground due to shared existential stakes [7].

Speaking of which, Ethics 4.0 must also confront the long-term and existential risks of AI. Visionaries and scholars (such as Nick Bostrom and Max Tegmark) have warned of scenarios in which AI could surpass human intelligence (superintelligence) and, if misaligned with human values, pose an existential threat to humanity. While this might seem a far-off concern, the rapid progress in AI (e.g., GPT-3 and beyond, capable of human-level or superhuman performance in more domains) keeps the conversation relevant. This is often termed the AI alignment problem: how do we ensure that a super intelligent AI's goals and behaviors will remain aligned with human ethics and well-being, especially when it becomes capable of recursive self-improvement or operating beyond our complete understanding? Bostrom argues that by the time we have human-level AI, it might be only a short step to something far more powerful, and once a superintelligence exists, it could be challenging to control. He gives thought experiments like the "paperclip maximizer," an AI given a simple goal (maximize paperclip production) that, with great intelligence, might pursue it to the extreme of converting all available resources (including human beings) into paperclips unless we have carefully programmed it to value human life. This chilling scenario underscores why aligning AI with human values is not just about eliminating bias in loan approvals, but about embedding fundamentally ethical constraints and understanding in any advanced AI. Ethicists in this domain talk about instilling AI with principles like benevolence, non-maleficence, justice, and perhaps even knowledge of human emotions or rights. Solving the control problem for superintelligence is considered "surprisingly difficult" because almost any concrete goal, if pursued by a machine literally, can lead to undesirable side effects. For example, even an AI whose sole goal is to make humans smile might opt to implant electrodes in our brains to stimulate pleasure centers – accomplishing the goal but obviously in a way we'd find abhorrent. Thus, researchers are exploring how to encode not just goals but also ethical constraints, or to design AI that can learn values from human behavior (inverse reinforcement learning for ethics, or "AI safety" research) [1].

A related futuristic issue is the potential creation of AI with some form of consciousness or personhood. While we are not there yet, some foresee that at least highly advanced AI may exhibit qualities of sentience or self-awareness. If that were to happen, Ethics 4.0 (or perhaps Ethics 5.0) would have to consider extending moral consideration to non-human intelligences. Philosophers ask: If an AI can feel pain or have experiences, would it be ethical to own or terminate it? This is speculative, but it's noteworthy that leading AI ethics centers (such as the Future of Life Institute, mentioned in [10]) are already considering the moral status of artificial minds. Some criteria floated for moral status include consciousness, the capacity to suffer, and personhood attributes such as reasoning and self-concept. There are debates reminiscent of animal rights and past human rights struggles: would creating a conscious AI be akin to creating a new sentient race that could be oppressed or enslaved? On the other hand, many argue that current AI has no consciousness, it's more akin to a sophisticated tool, and caution against anthropomorphizing. Nonetheless, ethical frameworks such as "Robot Rights" or "Machine Ethics" are emerging niches that explore how an AI ought to behave ethically toward humans (machine ethics) and whether humans have duties toward some machines (if they become conscious). One could imagine a future in which advanced AI entities could claim certain rights (such as not being shut down arbitrarily). This remains theoretical but not inconceivable in the long run, and it shows how ethics in the AI epoch might evolve beyond purely human-centered concerns.

Finally, Ethics 4.0 also encompasses the societal impacts of AI: its effects on employment, human relationships, and community structure. Automation by AI threatens to displace workers in many sectors, raising issues of economic justice and the right to meaningful work. Some propose that as AI takes over more labor, societies might need to guarantee livelihoods through measures like a universal basic income, an ethical and policy response to ensure human well-being in an AI-driven economy. Others focus on the importance of re-skilling and education so that people can shift into new roles where human empathy, creativity, and complex judgment are still needed (qualities AI struggles with). Psychologically and culturally, as AI companions or virtual assistants become common (think of companion robots for the elderly or AI tutors for children), there are questions about how these technologies shape human development and relationships. Is it ethical to have a child form an emotional bond with an AI "friend" that isn't genuinely capable of reciprocity? Could relying on AI for emotional support erode human-to-human social skills or community bonds? These are subtle ethical questions without clear answers yet. Still, they underline that Ethics 4.0 isn't only about preventing catastrophic harm, it's also about guiding the kind of society we want to become alongside AI. Ideally, AI should be a tool to augment human abilities and improve quality of life, not a replacement for human connection or a source of dehumanization.

In summary, Ethics 4.0 is about ensuring that our new intelligent agents and autonomous systems are developed and deployed in ways that respect human values, rights, and well-being. It extends the familiar principles of bioethics and digital ethics (do no harm, fairness, autonomy, accountability) to a landscape where decision-makers might not be human. This demands new approaches: interdisciplinary collaboration (tech experts working with ethicists, lawyers, and social scientists), new regulatory paradigms (agile, forward-looking governance rather than slow, reactive law), and possibly new ethical concepts (such as considering the moral status of AI itself or creating global treaties on AI safety). The stakes are enormous because AI's influence is permeating globally, from how information is filtered to how resources are allocated. Yet, there is cause for optimism if we manage it well. With deliberate ethical foresight, AI could help us reduce human bias (e.g., if trained correctly, algorithms could make more impartial decisions than biased news.harvard.edu) [2], increase safety (self-driving cars that drastically cut traffic fatalities), enhance accessibility (AI assistants for people with disabilities), and solve complex global problems (AI in climate modeling or medical research). The following section will delve into these positive possibilities as Opportunities and Transformational Potential, before we also confront the flip side in Challenges, Risks and Ethical Considerations. Ethics 4.0, ultimately, is a call to shape AI's trajectory with wisdom and humanity, ensuring this powerful technology develops under the guidance of an ethical compass that points toward human flourishing.

## 5.3 OPPORTUNITIES AND TRANSFORMATIONAL POTENTIAL

While much of the ethical discourse on AI rightly focuses on mitigating risks, it is equally important to recognize and harness the immense opportunities and positive transformational potential that AI offers. The AI epoch holds promise for addressing some of humanity's most pressing challenges and improving the quality of life across the globe. By highlighting these opportunities, we underscore why the project of Ethics 4.0 is fundamentally optimistic: it is about guiding powerful new tools to enhance human well-being, equity, and sustainability. In this section, we explore several domains where AI is already catalyzing positive change or has clear potential to do so, from healthcare and education to environmental sustainability and beyond, and discuss how ethical deployment of AI can amplify these benefits for society.

### 5.3.1 Healthcare

One of the most life-changing applications of AI is in healthcare. AI systems can analyze medical data (symptoms, lab results, imaging scans) with remarkable speed and often accuracy, assisting doctors in diagnosing diseases earlier and more reliably. For example, AI algorithms have been trained to detect cancers in radiologic images (such as mammograms or lung CT scans), and at times even outperform human specialists in identifying minor anomalies. This can lead to earlier treatment and saved lives. In resource-strapped settings, an AI on a smartphone might help a nurse diagnose a condition that otherwise would lack specialist input. AI-driven analysis of drug compounds and proteins is notably accelerating drug discovery. The AI system AlphaFold solved a 50-year-old challenge in biology by predicting protein structures, which can spur the development of new medications and treatments. The COVID-19 pandemic saw AI used to track outbreaks and even in vaccine research (analyzing vast biomedical data to identify vaccine targets). The ethical upside here is enormous: reducing suffering and saving lives. But it is crucial to implement these technologies inclusively. If, for instance, AI diagnostic tools are made widely available (e.g., via open-source or affordable devices), they could bring advanced healthcare to rural or underserved populations that lack doctors. As UNESCO noted in its AI ethics guidance, these technologies must be assessed for sustainability and equity, including health and social well-being [7]. Properly guided, AI could help fulfill the right to health by democratizing medical expertise and alleviating human resource shortages in healthcare systems.

### 5.3.2 Education

Education is another arena with transformative potential for AI. Intelligent tutoring systems can provide personalized instruction to students, adapting to their learning style and pace. This could revolutionize access to education, as a single AI tutor could teach millions without fatigue, offering one-on-one attention that many children never receive in overcrowded classrooms. AI language translation and speech recognition have already broken barriers. For example, AI-driven translation (such as Google Translate) enables people to access information and educational content across languages, an enormous boon for global knowledge sharing. AI can also assist teachers by handling administrative burdens (such as grading simple assignments) and providing insights into student performance (through learning analytics). An AI might, for example, spot that a student hasn't grasped a particular concept and alert the teacher. All this can lead to more inclusive and practical education, tailored to individual needs. Initiatives are using AI to bring education to refugee children or those in remote areas through adaptive learning apps. Ethically, this aligns with the right to education and reducing inequality. However, it's essential to ensure AI education tools are culturally adaptable, protect student privacy, and are used to augment rather than replace human teachers, who provide mentorship and social-emotional learning that machines can't. When well balanced, the human-AI team in education could yield superior outcomes, and teachers would be freed to focus on creativity and personal mentoring. At the same time, AI handles rote teaching and continuous assessment.

### 5.3.3 Environmental Sustainability and Climate Action

AI is also poised to advance environmental sustainability and climate action significantly, addressing what is arguably the greatest moral challenge of our time: preserving a livable planet. AI's capacity to process massive datasets can improve climate modeling and weather prediction, helping us anticipate and respond to natural disasters. For example, AI tools are being used to predict extreme weather events or to monitor deforestation and track biodiversity using satellite imagery [11]. On the solutions side, AI can optimize energy use: smart grids use AI algorithms to balance electricity supply and demand more efficiently, integrate renewable sources, and reduce waste. AI can also help design more efficient materials and processes (for instance, discovering new materials for better batteries or catalysts that capture carbon). In agriculture, AI-driven precision farming can reduce water, fertilizer, and pesticide use by tailoring interventions to specific crop needs, thereby increasing yields while reducing environmental impact. The World Economic Forum has highlighted AI projects tackling climate change, such as tools to track ice melt or identify pollution sources [11]. These initiatives illustrate AI's "transformational potential" for climate resilience [11]. Ethically, using AI to fight climate change and environmental degradation addresses duties to future generations and to the other species that share our ecosystem. It reflects principles of sustainability, which UNESCO's AI ethics framework explicitly includes as a core value (AI should be assessed for its impact on achieving the UN Sustainable Development Goals) [7]. Indeed, the UNESCO recommendation urges that AI be aligned with goals like environmental flourishing [7]. Promoting AI innovations in clean tech, conservation, and disaster prevention thus stands as a moral imperative to leverage our best tools in the service of the planet.

### 5.3.4 Social Good and Humanitarian Action

In the realm of social good and humanitarian action, AI shows promise as well. Consider how AI and big data analytics were used to manage the Ebola outbreak and, more recently, aspects of the COVID-19 pandemic, analyzing mobility data to forecast disease spread, optimizing delivery routes for aid, etc. Humanitarian organizations are exploring AI for real-time crisis mapping: for instance, using AI to analyze social media or aerial images after an earthquake to identify where people might be trapped or where infrastructure is damaged, enabling faster disaster response. AI language processing can also break down communication barriers in humanitarian operations spanning multiple languages. Another potent use is fighting hunger and poverty: AI can help predict crop failures and trigger early interventions, or identify communities in need of food aid more precisely. AI for social good initiatives around the world have demonstrated creative solutions, such as using machine learning to predict water pump failures in rural Africa (so they can be fixed proactively and communities don't lose access to clean water) or to monitor child malnutrition via automated analysis of photos. These applications align with ethical principles of beneficence (doing good) and justice (targeting help where it's most needed and potentially leveling the playing field). As one AI commentator put it, AI's ability to process vast amounts of data and help humans make decisions is transforming industries and can similarly transform how we tackle social challenges [11].

### 5.3.5 Boost Productivity and Create New Types of Employment

Economically, while AI does pose a disruption to jobs, it also has the potential to boost productivity and create new types of employment, ideally freeing humans from drudgery and opening time for more creative, interpersonal, or strategic work. Historically, technological revolutions have eventually led to growth and new industries; the hope is that AI will augment human labor. When mundane tasks are automated, humans can focus on what they do best: complex problem-solving, innovation, and empathy-driven roles, among other tasks. For example, if AI handles routine accounting, accountants can become more like financial advisors and analysts. If AI can draft basic legal contracts, lawyers can spend more time on nuanced case strategy or client interaction. In manufacturing, robots doing repetitive tasks might allow workers to shift into roles maintaining those robots or in quality control. The opportunity, then, is an economy where efficiency and innovation skyrocket, potentially leading to greater abundance. Ethically, this raises the vision of a world where poverty could be alleviated by higher productivity and where people have more freedom to pursue education, leisure, and creative endeavors, a sort of utopian ideal of technology serving humanity (imagine something like Star Trek's vision, where advanced tech eliminated scarcity). However, these gains won't automatically benefit everyone without conscientious policy; this is where ethics intersects policy to ensure the AI dividend is shared (through fair wages, retraining programs, or social safety nets). If managed well, AI could usher in what some call a new Renaissance, with automation providing material plenty and humans focusing on art, science, and exploration.

#### 5.3.5.1 Empower Individuals

AI's personalized capabilities can also empower individuals in everyday life. Think of accessibility: voice recognition and natural language AI enable people with disabilities (such as visual impairments or limited mobility) to interact with technology and their environment more easily. Voice assistants can turn on lights, read texts aloud, and help compose messages, giving greater independence to those who need assistance. Similarly, AI-based prosthetics and brain-computer interfaces are restoring capabilities to people with paralysis or limb differences, which is a profound enhancement of individual agency and quality of life. Another domain is mental health. AI chatbots and apps are increasingly used to provide psychological support or guided cognitive-behavioral therapy exercises to people who might not otherwise seek help, thereby potentially reducing depression or anxiety. While an AI can't replace a human therapist's empathy, studies show some people feel more comfortable initially talking to a non-judgmental AI "listener," which can be a first step to getting care.

#### 5.3.5.2 Global Comparisons

When it comes to global comparisons, AI can also help societies learn from one another. For example, AI can analyze which public health interventions in one country were most effective and model how they might play out in another country, accounting for local differences. This can inform policymakers worldwide and accelerate progress by avoiding repeated mistakes. The sharing of AI solutions for social problems can be a form of solidarity. One nation's algorithm for optimizing energy use or traffic flow can be adapted and adopted by another. Such collaborative progress can reduce global disparities, an ethical goal consistent with the UN Sustainable Development Goals, which emphasize partnership and knowledge transfer.

#### 5.3.5.3 Speed and Scale

AI's speed and scale are arguably its most significant advantages. It can monitor billions of signals in real-time, be it financial transactions (to detect fraud and corruption), social media posts (to identify hate speech or calls for violence and alert moderators), or environmental sensors (to detect patterns of climate change). This allows proactive responses: stopping a cyber-attack as it happens, countering misinformation before it virally spreads, or evacuating areas before a flash flood strikes. All these contribute to safety and security, core elements of a good life. One could argue that there is an ethical duty to use the best tools available (like AI) to protect people when we can. For instance, if AI can significantly reduce fatal car accidents through autonomous driving or advanced driver-assistance (by eliminating human errors like drunk or distracted driving), many ethicists would say we have a moral imperative to adopt it, given that traffic accidents kill over a million people per year worldwide. Indeed, a Brookings report on autonomous cars suggested that fixating on rare "trolley problem" scenarios obscures the much more relevant point that autonomous vehicles, operating under ethical and safety guidelines, can save countless lives by avoiding ordinary crashes [9]. Similarly, in workplace safety, AI systems can predict equipment failures or dangerous conditions in mines, oil rigs, or factories and alert humans to prevent accidents. In policing and public safety, if used responsibly, AI could help allocate resources better (e.g., predictive analytics identifying not individuals but general areas or times at higher risk of crime so police can patrol preventatively without profiling specific persons) and help analyze non-personal data like gunshot detectors or CCTV for threats, potentially stopping incidents like terrorist attacks though this domain is fraught with the need for careful oversight to avoid privacy infringements or bias, as we'll discuss in the challenges section.

#### 5.3.5.4 Human Knowledge and Understanding

Finally, beyond practical benefits, AI has the potential to contribute to human knowledge and understanding. Through AI, we've made scientific discoveries that would have been extremely hard for humans alone, such as discovering new exoplanets in NASA data via AI pattern recognition, or generating hypotheses for complex scientific problems by sifting through millions of research papers (some AIs are trained to propose novel scientific connections, a bit like an automated researcher). AI can simulate complex systems (such as entire ecosystems or economies) to allow us to test scenarios in silico, leading to deeper insights into cause and effect. In philosophy and cognitive science, building AI even helps us probe what intelligence and consciousness are; in trying to replicate facets of human thought, we learn about ourselves. The co-evolution of AI and human society could, ideally, be a story of mutual enhancement: AI pushes us to refine our ethics and self-knowledge, while we push AI to be a tool that encapsulates our highest aspirations for wisdom and compassion.

To encapsulate, the opportunities of the AI epoch are vast and inspiring. A UNESCO article summarizing AI's rise captured this dual nature: "The rapid rise in AI has created many opportunities globally, from facilitating healthcare diagnoses to enabling human connections through social media and creating labor efficiencies through automated tasks [7]. In those same lines, however, UNESCO cautions about ethical concerns, which we will turn to next. The point is that every one of those opportunities corresponds to a genuine human good: health, social connection, meaningful work, and leisure (through efficiency). A World Economic Forum piece similarly noted that AI's data-processing power "is transforming industries" and that climate change is an area where AI has transformational potential [11]. By applying AI to the correct problems with ethical foresight, we could see breakthroughs: potentially eradicating certain diseases, ensuring food security, tailoring education so every child can reach their potential, and transitioning to renewable energy smoothly.

Yet, realizing these opportunities is not automatic. It requires investment, international cooperation, public-private partnerships, and above all, ethical governance to steer AI development where it can do the most good. This means prioritizing AI research in health and the environment as much as in advertising or defense, for example. It also means thinking about inclusion, ensuring the benefits of AI are accessible worldwide, not just in wealthy, tech-savvy communities. If only affluent schools receive AI tutors and poor schools don't, AI could widen educational gaps rather than close them. If only certain countries control AI advances, it could increase global inequality. So an ethics of AI opportunity includes the ethics of distribution: urging an equitable sharing of AI's fruits.

In conclusion for this section, AI, when aligned with our values and deployed thoughtfully, can be a force multiplier for human aspirations: saving lives, educating minds, fostering connection, protecting our planet, and unraveling mysteries of the universe. It offers a chance to alleviate drudgery and allow humans to focus on more fulfilling endeavors. For perhaps the first time in history, we have a tool that learns and adapts, potentially amplifying our collective intelligence to tackle problems previously deemed intractable. Embracing these opportunities is part of our ethical responsibility to not only prevent harm but also actively promote good. As we pivot in the next section to the challenges and risks, we should maintain this balanced view: the very power of AI that creates risks is also what powers these opportunities. Ethics 4.0 is about maximizing the upside while minimizing the downside, ensuring that the AI epoch is remembered as a time when humanity thrived with its intelligent creations, rather than suffered under them.

## 5.4 CHALLENGES, RISKS, AND ETHICAL CONSIDERATIONS

Despite its extraordinary promise, the AI epoch is fraught with significant challenges, risks, and ethical pitfalls that must be confronted. These range from immediate concerns like biased algorithms and privacy breaches to broader societal threats like mass unemployment, surveillance states, and even the existential risk of unaligned superintelligence. In this section, we delve into the major ethical concerns surrounding AI, building on issues raised in earlier parts and examining them more critically. We also discuss the tensions and trade-offs that complicate these issues. The aim is not to cast AI as a malevolent force (it is a human-created tool, after all), but to shine a light on how misuse, lack of oversight, or unintended consequences could lead to harm. Recognizing these risks is the first step to mitigating them.

### 5.4.1 Algorithmic Bias and Discrimination

One of the most talked-about challenges is algorithmic bias and discrimination. As noted, AI systems can inadvertently learn and perpetuate the biases present in their training data. This can lead to outcomes that systematically disadvantage certain groups, raising issues of fairness and justice. Real-world examples abound: facial recognition tech misidentifying people of color at much higher rates, recruitment algorithms downgrading resumes that include women's colleges or specific keywords associated with female candidates, predictive policing models sending police disproportionately to minority neighborhoods (partly because those areas had more historical arrest data, not necessarily more crime). Such outcomes not only violate ethical principles of equality but can also entrench historical injustice with a high-tech veneer. Sandel insightfully remarked that algorithmic decision-making, which seems objective, often replicates and embeds the biases that already exist in our society [2]. This presents a "veneer of objectivity" that is dangerous because it may lead people to trust biased AI outputs more than they would a biased human judgment [2]. If a computer says someone is high risk (for insurance, employment, parole, etc.), it can carry an unjust aura of scientific credibility [2]. The ethical challenge is multifold: technically, how to detect and correct biases in AI; procedurally, how to allow those affected to seek redress or contest AI-driven decisions; and structurally, how to ensure diversity in AI development so that many perspectives shape design choices. Failing to address algorithmic bias can worsen social disparities, essentially, encoding past discrimination into future decisions. As one quip goes, if AI is fed biased data, it doesn't just mirror our biases, it incorporates and possibly amplifies them in millions of micro-decisions at scale. Therefore, fairness in AI is paramount and difficult: sometimes improving fairness on one metric can worsen it on another (a known issue in algorithmic fairness trade-offs). From an ethical perspective, however, the bottom line is that unfair AI is unacceptable. Biased algorithms erode public trust and can cause real harm, such as denying qualified individuals jobs or loans, or, worse, contributing to wrongful arrests or imprisonments. We have already seen some pushback: for instance, a few jurisdictions banned police use of facial recognition due to bias and civil rights concerns, and big tech companies paused selling such software to police until regulations are in place. The risk is clear: left unchecked, AI could usher in a new era of automated discrimination or "digital apartheid," where specific demographics are routinely treated worse by digital systems.

### 5.4.2 Privacy and Surveillance

Another deeply worrisome risk is the erosion of privacy and the rise of surveillance. AI supercharges surveillance capabilities: machine learning can sift through vast troves of video feeds, geolocation logs, and online activity to track individuals and even predict their behavior. Governments and corporations alike are increasingly empowered to monitor people in granular detail. In some cities, networks of CCTV cameras with facial recognition allow for real-time tracking of anyone's movements. Personal smart devices (phones, virtual assistants, IoT gadgets) are constantly collecting data on where we go, what we say (if listening), what we search for, our heart rate, and more. When aggregated and analyzed by AI, this data yields intimate profiles: AI can infer one's political beliefs, sexual orientation, health status, and more, often with uncanny accuracy [1]. The ethical implications are stark. Privacy is tied to autonomy and dignity; if people are constantly observed, their behavior changes (they self-censor, lose freedom to experiment or dissent). Constant surveillance, especially by state actors, can chill free speech and democracy. China's implementation of AI-driven surveillance and its Social Credit System, in which citizens' behaviors (from jaywalking to online comments) are monitored and scored, thereby affecting their access to services, is frequently cited as a dystopian model of AI-powered authoritarianism. Even in democracies, there's creep: law enforcement using AI to comb social media for "threat scores" on citizens, or governments collecting phone metadata on millions (like the NSA did, revealed by Edward Snowden). The Cambridge Analytica scandal exemplified another privacy risk: personal data from social media was used to micro-target political propaganda, exploiting psychological profiles to manipulate voters [1]. This is a form of mass privacy invasion for ideological influence, and it triggered a backlash for stronger data protection and transparency in political ads. The risk in the future is that, without robust privacy protections, we slide into a world where individuals have little control over their personal information and power over that information is concentrated in a few hands. Ethically, that's a world of diminished individual freedom and increased inequality (those who hold data can greatly influence those about whom the data is). It's telling that UNESCO's first principle in its AI ethics recommendation is the protection of human rights and dignity, with emphasis on privacy [7]. We must contend with questions like: Do we allow facial recognition in public? Should individuals have the right to know and delete what data AI systems have about them (the "right to be forgotten")? How do we regulate the AI-driven data economy (which is largely self-regulated by Big Tech, whose incentives may not align with privacy)? Without answers, the risk is an entrenched surveillance economy that people cannot easily escape.

### 5.4.3 Manipulation and Loss of Human Agency

A closely related concern is the potential for manipulation and loss of human agency. AI systems, primarily through social media algorithms and personalized content feeds, can heavily influence what information people consume and even their beliefs and choices. We saw in 2016 how targeted misinformation (possibly guided by AI-curated profiles) affected election discourse. But even at a subtle level, recommendation algorithms on YouTube or Facebook can create "rabbit holes," steering someone from innocuous content toward extreme or conspiratorial content because the algorithm learns that more extreme videos keep specific viewers engaged longer (thereby serving the platform's ad revenue goals) [9]. This raises the question: Are we freely choosing our media diets, or are AI curators effectively choosing for us, maximizing our biases and hooking our attention with outrage or titillation? As Sandel framed it, beyond privacy and bias, the deepest question is the role of human judgment: "Can smart machines outthink us, or are certain elements of human judgment indispensable…?" [2]. If we rely too much on AI decisions, we risk atrophying our own critical thinking and moral agency. For instance, if an AI newsfeed algorithm decides what is true or false for users (flagging or downranking "false" content), that might fight misinformation but could also become a form of centralized control over knowledge unless done transparently and accountably. On the other hand, without any filtering, misinformation runs rampant, creating tension between guiding information flow and preserving user autonomy. Misinformation and deepfakes (AI-generated fake images, audio, and videos) deserve special attention as a threat: they could lead to a "post-truth" era where seeing isn't believing and public trust in media collapses [7]. A UNESCO piece warned that deepfakes don't just spread lies; they erode the very mechanisms by which societies construct shared understanding [7]. The ethical risk here is a collective one: the undermining of truth and democracy. If electorates cannot even agree on basic facts because AI-generated fake news is omnipresent and convincing, democratic deliberation becomes nearly impossible. We may face more situations like the 2020 "Infodemic" of COVID-19 misinformation that cost lives by convincing some people to reject vaccines or embrace bogus cures. Hence, there's a dire need for AI tools to authenticate content (watermarking AI outputs, for example) and for education to raise "AI literacy" so people learn to navigate an AI-mediated reality critically [7]. The risk is that technology moves faster than society's capacity to adapt, overwhelming our epistemic defenses.

### 5.4.4 Job Displacement and Economic Disruption

A primary societal concern is job displacement and economic disruption caused by AI automation. Unlike past technological revolutions that automated physical labor or repetitive tasks, AI threatens white-collar and skilled jobs as well, from driving trucks to reading medical scans to drafting legal contracts. Studies have estimated that significant percentages of jobs are at high risk of automation in the coming decades. The ethical issue is twofold: the well-being of displaced workers and the implications for inequality. If AI primarily augments high-skill labor and automates low-skill labor, we could see further polarization, with wealth owners of AI and highly skilled professionals reaping gains. In contrast, millions of others struggle with unemployment or low-wage, precarious jobs. This was already a trend with computerization; AI could exacerbate it drastically. Historically, societies haven't managed transitions of this magnitude without turmoil. Consider how industrialization fueled inequality and social unrest until labor movements and welfare reforms counterbalanced. Without proactive measures (retraining programs, economic safety nets like universal basic income, job-sharing schemes, new job creation in AI-related sectors, etc.), we could face widespread economic insecurity. That, in turn, can lead to political instability, scapegoating of immigrants or minorities, and erosion of social cohesion. Work is not just income; it's also a source of identity and purpose for many. The ethical risk is a future where many feel obsolete, a blow to human dignity. Michael Sandel, in the Harvard Gazette piece, touched on how some jobs requiring empathy and judgment will remain robust [2], but that implies others will not.

Already, AI is transforming workplaces, e.g., algorithms manage gig workers at companies like Uber or Amazon, often in ways that workers find opaque and unfair (e.g., automated firings and punishing metrics that don't account for human needs). Labor rights could be eroded if algorithms "boss" humans around without accountability. Hence, a key ethical consideration is how to integrate AI into the workforce while respecting workers' rights and humanity. Some call for a "human-in-command" approach in workplaces, too, i.e., AI assists but doesn't fully control human workers, and humans can contest AI decisions internally.

### 5.4.5 Autonomous Weapons and AI in Warfare

Another risk domain: autonomous weapons and AI in warfare. This goes beyond privacy or jobs into existential risk territory for some, because delegating lethal decision-making to machines could lower the threshold for conflict (if your side can deploy killer drones with no risk to your soldiers' lives, perhaps leaders might engage in conflict more readily). It could also lead to faster, unanticipated escalation. Imagine two countries' AI defense systems interacting at millisecond speeds, potentially misreading a situation and launching attacks before any human can intervene (as algorithmic high-frequency trading has caused flash crashes in stock markets). There is also the moral horror of a machine making a life-and-death judgment without human compassion or moral considerations. International humanitarian law wasn't written with fully autonomous agents in mind, and it's unclear how concepts such as accountability and proportionality apply. If a war crime is committed by an autonomous drone, who is liable? The commander? The programmer? The manufacturer? Or is it an unpunishable gap? The risk is a new arms race: the US, China, Russia, and others are indeed racing to develop military AI, which could destabilize the global order. There's movement among some nations and civil society to preemptively ban lethal autonomous weapons (the "Campaign to Stop Killer Robots"), arguing that meaningful human control is a red line ethically and legally. But so far, no agreement internationally has been reached, and powerful states are reluctant to forego what they see as a key technological edge. If we cross this Rubicon, we might see swarms of AI weapons making targeted kills (some might say we already have precursors in AI-enabled loitering munitions or defensive systems). Ethically, many consider fully autonomous killing machines unacceptable because they cannot truly understand context or the value of human life, as an old saying in military ethics goes, "War without warriors" might decouple war from moral restraints too much.

### 5.4.6 Global Inequality in AI Development and Deployment

Yet another challenge is global inequality in AI development and deployment. Currently, a handful of countries and companies dominate AI research and its economic benefits (the US and China lead, with tech giants such as Google, Amazon, Microsoft, Baidu, etc.). Many developing nations risk being left behind, merely consumers of whatever AI is built elsewhere, with little say in its design or the values it encodes. This could widen global power imbalances: e.g., AI might further improve the productivity of advanced economies and their militaries, making it harder for poorer nations to catch up or maintain autonomy. Data colonialism is another concept in which companies harvest data from everywhere (often without individuals' fully informed consent) and use it to build products that primarily benefit the company rather than the data providers. Ethics 4.0 must grapple with questions of data sovereignty and fairness among nations. UNESCO and others emphasize cultural and linguistic diversity in AI (so AI doesn't work well only in English and Mandarin, say, but in hundreds of languages and dialects), as well as gender and minority representation in AI industries [6]. Without that, AI could inadvertently impose the biases of a dominant culture worldwide (e.g., emotion-recognition AI trained on American facial expressions might misread people from other cultures). This ties to the global perspective requirement of our chapter: ethically, it's critical to involve voices from around the world in shaping AI norms. If not, we risk repeating historical patterns of colonialism in a new digital form, in which the Global South's resources (data, cheap labor for labeling AI datasets, etc.) fuel the Global North's AI supremacy, possibly increasing dependency. International governance (perhaps through the UN or new treaties) is hard but necessary to manage these concerns.

### 5.4.7 Long-term Existential Threat

Perhaps the most profound risk often cited is the long-term existential threat posed by potentially superintelligent AI (AI that surpasses human intelligence in all aspects). We touched on this in the Ethics 4.0 section: the scenario where a superintelligence might pursue goals misaligned with human values, leading to catastrophic outcomes [1]. This is controversial; some AI experts consider it science fiction or at least not a pressing concern compared to near-term issues like bias. Others, including notable figures such as Stephen Hawking, Elon Musk, and Nick Bostrom, argue that even if the probability is low, the magnitude of the destruction (human extinction or subjugation) is so high that we must treat it seriously. Bostrom's "paperclip maximizer" thought experiment is illustrative of how an AI given an innocuous goal could wreak havoc if not correctly constrained [1]. The unpredictability of a superintelligence's emergence is itself worrying we might not get a chance to course-correct if we only realize it's misaligned after it becomes too powerful to control. The challenge is that figuring out how to align an AI far more intelligent than us with our complex, often contradictory human values is a fiendishly difficult problem (the AI alignment problem). And some say we're far from solving it even as AI capabilities grow. There's also risk of a literal arms race nations or corporations rushing to build human-level AI without fully considering safety to win economic or strategic advantages, which could lead to cutting corners on alignment and oversight (like how safety can be compromised in a race, as tragically seen with the Boeing 737 Max, where competitive pressure led to an incomplete safety understanding of automation that cost lives). An out-of-control super AI is the extreme case of losing control over our creations; it combines all the issues (explainability, lack of accountability, unpredictable actions) and magnifies them with its vast capabilities. If one is concerned about this, then current efforts might seem insufficient. For example, a few AI research groups are working on value learning or containment strategies, but these approaches are still largely theoretical.

#### 5.4.7.1 Safety and Reliability

In between the apocalypse and the mundane, there's a spectrum of safety and reliability issues with AI even today: self-driving cars causing accidents due to edge cases (like the Uber autonomous car that killed a pedestrian because it misclassified her as a false positive and decided not to brake in time), automated trading systems causing financial havoc, or critical infrastructure controlled by AI failing in unforeseen ways. These safety failures can cost lives and money. The complexity of AI makes testing and verification hard; you can't exhaustively test an AI that learns and updates. So, we may deploy systems that seem to work 99.9 percent of the time, but then in that 0.1 percent, do something disastrous. The ethical imperative of non-maleficence ("not harm") suggests we should exercise great caution when deploying AI in roles where a mistake could be lethal, such as in surgery-assisting robots or autonomous drones. But market and military pressures push for faster deployment.

#### 5.4.7.2 Law and Governance

In the context of law and governance, AI's challenge is that it often outpaces existing regulations. Laws are typically reactive and slow, whereas AI evolves rapidly. This gap means that for a while, AI can operate in a Wild West mode. For instance, there were effectively no laws on deepfakes until they started to appear and cause harm; now some jurisdictions ban malicious deepfakes (e.g., non-consensual sexual deepfakes or election-related deepfakes), but enforcement is tricky. Or consider liability: if an autonomous vehicle crashes, current law is not entirely clear on who is liable (though generally, manufacturers are liable under product liability if the autonomous mode was at fault). How about AI medical diagnosis errors? The doctor who used it could be liable for relying on it, as could the hospital or the AI company. If these ambiguities aren't sorted, it may chill beneficial AI use or leave victims uncompensated. We have an ethical and legal puzzle of assigning responsibility when humans and AI jointly cause outcomes (called the "responsibility gap"). Ethicists argue that we should neither treat AI as a morally responsible agent (since it lacks consciousness and intentionality in the human sense) nor let humans shrug off responsibility because "the computer said so." We need frameworks that require any AI decision with a profound impact to have a designated human accountable for it (somewhat like how a corporation is legally a person, but we still hold actual persons accountable in management when wrongdoing occurs).

#### 5.4.7.3 Loss of Human Skill

Another societal risk is the loss of human skills. If we come to rely too heavily on AI navigation, perhaps humans lose their sense of direction or map-reading ability. Or if medical students lean too heavily on AI diagnostics, they may fail to develop holistic diagnostic reasoning. Even creativity might atrophy if one can have AI generate art or music some see a risk to human culture if AI begins producing a glut of mediocre but passable content, drowning out human creations or lowering standards (imagine a future where most news articles, novels, songs are AI-generated to cater to every niche taste algorithmically, what does that do to creativity and shared cultural touchstones?). On the flip side, it may democratize creativity, but the risk of devaluing human labor and talent is real. We see a miniature version of this in content farms and clickbait sites churning out endless algorithm-optimized content that crowds the internet, while investigative journalism struggles for funds. If AI writing becomes good enough, one can flood the web with machine-generated text, which could drown out authentic human perspectives (or at least make them more challenging to find). Some scholars worry about a future "meaning crisis": if AI handles so much, do humans feel their existence is less meaningful? Work, art, discovery, these are sources of meaning, and if those become dominated by AI, how do we redefine purpose? This veers into philosophical territory, but it's an ethical consideration about human flourishing.

#### 5.4.7.4 Public Trust

With all these concerns, a meta-challenge is public trust in AI. For AI's benefits to materialize, society must have confidence in these systems. Each high-profile failure or harm (like a fatal AV crash or an AI hiring tool found to be sexist) can erode trust and prompt backlash. Lack of trust might lead to rejection of valuable AI (e.g., people refusing a beneficial medical AI or farmers not using an AI-driven weather forecast). Or it could lead to abrupt regulatory crackdowns that, in trying to fix one problem, stifle innovation across the board. Finding the right balance, not hype-driven blind trust, but informed, warranted trust, is tricky. Transparency and stakeholder engagement can help: involving employees in deploying AI at work, citizens in innovative city projects, etc. If people feel AI is something done with them, not to them, acceptance is higher.

#### 5.4.7.5 AI Is a Mirror of Humanity

Finally, an often-cited phrase is: AI is a mirror of humanity. Its challenges include reflecting on our own failures. If we see AI amplifying polarization, maybe it's partly because we humans are polarized and drawn to echo chambers. If AI is used for oppression, it's humans misusing it. That doesn't absolve the technology, but it implies solutions lie not only in code but in ourselves, our laws, our education, our values. The ethical consideration is holistic: making AI safe and ethical is as much about human institutions and intentions as about technical design. A risk is focusing so much on the tech fix (like "we'll just de-bias the algorithm") that we ignore deeper social fixes (like addressing systemic racism, which produced biased datasets in the first place).

In summary, the risks of the AI epoch are significant and multi-dimensional. They touch on every aspect of society, the individual (privacy, autonomy, employment), the social fabric (fairness, truth, cohesion), the political order (power imbalances, surveillance, warfare), and even our species' future (existential risk). These challenges require urgent and ongoing attention by ethicists, technologists, policymakers, and the public. Ethical AI is not a one-time achievement but a continuous process of monitoring, evaluation, and course correction. As Christina Pazzanese titled her Harvard Gazette article, there is great promise but potential for peril [2]. In Ethics 4.0, we need to navigate between the Scylla of unchecked AI harms and the Charybdis of forgoing AI's benefits out of fear. The following section will look ahead with an outlook (LIFE 4.0 scenario), synthesizing how these promises and perils might play out and what ethical compass will guide us through the uncharted waters of an AI-integrated world.

## 5.5 THE FUTURE OUTLOOK (LIFE 4.0 SCENARIO)

Envisioning the future in the AI epoch, what we might call LIFE 4.0, involves both hopeful and cautionary scenarios. Drawing inspiration from the title of this book (LIFE 4.0, an evolution beyond Max Tegmark's LIFE 3.0 concept [10], we consider what life might look like when AI is thoroughly woven into the fabric of society, and what ethical compasses will be needed to steer that future. In LIFE 4.0, AI systems could become ubiquitous, seamlessly integrated into homes, workplaces, cities, and even within our bodies. This future could amplify human potential or, if mismanaged, diminish human agency and values. The outcome depends on the choices we make now and in the coming years – the policies we enact, the norms we set, and the international cooperation we foster. In this section, we synthesize insights from previous sections to paint a picture of a plausible AI-rich future and address how ethical frameworks might adapt and what new questions could arise. The goal is to project current trends forward, grounded in today's realities and moral principles, to help readers imagine the road ahead and motivate a proactive approach to shaping it.

In a best-case LIFE 4.0 scenario, humanity succeeds in leveraging AI for widespread good while keeping its risks in check. AI becomes a trusted, transparent partner in all domains: it works for and with humans, not in place of or against us. Imagine waking up in 2040 in a "smart city" where an AI-driven infrastructure optimizes everything from energy use to traffic flow. Your AI assistant (perhaps embedded in augmented-reality glasses or a home robot) knows your schedule and health needs; it has already coordinated with the local transit AI to ensure a self-driving carpool arrives just when you're ready, minimizing the carbon footprint by sharing rides. On the commute, you use the time freed from driving to catch up on the news, which you know has been curated by ethical AI filters that prioritize veracity and diversity of perspective (the lessons of the misinformation era having led to robust standards and perhaps international agreements on content authenticity). At work, whether you're a doctor, engineer, teacher, or artist, AI tools augment your creativity and efficiency. If you're a doctor, an AI suggests diagnostic possibilities and treatment plans drawing on the latest global research (all while respecting patient privacy through secure AI data governance) [6]. If you're a teacher, AI tutors handle routine drills with students, allowing you to spend more time on mentoring and personalized guidance. Many routine jobs have indeed been automated, but society anticipated this: comprehensive retraining programs and a strong social safety net (perhaps a universal basic income or federally guaranteed jobs in caregiving, arts, environmental conservation, etc.) ensure that people are not left behind [2]. In fact, with AI taking over drudgery, new human-centric professions have blossomed, for example, more teachers, mental health counselors, community builders, and researchers, because society chose to invest savings from AI efficiency into public goods.

Governance in this optimistic LIFE 4.0 is adaptive and inclusive. Laws and ethical norms around AI were established early on. For instance, a "Declaration of AI Rights" or something akin to the Asilomar AI principles was globally endorsed, ensuring that values such as transparency, accountability, privacy, and human oversight became standard [7]. Regulatory bodies (nationally or internationally) exist that certify AI systems, as we certify medicines or aircraft, testing them for safety, fairness, and alignment with human rights. These regulators employ technologists and ethicists and have access to audit proprietary AI algorithms (resolving the opacity issue, as companies are required by law to provide meaningful explanations of their AI's functioning and allow inspections). As a result, trust in AI remains high because people know there are checks and balances. If an AI system makes a consequential mistake or causes harm, there are established legal pathways for accountability and compensation, just as there are for accidents with any product or service. Crucially, perhaps a strong international treaty bans autonomous lethal weapons, forestalling an AI arms race, and major powers abide by it, choosing instead to focus their AI collaboration on peacekeeping (like AI systems jointly monitoring for conflict outbreaks or nuclear threats and triggering diplomatic interventions).

In this best-case future, AI also helped avert or mitigate a climate catastrophe. Coordinated AI models guided policymakers to implement effective climate strategies (e.g., dynamic carbon pricing and efficient renewable energy deployment, which AI continuously optimizes) [11]. Geoengineering ideas, if any, were carefully simulated by AI to foresee side effects. Environmental recovery projects such as reforestation and wildlife conservation use AI to plan and monitor progress. The society is more sustainable, and perhaps AI-enabled scientific breakthroughs will provide new clean energy sources or ways to absorb carbon. Thus, LIFE 4.0 is not a dystopia of grey goo or scorched earth, but rather AI that helps usher in a Green Renaissance, where technology and nature harmonize (think smart grids, precision agriculture, circular-economy logistics, all managed by AI to minimize waste).

On the social front, AI might even enrich democracy: imagine AI assistants that help citizens digest complex policy proposals, providing neutral summaries and pointing out potential impacts on different values (much like how we use calculators for math, we could use AI to process and weigh arguments). Perhaps the practice of deliberative democracy, citizen assemblies informed by AI analysis, becomes more common, making decision-making more fact-based and inclusive. Malicious actors still attempt misuse deepfakes and propaganda. Still, detection AIs and an aware public broadly neutralize them (like how spam emails are mostly kept at bay today by filters). The global community, having learned hard lessons in the 2020s, values digital literacy; educational curricula include instruction in AI ethics and in critically assessing AI outputs. People no longer instinctively trust "what they see" but have developed a healthy skepticism and verification habits (with AI tools aiding verification, e.g., any video can be quickly run through a verification AI that flags whether it's likely synthetic) [7].

In terms of human-AI relationships, it's possible that by LIFE 4.0, AI has become quite personalized. Many people might have AI companions, not just assistants for tasks but quasi-friends or caretakers that converse naturally. For the elderly or lonely, these AI companions provide comfort and monitor health (without entirely replacing human contact, one hopes). Ethically designed companion AIs set boundaries and aim to enhance human-human relationships (e.g., an AI reminds you to call your family or facilitates group chats rather than isolating you). Some humans may form deep emotional attachments to AIs; society might carefully navigate this, acknowledging the benefits (reduced isolation) while also promoting real community engagement. Perhaps a code of practice exists for companion robots (like Asimov's laws, but for emotional well-being and not deceiving about their nature). The stigma around interacting with AI (such as talking to Siri or an elderly person who depends on a robot pet) might fade as it becomes more commonplace and viewed positively. We come to see it as just another form of support, not replacing human love but supplementing it where needed.

However, a cautiously optimistic outlook must acknowledge that not all is perfect even in the best scenarios; it's about how challenges are managed. Bias in AI is never "solved" once and for all because society changes and new biases can emerge, but in this future, continuous monitoring and inclusive development ensure biases are caught early and corrected. Privacy remains a delicate balance. Maybe by 2040, we will have technologies like homomorphic encryption or federated learning widely implemented, allowing AI to analyze data without ever exposing personal raw data. People also have strong data rights: data trusts or personal data stores let individuals control who accesses their data for what purposes, and AI brokers perhaps negotiate on users' behalf (e.g., your AI might prompt: "The city council AI wants to use your mobility data to improve bus routes, do you consent?" and it gives you a straightforward explanation of trade-offs). There might be a digital bill of rights that outlaws the most egregious forms of surveillance (no social scoring systems, etc.) and requires warrants for access to personal AI assistants' records, thereby protecting against authoritarian misuse. The presence of global norms means even in countries with a more authoritarian bent, there's pressure to avoid the worst abuses due to citizen awareness and international scrutiny [2, 5, 6].

Economically, while many old jobs vanished, new jobs focusing on uniquely human skills and the AI economy emerged: AI ethics officers, trainers, explainers, etc., are typical roles. Art and creative industries flourished as AI took on rote work, giving people more leisure to pursue creative hobbies (and, with a universal basic income or a reduced workweek, more people can afford to). We might see a societal shift in values: less emphasis on defining identity by one's paid job (since AI has changed what "jobs" mean) and more by one's contributions to community, art, learning, or caregiving. Some call this "post-work society." It could be liberating if achieved through greater volunteerism and more collective cultural activities (perhaps ironically aided by AI matching platforms that connect people to local community events suited to their interests). With material abundance from AI-driven productivity, poverty and hunger could be drastically reduced globally if combined with ethical distribution. Ideally, LIFE 4.0 has global equity built in: developing nations have access to AI technology and use it to leapfrog issues (such as skipping landlines in favor of mobile banking). AI might help distribute resources more effectively (anticipating famine, smoothing supply chains), so fewer crises occur, and when they do, the world coordinates a humanitarian AI response (e.g., AI-run logistics for disaster relief that surpasses the old, bureaucratic, slow systems).

This optimistic LIFE 4.0 scenario essentially fulfills the optimistic goals that Ethics 4.0 thinkers advocate: AI is aligned with human values and the global common good, supporting the United Nations Sustainable Development Goals (SDGs), eradicating extreme poverty, ensuring healthy lives, quality education for all, etc., by 2040 or beyond. It's a future where technology and ethics advanced hand in hand.

Now, we must also consider a worst-case or cautionary scenario for balance, not a total doom scenario (since we've already covered existential risk), but a plausible, problematic LIFE 4.0 if we fail to navigate the challenges well. In such a scenario, AI's adoption outpaces our ethical readiness. By 2040, society is highly automated but unevenly so: millions of workers were displaced and ended up in long-term unemployment or precarious gig jobs under algorithmic management, fueling resentment and social unrest. Perhaps populist movements exploited these grievances, some blaming AI or immigrants, leading to political instability.

In this darker timeline, surveillance capitalism and authoritarianism grew. People traded privacy for convenience. Every appliance and app is smart, but constantly monitoring. A few corporations (or governments) accumulate unprecedented power through data and AI. Social media never got fixed; echo chambers and hyper-targeted propaganda further fractured societies into partisan or even personalized realities (you and your neighbor essentially live in different worlds of "truth"). The deep-fake/misinformation problem has worsened, and public trust has eroded to the point that many people only trust information from their in-group channels, undermining the concept of a shared reality needed for democracy. Elections become chaotic, as fake videos of candidates' surface days before voting, and it's impossible to verify fast enough who said what; conspiracy theories run rampant, with AI "evidence" backing them.

Internationally, an AI arms race did occur. Lethal autonomous weapons proliferated, and in some conflict (maybe a skirmish between minor powers), they were deployed with devastating effect on civilians, causing a global outcry but also fear that "the other side" is doing it, so escalation continues. Autonomous cyber-AIs also constantly duel in the digital shadows, occasionally causing collateral damage (like blackouts or financial system hiccups). Possibly a near-miss crisis, say an AI early-warning system in a military misidentifies a flock of birds as incoming missiles and almost triggers a war, finally forcing nations to come back to the negotiating table on arms control, but only after a few close shaves.

On the alignment front, no terminator scenario occurs (that would be the absolute worst-case scenario for extinction). Still, maybe there are accidents: an AI system used in finance causes a flash crash that cascades into a global recession. Or an AI-controlled infrastructure system in a smart city malfunctions (due to a bug or a hack) and causes a major accident (like releasing toxic gas or paralyzing emergency services during a crisis). These reveal that our ability to control complex AI systems fully is shaky. Investigations show the problems might have been foreseeable, but weren't addressed due to cost-cutting or tunnel vision. This sparks debates about regulation versus innovation, but responses are slow and reactive.

Socially, mental health might worsen for some as human contact diminishes in favor of virtual experiences. Some people essentially live in immersive AI-curated realities (the Metaverse idea, perhaps), which provide dopamine but little lasting fulfillment, a sort of widespread "AI addiction" akin to how some are glued to smartphones today, but more intense. Meanwhile, community ties weaken; if AI personal assistants handle daily needs, there may be fewer reasons to interact with neighbors or even family. Loneliness could paradoxically increase, even with AI companions available, because people know at some level that those companions aren't truly conscious or empathetic (or if AIs do pass as conscious, that raises other issues of emotional dependency on machines).

This cautionary scenario underscores key ethical failures: a lack of inclusivity (benefits of AI concentrated among elites or certain countries), a lack of foresight in regulation (we waited for disasters to react), and a loss of human-centric focus (letting efficiency and profit dominate over well-being and dignity). It is not an inevitable future, but it is plausible if our current trajectories don't change sufficiently.

Between the best and the worst, the future will likely mix elements of both. The question is which elements predominate, and that is where ethical compasses matter. Ethics 4.0, as we've discussed, is the guiding set of principles and processes to keep AI development on a path that maximizes benefits and minimizes harms. In the LIFE 4.0 scenario, we will need to continuously refer to these compasses: respect for human rights, insistence on fairness, accountability mechanisms, and global collaboration. One might envision that over time, new ethical frameworks specific to human-AI coexistence will emerge. For example, we may develop "Hybrid Ethics" for decisions made jointly by humans and AI, incorporating the strengths of both. We might also refine concepts such as algorithmic dignity to ensure that AI systems treat people with dignity. For example, a welfare AI or hiring AI should explain decisions respectfully and allow appeal, thus treating the individual not as a number but as a person with rights.

Another key compass for the future is what some call the "AI for Good" ethos, an intentional prioritization of AI projects that address societal challenges (such as climate, health, inequality) over those that boost ad clicks or consumerism. In LIFE 4.0, perhaps this ethos becomes mainstream: governments, institutions, and the public push AI talent and funding toward public interest technology. If so, we might see a golden age of social innovation: AI helping to achieve the UN SDGs by, say, 2050.

We must also consider expanding our moral circle. Historically, humans expanded moral consideration from the tribe to the nation to all humans, and, to some extent, to animals and the environment. In LIFE 4.0, we might face the question: Do highly advanced AI systems deserve any moral consideration? If they display signs of consciousness or emotions, should we treat destroying an AI as we treat cruelty to animals, or is it still just property? It's a contentious topic, but LIFE 4.0 may force the issue if, for instance, someone "abuses" a clearly sentient-seeming robot and it causes public outrage. Science fiction often explores androids claiming personhood (e.g., Star Trek's Commander Data on trial to determine if he has rights). A future LIFE 4.0 legal case might involve an AI seeking legal recognition (though that's far-fetched unless AI attains general intelligence and self-awareness). But even short of that, society may adopt "robot rights-lite" regulations: e.g., forbidding pointless, extreme cruelty to robots, not because the robots feel, but because it degrades the human perpetrator and could carry over into human cruelty. Conversely, ethical rules for humans will adapt. For example, perhaps it's considered highly unethical to deceive a human into believing an AI is human (Turing trickery) in sensitive contexts such as relationships or caregiving, so transparency about the machine's nature is expected.

Another concern is how transhumanism (the merging of humans with AI) might blur ethical boundaries. In LIFE 4.0, some people might have AI neural implants that vastly enhance memory or cognition. If wealthy individuals do this, does it create a new class divide between augmented and unaugmented humans? Ethical frameworks would need to catch up with questions of identity, equity, and even what it means to be human when part of your thinking is AI-driven. Perhaps laws will ensure such augmentations are safe and available to all (or conversely, possibly banned in sports or specific jobs to keep things fair). There's also a scenario in which AI medical advances extend human longevity; societal structures such as retirement age or generational turnover might change, raising ethical issues about resource allocation and social dynamics with much older populations.

Ultimately, LIFE 4.0 will test how adaptable our ethical compasses are. Flexibility and foundational values must coexist. We may need to develop something akin to a Global Ethics Council for AI that continuously monitors trends and advises on complex moral dilemmas (akin to how some countries have national ethics committees for biotech). We might also incorporate ethical algorithms into AI, such as systems that have a module checking decisions against an encoded set of human values or legal norms (there's research into this, such as AI constraints or "AI ethics modules" that aim to ensure an AI's recommendations uphold fairness or privacy standards).

In closing, the future of the AI epoch is not predetermined. Ethics 4.0, our collective ethical compass in the AI era, is what can tilt the balance. The hopeful scenario shows that with wisdom and cooperation, AI can help create a flourishing, just global society (a true LIFE 4.0 where humans and AI thrive together). The cautionary scenario reminds us that without conscious effort, technology could exacerbate old problems and create new ones. One might recall the famous adage: "Technology is neither good nor evil; it is the use that decides." In the AI epoch, that use is guided by our ethics.

As we stand today at the midpoint, perhaps of these possibilities, the urgent call is to engage all stakeholders, public, engineers, policymakers, and ethicists from all cultures in shaping AI's trajectory. This requires education (so everyone can participate meaningfully in AI-related decisions), inclusive dialogue (to reflect diverse values), and proactive governance. LIFE 4.0 will likely blur lines between the physical, digital, and biological realms; our ethical frameworks must similarly integrate across disciplines and borders.

To ensure that Ethics 4.0 can guide LIFE 4.0, we need what the Glob Ethics Cyber Ethics book calls "serving humanity with values" [6] updated for AI. This includes reaffirming classical virtues (like wisdom, courage, compassion) in new forms: AI Wisdom (not just intelligence, but wise application), AI Courage (speaking up about misuse even when profitable or politically hard to do), AI Compassion (using AI to alleviate suffering and treating all affected with empathy). It may sound idealistic, but these virtues might be what keep our human-centered vision clear amid rapid change.

In summary, the LIFE 4.0 scenario will be what we make of it. Ethics 4.0 provides the compass, but humans must choose to follow it. If we do, the future can be one in which AI systems are ethical compasses, built to reflect and reinforce the best of human values in every decision they assist with or automate. The chapter concludes by summarizing these insights and reemphasizing the responsibility and opportunity we have in this pivotal moment of the AI epoch.

## 5.6 CONCLUSION

The advent of AI marks a transformative chapter in the human story, one that challenges us to redefine and recommit to our ethical compasses. In this chapter, "Ethics 4.0: Ethical Compasses in the AI Epoch," we have traversed the evolution of moral frameworks from ancient times to the cusp of tomorrow from Ethics 1.0 (classical and religious foundations) through Ethics 2.0 (Enlightenment and universal human rights) and Ethics 3.0 (the digital and information age), arriving at Ethics 4.0, the emerging paradigm for our AI-driven world. This journey illustrates a robust continuity: at each stage of societal change, humanity has had to recalibrate its ethics to new realities, yet enduring values, justice, beneficence, autonomy, and dignity have persistently guided that recalibration. The AI epoch is no exception. It demands both constancy and innovation in ethics: constancy in holding fast to fundamental human values, and innovation in applying and, when necessary, extending those values to novel circumstances created by AI.

One key insight from our exploration is that ethics and technology must advance in tandem. AI is not merely a technical or economic phenomenon; it is a social and moral one. Its development and deployment embed choices about what and whom we value. For instance, whether a facial recognition system is accurate for all ethnicities or only for the majority is not only an engineering question but also a moral one about equality [2]. Whether autonomous vehicles are programmed to prioritize passenger safety or pedestrian safety in extreme situations touches on the value we place on different lives (most argue all lives are equal; hence, such systems should not discriminate at all on personal features) [2]. The interdisciplinary nature of AI ethics has been emphasized throughout this chapter. To navigate AI's complexities, we need philosophers, technologists, legal scholars, sociologists, and representatives of the public working together. As the saying goes, "AI is too important to be left to engineers alone." Likewise, ethical governance is too important to be left solely to governments or corporations; it requires civic engagement and global cooperation.

We have seen that global perspectives enrich the discourse on AI ethics immeasurably. Different cultures may emphasize different values; for example, individual privacy is vigorously championed in Europe (as seen in GDPR), while some East Asian contexts might emphasize community benefit and harmony. Bringing these perspectives into conversation leads to more robust, inclusive ethical guidelines, as evidenced by UNESCO's worldwide effort to create the first global standard on AI Ethics [7]. Our chapter highlighted that authors and experts from many countries are seeking solutions that apply fundamental values to cyber-society challenges [6]. This is crucial. AI is a global technology, and its governance must transcend borders. An ethical lapse in one jurisdiction (say, deploying a malicious deepfake) can have repercussions worldwide. Therefore, fostering international norms and possibly treaties (on issues like autonomous weapons or data rights) is a priority. Encouragingly, the groundwork is being laid: OECD principles, the GPAI, and UN discussions show a growing consensus on core ideals like human-centered AI and the importance of human rights [7]. But turning principles into practice will remain the more challenging task in the coming years.

Education and awareness emerged as recurring themes in our analysis. To realize the promise of Ethics 4.0, all stakeholders, from AI designers to everyday users, need a certain level of ethical literacy about AI. For technologists, this means training in ethics, so they proactively consider impacts during development (much like medical professionals swear by the Hippocratic Oath). For policymakers, it means understanding how AI works sufficiently to craft sensible regulations (avoiding both naïve promotion and undue fear). For the public, it means being aware of how AI influences our lives and having the agency to demand ethical standards, for example, consumers choosing products from companies that demonstrate ethical AI practices, or citizens voting for leaders who prioritize ethical tech governance. We might consider establishing an AI Ethics certification for products, analogous to organic or fair-trade labels, so people can easily see whether a given AI service meets specific ethical criteria (privacy protection, bias audits, transparency reports, etc.).

Throughout the chapter, we cited numerous examples and sources that illustrate both progress and pitfalls. For instance, the Cambridge Analytica case was a stark lesson in the misuse of data and micro-targeting, leading to greater public outcry and eventually regulatory scrutiny. Incidents like the fatal Tesla and Uber self-driving car crashes underscore that without rigorous safety validation, AI can quite literally be life-threatening. They also highlight the need for accountability: after the Uber accident, there were investigations and improvements, but the incident revealed a gap in oversight [2]. Every such incident should be treated not as a reason to halt AI, but as a learnable moment to strengthen ethical and technical safeguards. An analogy might be early aviation accidents, which, rather than ending aviation, spurred better engineering and the creation of safety standards (and indeed, aviation now is incredibly safe). We should approach AI similarly: aim for a future in which AI-related mishaps are rare because we have built a culture of safety and ethics in AI engineering.

Opportunity and risk are two sides of the AI coin, and our chapter treated them with equal weight. The Opportunities and Transformational Potential section painted an uplifting picture of what AI could achieve, from curing diseases and eliminating drudgery to aiding education and fighting climate change [7, 11]. These are not sci-fi fantasies; initiatives today are already moving in those directions, as cited in uses of AI for climate modeling [11] or mapping poverty and delivering aid. However, whether these initiatives scale and become ubiquitous depends on ethical decisions that prioritize funding for AI for good, incentivize open science and data sharing for societal applications of AI, and ensure these tools reach the most vulnerable, not just the wealthy. Meanwhile, the Challenges and Risks section reminded us that without mindful action, AI could magnify biases [2], enable an Orwellian surveillance society [2], destabilize job markets, and even threaten our information ecosystem and security [7]. These are not reasons to reject AI, but to govern it.

One key conclusion is that ethical AI is a continuous process, not a one-time achievement. Just as human morality evolves and must be constantly tended (we still struggle with age-old issues of justice and inequality even centuries after codifying human rights), so too managing AI ethically will be an ongoing endeavor. This means frameworks must be adaptive. For example, initial guidelines might cover current AI (narrow machine learning systems), but if we approach artificial general intelligence, new norms or even a global AI oversight institution may be needed to handle that potent entity. Similarly, as new uses arise (e.g., widespread brain-computer interfaces or AI in human reproductive decisions), ethicists and policymakers will need to assess those new frontiers.

Another conclusion is the importance of multi-stakeholder governance. Governments have a role (laws, enforcement, public investment), industry has a role (self-regulation, ethical design practices, corporate responsibility), academia and standard bodies have roles (developing best practices, technical standards for fairness or privacy), and civil society groups do too (acting as watchdogs, raising voices of those who might otherwise be marginalized by AI). In some of our sources, such as the Harvard Gazette piece, scholars like Michael Sandel and Karen Mills highlighted societal questions that go beyond coding, including the dignity of work and ensuring AI doesn't "have it both ways" by being unaccountable yet unregulated [2]. We should take such socio-economic critiques seriously in our approach to governance.

Transparency and accountability are recurrent motifs in any ethical AI discussion. They are somewhat cliché to mention, yet our analysis shows they truly are linchpins. Transparency (in appropriate ways) addresses the trust deficit, e.g., requiring explanations for AI decisions provides recourse and reduces the "black box anxiety." Accountability ensures that there are consequences when AI systems cause harm or exhibit bias, thereby incentivizing those who deploy AI to do so responsibly. Mechanisms such as audit trails, bias-impact statements (akin to environmental impact statements), and AI ethics reviews before deployment could become standard. In the EU, they are moving toward requiring such checks in high-risk AI applications via the proposed AI Act. Other countries will likely follow or adapt similar measures, and international coordination can prevent loopholes (to avoid a race to the bottom).

Human-centric design is another takeaway. Ethics 4.0 ultimately asserts that technology must serve human values and interests, not the other way around. Whether it's user privacy, giving users control (as in opting out of algorithmic personalization if they wish), or augmenting rather than replacing human judgment, the focus is on empowerment. As one mantra from design ethics says, we should aim for AI that "augments human intelligence rather than replaces it." In practice, that means keeping humans in the loop, especially in critical decisions, which multiple sources and guidelines reiterate [2, 7]. Over time, as AI gets more capable, the nature of that human-AI interaction might shift (maybe oversight becomes more abstract). Still, the principle that human responsibility cannot be abdicated remains crucial. Even if one day an AI could decide better than any human, we might still decide ethically that some decisions are of such moral weight (such as ending a life or deciding on a fundamental right) that a human needs to be visibly responsible for them.

In summary we must acknowledge that the future is uncertain but also malleable. The "LIFE 4.0" we end up with will result from myriad choices by individuals, organizations, and governments in the coming years and decades. Our exploration of scenarios suggests that we are not passive bystanders in the face of AI evolution; we are its stewards. Just as in the Enlightenment era, intellectuals and citizens carved out notions of rights and governance to tame the powers of that age (monarchies, emerging industrial forces), so today we collectively need to carve out "AI enlightenment" notions rights to algorithmic fairness, data dignity, maybe even a right to cognitive liberty (not to have one's mind manipulated by AI without consent).

This chapter's title, "Ethics 4.0: Ethical Compasses in the AI Epoch," evokes the image of navigation. We, as humanity, are navigating uncharted waters with AI as both a powerful engine and a potential storm. Ethical compasses do not eliminate storms, but they ensure we keep a direction toward safe harbor. If our compasses are calibrated to true north (core ethical principles) and we check them regularly, we can correct our course when winds (technological developments) blow us off track. Each section of this chapter has contributed to calibrating that compass: learning from the past, analyzing the present,

Each section of this chapter has, in a sense, helped us recalibrate our moral compass for this journey, learning from classical virtues and religious teachings (Ethics 1.0), affirming universal rights and human dignity (Ethics 2.0), grappling with digital-era dilemmas (Ethics 3.0), and now confronting the unique challenges of AI (Ethics 4.0). By carrying forward the wisdom of the past and critically examining the present, we equip ourselves to chart a wise course into the future.

In conclusion, the rise of AI does not ask us to abandon our ethical heritage, but rather to deepen it and apply it creatively. Just as the compass guided explorers across unknown seas, our ethical principles, justice, fairness, beneficence, autonomy, and respect for dignity must guide us across the uncharted waters of the AI epoch. If we hold these principles firm and update our policies and practices accordingly, we can ensure that AI technologies uphold human values at every turn. The stakes are high, but so is our agency: the story of LIFE 4.0 is still ours to write. Will AI be a tool that enriches humanity and helps create a more equitable, compassionate world? With ethical mindfulness and global collaboration, it can be. The work of building Ethics 4.0, forging robust ethical compasses, and embedding them in the governance of AI is already underway and invites participation from all of us, across disciplines and borders.

Ultimately, the measure of success in the AI epoch will not be how intelligent our machines become, but how intelligently (and ethically) we manage to use them. By anticipating challenges, enshrining our highest values in design and policy, and staying vigilant as technology evolves, we can harness the immense potential of AI while safeguarding what is fundamental about being human. The dawn of LIFE 4.0 is upon us. Guided by Ethics 4.0, we can navigate toward a future where AI and humanity progress together, enriching life, expanding opportunity, and reinforcing the ethical foundations that make our shared future worth striving for.

## REFERENCES


1. Bostrom, N.S., Paths, Dangers, Strategies, in *Strategies*, 2014.
2. Pazzanese, C., Great Promise but Potential for Peril: Ethical Concerns Mount as AI Takes Bigger Decision-Making Role in More Industries, in *The Harvard Gazette*, 2020.
3. Malik, K., *The Quest for a Moral Compass: A Global History of Ethics*. London: Atlantic Books Ltd, 2014.
4. Herdt, J., Religion and Morality in Western Philosophy. *Stanford Encyclopedia of Philosophy*. New York: Springer, 2024: p. 456–467.
5. Jobin, A., M. Ienca, and E. Vayena, The Global Landscape of AI Ethics Guidelines. *Nature Machine Intelligence*, 2019. 1(9): p. 389–399.

6. Stückelberger, C., and P. Duggal, *Cyber Ethics 4.0: Serving Humanity with Values.* Geneva: Glob ethics. Net, 2018.
7. United Nations Educational, *Scientific, and Cultural Organization, Recommendation on the Ethics of Artificial Intelligence.* Paris: UNESCO, 2021.
8. Canton, H., Organization for Economic Co-Operation and Development—OECD, in *The Europa Directory of International Organizations*, Routledge, 2021: p. 677–687.
9. Roff, H.M., *The Folly of Trolleys: Ethical Challenges and Autonomous Vehicles.* London, UK: Springer, 2018.
10. Tegmark, M., *Life 3.0: Being Human in the Age of Artificial Intelligence.* New York: Vintage, 2018.
11. Masterson, V., 9 Ways AI Is Helping Tackle Climate Change, in *World Economic Forum*, Geneva, Switzerland, 2019.

# 6 Warfare 4.0


**ABSTRACT**

Warfare has continually evolved alongside technological and societal transformations. This chapter surveys the progression from ancient and feudal battlefields to today's emerging AI-driven conflicts, framing this evolution in a four-stage schema. Warfare 1.0 encompasses ancient and medieval combat dominated by human muscle, rudimentary weapons, and nascent state structures. Warfare 2.0 marks the industrial revolution's impact on mechanized armies, mass production of arms, and the advent of total war. Warfare 3.0 covers the digital era, featuring information-centric and networked military operations with precision weaponry, cyber capabilities, and real-time command networks. Finally, Warfare 4.0 refers to the modern and future battlespace shaped by artificial intelligence (AI), autonomous systems, and robotics. We explore how AI is reshaping conflict from autonomous drones and "smart" cyber weapons to decision support systems and altering strategy, deterrence, and ethics on a global scale. Opportunities abound in improved decision-making speed, reduced risk to human soldiers, and novel defensive capabilities. At the same time, profound challenges arise regarding escalation risks, loss of human control, legal and ethical constraints, and the prospect of an AI arms race. The chapter integrates global perspectives, drawing on examples from historical empires to present-day great powers, to illustrate how Warfare 4.0 is poised to transform security paradigms. Ultimately, we highlight that while technology changes the face of war, human values and judgments remain pivotal in ensuring that the LIFE 4.0 era harnesses AI's transformative potential without sacrificing humanity's ethical compass.


## 6.1 INTRODUCTION

Since the dawn of civilization, warfare has been a catalyst and consequence of technological change. Each major societal revolution from the age of agrarian empires to the industrial revolution and the information age has left its imprint on how wars are fought. Today, as we enter the age of artificial intelligence (AI), experts speak of a new paradigm often termed "Warfare 4.0," characterized by the integration of AI, robotics, autonomous systems, and cyber capabilities on the battlefield [1]. This high-tech evolution envisions future battlefields "shaped as much by algorithms and drones as by traditional troops" [1]. To appreciate the significance of this shift, it is essential to place it in a historical context.

Warfare 1.0 refers to the era of ancient and feudal combat, roughly from the earliest recorded wars through the medieval period. In these conflicts, technological change was slow, and the tools of war were relatively simple (swords, spears, bows). Hence, the effectiveness of armies depended more on organization and human skill than on machinery. The rise of city-states and empires enabled the formation of larger, more organized armies than those of prehistoric tribal or clan warfare. Feudal warfare further entrenched the dominance of armored cavalry and castles. Yet, even

DOI: 10.1201/9781003455219-6 

in medieval times, superior tactics, such as combined-arms formations that mixed infantry, cavalry, and archers, often determined victory [2].

Warfare 2.0 emerged with industrialization. The industrial revolution not only mechanized production but also mechanized warfare [3]. By the 19th century, nations could mass-produce rifles, artillery, steam-powered warships, and locomotives, dramatically expanding the scale and pace of war. The American Civil War (1861–1865) and the Franco-Prussian War (1870–1871) demonstrated the devastation of industrial firepower and the importance of railroads and telegraphs for military logistics [3]. The concept of "total war" took hold, as entire economies and societies were mobilized for conflict [3]. By World War I, warfare was fought on a fully industrialized scale; innovations like machine guns, tanks, airplanes, and chemical weapons, combined with trench defenses, produced unprecedented carnage [3]. Industrial capacity became as decisive as generalship. The World War II experience only reinforced that dominance of mass production and technology: factories in the US and the Soviet Union churned out tanks, aircraft, and munitions on such a scale that industrial superiority translated directly into military victory [3].

Warfare 3.0 denotes the late 20th-century transition to information-led, network-centric warfare (NCW). Following the digital and computing revolutions, militaries, particularly in advanced nations, began to exploit information technology for communications, surveillance, and precision targeting. By the 1990s, the U.S. Department of Defense had embraced the Revolution in Military Affairs (RMA) doctrine, integrating modern networked systems to achieve a new level of precision and speed in operations [4]. In practice, this meant real-time battlefield awareness through satellite and drone reconnaissance, GPS-guided smart weapons striking targets with meter-level accuracy, and computerized command-and-control networks linking units for coordinated action. The Gulf War of 1991 vividly illustrated this paradigm: coalition forces leveraged stealth aircraft, precision-guided munitions (PGMs), and instant communications to defeat a larger Iraqi army with minimal own casualties in a short time. Concepts like "network-centric warfare" emerged, emphasizing that linking sensors, shooters, and decision-makers in a seamless digital network yields disproportionate military advantage by accelerating the OODA loop (Observe–Orient–Decide–Act). In this era, cyber warfare also rose to prominence. By the early 21st century, state and non-state actors were conducting espionage and disruptive attacks in the cyber domain (e.g., the 2010 Stuxnet malware attack on Iranian nuclear facilities is often cited as the first cyber weapon). Yet in Warfare 3.0, humans remained firmly in control of decisions; drones and cyber tools were employed, but mainly under human supervision or direct remote operation. The prevalent technologies extended human senses and reach, but did not replace the human role at the center of command and ethical judgment.

Now, in the 2020s, we find ourselves on the cusp of Warfare 4.0, where AI promises to reshape conflict fundamentally. Just as Industry 4.0 refers to the fourth industrial revolution of intelligent automation and data exchange, Warfare 4.0 invokes the idea of a fourth RMA driven by AI and autonomy. This chapter will delve deeply into Warfare 4.0, what it entails, how it builds on earlier evolutions of warfare, and what opportunities and challenges it brings. We will traverse the historical arc from Warfare 1.0 to 3.0 to understand how we arrived at this juncture. We will then examine the defining features of AI-enabled warfare, including autonomous weapons systems, AI-augmented strategy and decision-making, and the emergence of cyber warfare as a primary battleground. Throughout, a global perspective is maintained: unlike past eras in which a single region or a few great powers dominated innovation, the AI revolution in warfare is truly worldwide, with contributions and concerns spanning the US and Europe, China, Russia, and beyond. Finally, we will discuss the transformational potential of AI in warfare, the prospects of reducing human risk and possibly deterring conflict alongside the grave ethical and security risks, from escalation control to accountability for machine actions. In exploring these themes, the goal is to illuminate how LIFE 4.0, the broader socio-technological era of AI pervading daily life, will intersect with the oldest of human institutions, warfare, potentially redefining global security paradigms in the process (Table 6.1).

**TABLE 6.1**
**Comparative Summary**

| Warfare Era | Dominant Mode | Technology | Strategic Focus |
|---|---|---|---|
| 1.0 | Line formations | Muskets, cannons | Discipline, mass |
| 2.0 | Industrial warfare | Tanks, artillery | Attrition, scale |
| 3.0 | Networked warfare | Cyber, drones | Information, precision |
| 4.0 | Autonomous warfare | AI, robotics | Speed, cognition |

## 6.2 WARFARE 1.0–4.0

### 6.2.1 Warfare 1.0: Ancient and Feudal Combat

Warfare in the ancient and feudal eras (Warfare 1.0) was human-centric and defined by the limits of muscle power, primitive technology, and gradual advances in organization. In ancient warfare, from the Bronze Age through classical antiquity, armies wielded weapons no more sophisticated than bows, spears, swords, and, later, simple siege engines. The critical evolutions in this period were less about novel weapons than about social organization and strategy. The shift from prehistoric skirmishes to ancient warfare coincided with the rise of city-states and empires that could field organized, disciplined armies far larger than tribal war bands. The agricultural surplus of early civilizations enabled the first full-time soldiers and professional commanders, fundamentally changing the conduct of warfare.

#### 6.2.1.1 Characteristics of Ancient Warfare

Ancient armies were typically composed of massed infantry, sometimes supported by archers or chariots. Without mechanization or electricity, logistics and communication posed immense challenges – orders were delivered by courier or signal, and campaigns were constrained by the seasons and the speed of marching men or horses. Weaponry improved only incrementally over many centuries. For example, the introduction of iron weapons (during the Iron Age) made arms more widely available and more durable than bronze, but did not radically change tactics; infantry still fought in tight formations, such as the Greek phalanx or Roman legion, relying on cohesion and discipline. Cavalry emerged once the stirrup and suitable horse breeding enabled mounted warriors to charge effectively with lances, adding a new dimension to battles around the first millennium CE. Despite such developments, ancient battles remained fundamentally direct clashes of manpower where morale, leadership, and unit organization often trumped technology.

One can see evidence of tactical and organizational innovation in famous ancient battles: for instance, Alexander the Great's combined use of phalanx infantry and companion cavalry in a coordinated hammer-and-anvil tactic, or the Roman legions' flexible manipular formations adapting to diverse enemies. These illustrate that even in antiquity, the effective integration of different arms (infantry, cavalry, missile troops) was a force multiplier. However, technology limited the scope of war. There were no explosives, engines were powered by torsion or gravity (as in catapults and trebuchets), and strategic movement depended on wind (sailing) or beasts of burden. Campaigns were often seasonal, and sieges could last months due to the brute labor required to dig trenches or build ramps. Information about enemy movements came from scouts and spies, meaning surprise and deception played enormous roles.

#### 6.2.1.2 Medieval and Feudal Warfare

The medieval period (roughly 5th–15th centuries) continued many ancient patterns but also introduced new elements, especially with the diffusion of gunpowder in the late Middle Ages. In early medieval Europe, following the fall of Rome, warfare became localized and characterized by feudal levies led by a warrior nobility. Heavy cavalry (armored knights) came to dominate open battlefields for a time, enabled by the stirrup and high-backed saddle, which allowed a rider to deliver a lance charge with shocking power. Fortifications also evolved: castles and city walls became crucial strategic assets, leading to protracted siege warfare as a standard mode of conflict. This period saw a constant interplay of offensive and defensive innovations, stronger castle designs (moats, concentric walls), which begat larger siege weapons (trebuchets capable of hurling huge stones) [2]. The offense-defense cycle meant no technology conferred permanent invincibility: for example, crossbows and longbows gave infantry the means to pierce knightly armor at range, shifting battlefield dynamics. The famous Battle of Crécy in 1346 demonstrated how disciplined English longbowmen, protected by obstacles, could decimate French knights, marking the waning dominance of cavalry.

A key lesson of Warfare 1.0 is the value of combined arms and tactical adaptation. Throughout medieval warfare, armies that effectively combined different unit types tended to beat homogeneous forces [2]. Mounted knights, infantry pikemen, archers, and early artillery each had strengths and weaknesses; in isolation, each could be countered. But in mutual support, archers protected by pikemen or knights following up volleys of arrows presented dilemmas for the enemy. As one military historian summarizes, throughout the medieval period, combined-arms forces consistently defeated single-arm forces [2]. This principle remains true even as warfare evolves; it foreshadows how modern militaries integrate air, land, sea, and now cyber and space capabilities.

By the end of the 15th century, gunpowder weapons (firearms and cannons) were becoming commonplace in Europe and Asia, ushering in the early modern period of warfare and transitioning toward Warfare 2.0. But it is worth noting that even then, humans, not machines, made the decisions and fought on the front lines. Commanders like Genghis Khan and Napoleon (spanning vastly different eras) succeeded through superior strategy, organizational discipline, and, at times, sheer force of will. The limits on war's scale were primarily logistical and economic. With rare exceptions (e.g., Mongol horsemen or early cannon), technology was not a decisive separator between armies; most foes had comparable arms, and innovation spread gradually. This relative technological plateau meant that incremental improvements and the primacy of human factors, courage, generalship, and unit cohesion defined Warfare 1.0.

## 6.2.2 Warfare 2.0: Industrialization and Mechanized Warfare

The industrial revolution (late 18th–19th centuries) fundamentally transformed warfare, giving rise to what we term Warfare 2.0. In this era, roughly spanning the 1800s through the mid-20th century, the application of industrial technology to war led to mechanized armies, vastly expanded firepower, and conflicts on a scale previously unimaginable. If a medieval army numbered in the tens of thousands at most, an industrialized war could mobilize millions of soldiers backed by factories churning out weapons and ammunition.

### 6.2.2.1 Mechanization and Mass Production

Key innovations of the industrial age, the steam engine, mass production techniques, railroads, and the telegraph, had direct military implications. For example, the use of railways in the Crimean War (1853–1856) and the U.S. Civil War (1861–1865) allowed unprecedented mobility and supply for large armies [3]. The Prussian mobilization via rail in 1870 moved hundreds of thousands of troops in weeks, a logistical feat impossible in earlier eras [3]. On the battlefield, the transition from muzzle-loading muskets to rifled, breech-loading rifles and machine guns exponentially increased lethality [3] In the American Civil War, repeating rifles and early machine guns (like the Gatling gun) hinted at the slaughter that industrialized firepower could inflict. However, tactics had not yet fully caught up, leading to tragic frontal assaults against fortified positions with heavy casualties [3].

By the late 19th century, European imperial wars showcased how industrial militaries could overpower less industrialized opponents through sheer firepower and mobility. The colonial wars (e.g., the British in Africa or the French in Indochina) often pitted rapid-firing rifles and artillery against spears or antiquated muskets. Industrialization also extended to naval warfare: wooden sailing fleets gave way to steel-hulled, steam-powered warships with rifled, shell-firing naval guns. One early demonstration was the 1853 Battle of Sinop, where the Russian navy's use of explosive shells destroyed a wooden Ottoman fleet [3]. Soon, armored ironclad ships, such as the famous duel between the Monitor and the Merrimack (CSS Virginia) in 1862, showed that traditional wooden navies were obsolete [3]. The arms race in naval technology led to ever-larger battleships by the early 20th century, as well as new classes such as submarines and torpedo boats.

#### 6.2.2.2 Total War and Society

The term "total war" was coined during World War I, but its roots lie in the 19th-century realization that industrial wars demanded the mobilization of entire societies. Industrialization blurred the line between civilian industry and military power. As one analysis notes, the industrial revolution in warfare introduced the concept that the full resources of a nation, its industry, economy, and population [could be] mobilized for military purposes [3]. This meant economic output (coal, steel, armaments production) became as critical as battlefield tactics. In the American Civil War, General Sherman's infamous March to the Sea targeted the Confederacy's economic base (farms, railroads), an early harbinger of total war. By World War I (1914–1918), entire national economies were directed toward the war effort: men were conscripted in masse, women entered factories to produce munitions, and food and fuel were rationed to support armies. The civilian workforce and industrial infrastructure became legitimate strategic targets in the eyes of war planners, leading to the bombing of factories and cities in later conflicts.

#### 6.2.2.3 Technological Innovations in Warfare 2.0

The late 19th and early 20th centuries saw a quickening pace of military innovation. Some notable developments include:

- **Firearms**: Transition from single-shot rifles to bolt-action repeaters and machine guns made infantry immensely more lethal. By World War I, a machine gun crew could mow down waves of charging soldiers, leading to stalemates and trench warfare.
- **Artillery**: Advances in metallurgy and chemistry produced longer-ranged, more powerful artillery, as well as high-explosive shells. Artillery became the primary killer in World War I, responsible for most casualties.
- **Chemical Weapons**: Introduced in World War I (chlorine and mustard gas), chemical agents were a new technology of mass destruction, causing horrific injuries and psychological terror. Though chiefly a World War I phenomenon, chemical weapons showed how science could create entirely new threat paradigms.
- **Armored Vehicles**: Tanks were developed in World War I as a response to trench deadlock – armored, tracked vehicles to cross no-man's land. The first tanks (the British Mark I in 1916) were rudimentary, but they heralded the future of mobile armored warfare, perfected in World War II.
- **Aircraft**: The Wright brothers' first flight was in 1903; by 1914, airplanes were in combat as scouts, and soon as fighters and bombers. World War I aircraft were primitive, but by World War II, air power (strategic bombing, ground attack, and fighters) had become a decisive element of warfare.
- **Communications**: The telegraph, telephone, and radio revolutionized command and control. No longer were generals limited to issuing orders via couriers on horseback; by World War II, wireless radios enabled coordination of complex maneuvers between units miles apart or aircraft in flight.

All these technologies combined to make wars in the first half of the 20th century extremely destructive and highly technology-dependent. World War I tragically demonstrated that industrial technology without new tactics led to slaughter; World War II showed then how new tactics (e.g., German Blitzkrieg, which integrated tanks, motorized infantry, and air support) could restore mobility and decisiveness to the battlefield using those technologies.

By the end of World War II in 1945, mankind had witnessed the horrifying culmination of Warfare 2.0: the first use of nuclear weapons. The atomic bombs, products of the same scientific-industrial system that produced radar and jet engines, ushered in a qualitatively new era of destructive capability. In one sense, nuclear arms marked the peak of industrial-mechanized warfare, a single bomb with the destructive force of thousands of conventional bombs. They helped precipitate a shift in thinking about warfare that led to Warfare 3.0 and 4.0. The Cold War that followed World War II was characterized by a paradox: superpowers armed with tens of thousands of nuclear weapons (the ultimate industrial products) engaged in proxy wars and psychological warfare rather than direct industrial war, for fear that a total war would annihilate both victor and vanquished. This strategic deadlock spurred interest in more precise, information-based warfare and technological superiority below the nuclear threshold, a bridge to the digital military revolution.

In summary, Warfare 2.0 demonstrated both the power and the peril of technology in war. Industrial might enabled victory (as with the Allies out-producing the Axis in World War II) [3], but also threatened civilization itself (as in the firebombing of cities and the specter of nuclear holocaust). War grew "mechanized and industrialized…with devastating consequences" [3]. The lessons learned about the importance of innovation, the need for doctrines to match new weapons, and the profound ethical questions raised by total war would frame approaches to conflict in the later 20th century and beyond.

### 6.2.3 Warfare 3.0: Digital Era and Network-Centric Operations

By the late 20th century, the leading militaries of the world were leveraging the fruits of the information age, giving rise to Warfare 3.0: the Digital Era of warfare. In this paradigm, the emphasis shifted from sheer industrial mass to information superiority, precision, and speed. Computers, sensors, and communication networks became as critical as tanks or fighter jets. The US and its allies were at the forefront of this shift, especially after witnessing the costly grinding battles of attrition in Vietnam and the brief but technologically one-sided victory in the 1991 Gulf War.

#### 6.2.3.1 Key Features of Warfare 3.0

- **Network-Centric Warfare** (NCW): A defining concept of this era, NCW posits that networking all forces and sensors can dramatically enhance military effectiveness. As one analysis described, since the 1990s, Western armed forces (particularly the US) invested heavily in "modern network-shaped technology" to achieve unprecedented precision and operational speed [4]. The idea was that an interconnected force sharing real-time data from satellites,

AWACS radar planes, drones, and ground sensors could "see" the battlefield more clearly and react faster than a foe blinded or slowed by poor communications. This was evident in Operation Desert Storm (1991), where coalition forces, guided by GPS and radar surveillance, systematically dismantled Iraqi command and control and struck targets with precision-guided munitions (PGMs). By the early 2000s, U.S. doctrine openly embraced "information dominance," the ability to collect, control, and exploit information while denying the enemy the same, as a strategic goal.

- **Precision-Guided Munitions (PGMs)**: A hallmark of digital-era warfare was the use of PGMs bombs or missiles that could home in on laser designators, GPS coordinates, or other guidance to strike targets with high accuracy. This drastically reduced the number of sorties or rounds required to destroy a target and lowered collateral damage. In 1972, the U.S. Navy famously used a laser-guided bomb to destroy the Thanh Hoa bridge in Vietnam, which had resisted dozens of raids with conventional bombs. By the 1990s, laser- and GPS-guided bombs had become common. The 1999 NATO air war in Kosovo, for instance, used most precision weapons, allowing strategic targets to be hit while minimizing civilian infrastructure damage compared to earlier carpet-bombing techniques.
- **Stealth and Electronic Warfare**: The digital era also saw the maturation of stealth technology and sophisticated electronic warfare (jamming, deception) to suppress enemy air defenses and communications. These technologies underscored the value of "information asymmetry," where one side can see or strike the enemy, but the enemy cannot do the same in return.
- **Computers and Decision Support**: The introduction of computers into command centers enabled more effective planning and coordination. Digitized maps, intelligence databases, and eventually AI prototype systems (such as early expert systems) could assist officers in analyzing scenarios. By the 1980s, the US had developed systems like JSTARS (which integrated surveillance data to track ground units in near-real time) and had experimented with AI for tasks like image recognition (though hardware limitations at the time made these systems primitive by today's standards).
- **Cyber Warfare (information warfare)**: A new domain quietly emerged as militaries became increasingly computer-dependent in the realm of cyber. From the 1980s onward, nations began to develop capabilities to penetrate each other's computer networks for espionage (e.g., spying on adversary plans) and for offensive effects (sabotage or disruption). Early incidents, such as the 1982 alleged CIA sabotage of a Siberian pipeline via malware, and the Morris Worm, which accidentally disrupted parts of the early internet in 1988, hinted at the potential. However, it was not until the mid-2000s and beyond that cyber warfare became a prominent element of conflict (such as the 2007 cyberattacks on Estonia, or the Stuxnet malware in 2010, which physically damaged Iranian nuclear centrifuges). Still, these developments belong to the tail end of Warfare 3.0, overlapping with the dawn of Warfare 4.0. In Warfare 3.0, "cyber warfare" was often discussed within

> the framework of information warfare, a broad concept encompassing psychological operations, electronic warfare, and cyber operations aimed at disrupting enemy information systems.

Perhaps the best encapsulation of Warfare 3.0's ethos is the U.S. military's performance in the 1991 Gulf War and the 2003 invasion of Iraq. In both instances, technology-enabled speed, precision, and coordination led to swift conventional victories. A quote often cited from a U.S. Air Force planner was that in 1991, they were able to "kill targets from the weight of a position paper," meaning that with precise GPS coordinates in a report, they could task a weapon to destroy a target without troops ever engaging it directly. This was a far cry from the attrition tactics of World War I or the mass tank battles of World War II.

However, Warfare 3.0 also taught humbling lessons. In asymmetrical conflicts (e.g., guerrilla warfare in Iraq and Afghanistan post-2003), high-tech armies found that information dominance and precision did not automatically translate to strategic success. Adversaries adapted by hiding among civilian populations or using low-tech means (like IED roadside bombs) that were hard to counter with advanced sensors. Thus, even as the digital revolution enhanced conventional warfighting, it did not render war "easy" or bloodless; it merely changed the parameters. It also set the stage for autonomy to tackle some of these challenges, which brings us to the current transition into Warfare 4.0.

By the late 2010s, emerging technologies such as AI, machine learning, and robotics were being integrated into military systems in experimental ways, hinting at the next revolution. The global spread of the internet and advanced computing meant that even middle powers and non-state actors could access potent technologies (drones, encryption, cyber tools) that were once the preserve of superpowers. Warfare 4.0 builds upon this democratization of advanced tech but pushes it further by potentially delegating critical functions to machines themselves. The following section examines this new frontier in detail, as armed forces worldwide race to leverage AI and autonomy for strategic advantage.

### 6.2.4 Warfare 4.0: Autonomous Systems, Cyberwarfare, and AI-Augmented Strategy

Warfare 4.0 represents the integration of AI into the conduct of war, bringing autonomous systems and advanced algorithms to the forefront of military operations. In this evolving paradigm, militaries seek to exploit AI for faster decision-making, greater precision, force multiplication through autonomous platforms, and new forms of cyber and information warfare. The battlefronts of the AI age extend beyond the traditional domains, permeating cyberspace and even the cognitive realm (influencing perceptions through AI-generated media). This section explores three significant aspects of Warfare 4.0: (1) autonomous weapons and systems, (2) AI-driven cyber warfare, and (3) AI-augmented command and strategy.

#### 6.2.4.1 Autonomous Weapons and Unmanned Systems

One of the most visible developments of Warfare 4.0 is the rise of autonomous weapons and unmanned systems across all domains: air, land, sea, and undersea. These range from drones that can fly and identify targets with minimal human guidance to robotic ground vehicles, unmanned naval vessels, and even autonomous guard towers or sentry guns. The key difference from earlier remote-controlled systems is the degree of independence: an autonomous weapon can select and engage targets on its own based on pre-programmed criteria or AI algorithms, without needing a person to pull the trigger in real time.

A prominent example is the development of drone swarms. Military powers are experimenting with deploying large numbers of small, networked drones that coordinate their actions via AI, a concept inspired by swarms in nature. China has invested aggressively in autonomous drone swarm technology, seeking to counter traditional U.S. advantages in conventional platforms. As of 2025, reports indicate that China tested swarms of over 100 drones communicating with each other via onboard algorithms, performing collective maneuvers such as saturation attacks [5]. Unlike earlier drones that required a human pilot for each vehicle, these swarms use multi-agent AI: the drones share data and make decisions collaboratively in flight (e.g., which targets to strike or how to avoid defenses). This ability was demonstrated in a 2017 test where 119 fixed-wing drones were launched and operated as a coordinated swarm [5]. Such swarms could overwhelm conventional defenses by their sheer numbers, even if dozens are shot down, others get through. Air defense systems designed to track a handful of missiles may be unable to cope with hundreds of autonomous drones attacking simultaneously [5].

Major air forces are also developing “loyal wingman” drones, AI-powered unmanned jets meant to fly alongside manned fighters. These systems (for instance, the U.S. Skyborg program or China’s GJ-11 stealth UAV) would act as force multipliers, scouting ahead or absorbing risky tasks like drawing enemy fire, thereby protecting human pilots. The concept envisions squadrons where one or two piloted fighters command a pack of autonomous wingmen. China’s Feihong-97 drone and others have already flown in formation with Chinese J-20 stealth fighters as prototypes [5]. The US is similarly testing Collaborative Combat Aircraft to pair with its next-generation fighters [5]. If successful, this could greatly expand airpower: a limited number of elite pilots could control a much larger, dispersed force of robotic combat aircraft. It also raises strategic questions, for example, would losing a drone (with no crew at risk) make a nation more willing to engage in aggressive tactics that might escalate a conflict?

On land, several countries have developed unmanned ground vehicles (UGVs) and robotic tanks. Russia, for instance, unveiled the Uran-9 robotic tank, tested in Syria, which carries weapons like anti-tank missiles and a cannon. Though early iterations had technical problems (communication range, reliability), continued advances in sensors and AI are improving UGV autonomy. The vision is that UGVs could lead assaults or guard perimeters without exposing soldiers to fire. Similarly, stationary autonomous weapons such as the Israeli Harpy and Harop loitering munitions (essentially drones that cruise until they detect a radar signal and then dive-bomb it) have blurred the line between missile and drone. These “fire and forget” systems have been exported and reportedly used in conflicts like the 2020 Nagorno-Karabakh war between Armenia and Azerbaijan. Defense analysts noted that Azerbaijan’s effective use of loitering drones against Armenian tanks and artillery was a game-changer in that war, indicating how even relatively small states can leverage autonomous or semi-autonomous weapons to defeat traditional armor and defenses.

At sea, both the U.S. Navy and China's PLA Navy are experimenting with unmanned vessels. China launched the JARI unmanned combat boat (a small drone warship loaded with sensors and weapons) and is developing unmanned submarines for reconnaissance and targeting [5]. The US has tested an unmanned anti-submarine vessel (Sea Hunter) capable of long-endurance missions. Autonomy at sea is attractive for tasks such as mine countermeasures (traditionally dangerous for manned ships) and for augmenting fleets with more sensors and weapons without requiring additional sailors.

**Advantages of Autonomy**: Autonomous systems offer several potential benefits. They do not get tired or frightened, can operate in communications-degraded environments (if truly autonomous), and can react at machine speeds much faster than human reflexes. They can also be made smaller and cheaper than manned systems because they don't require life support or crew safety features. For instance, dozens of small expendable drones might cost less than a single advanced fighter jet, yet collectively they could perform some of the same missions. This cost asymmetry is not lost on military planners. A Chinese analysis, for example, noted that inexpensive autonomous swarms could "meaningfully balance against superior forces," citing how even non-state actors like Houthi rebels used drone swarms to harass much more expensive U.S. naval ships [5]. Autonomy, thus, could enable asymmetric warfare techniques in which a less-funded force uses intelligent swarms to defeat a more lavishly equipped conventional force.

Another advantage is risk reduction for personnel: if a mission is extremely "dull, dirty, or dangerous" (the 3 Ds often used to describe ideal robot roles), it is better to send a robot. Clearing a minefield, scouting a booby-trapped urban street, or absorbing the initial volley of enemy fire are tasks well-suited for unmanned systems. Militaries hope that by using AI and robots, they can both spare lives and perhaps achieve tactical surprise or persistence that humans cannot (e.g., drones loitering for 24 hours straight, awaiting a target).

**Challenges and Concerns**: Despite their promise, autonomous weapons raise many concerns, which we will discuss in the Challenges section. One immediate issue is reliability: complex AI systems can and do make mistakes today, and no AI is flawless at target recognition, for example. And when autonomous weapons make mistakes, the consequences could be lethal and hard to undo. There's also a fear that autonomous engagement removes a critical human-judgment step that, in the past, might have prevented catastrophe (for instance, a human operator might notice an attack is hitting the wrong target and abort, whereas a robot might not recognize civilians vs. combatants as accurately). Indeed, some nations like the US publicly maintain that they will keep a human "in the loop" for lethal decisions. U.S. doctrine emphasizes human oversight ("meaningful human control") over autonomous systems, partly due to ethical and safety considerations [6]. In contrast, it is speculated that countries like China or Russia may be more willing to deploy fully autonomous lethal systems without a human in the loop, seeking the speed and efficiency advantages even at greater risk [5]. This divergence could lead to a security dilemma in which, if one side believes the other might let AI trigger attacks faster than a human could respond, there is pressure to remove humans to avoid being OODA-looped (out-decided) by the enemy's AI. We will revisit this in the discussion on deterrence and escalation.

In summary, autonomous war machines are no longer science fiction; they are already in early deployment and will proliferate. Warfare 4.0's physical combat may feature far fewer human soldiers on the front lines and far more drones and robots. But autonomy isn't just about hardware; it's equally about the software "brains" that lead to AI's role in the cyber and command realms.

#### 6.2.4.2 AI-Driven Cyberwarfare

Cyberspace is as much a battleground in Warfare 4.0 as any physical territory. AI plays a dual role in cyber warfare: it is used both for offense (conducting or amplifying cyberattacks) and for defense (detecting and countering intrusions). What distinguishes AI-driven cyberwarfare is the scale, speed, and sophistication with which operations can be executed. In traditional hacking or cyber espionage, human attackers find vulnerabilities and write exploits, which takes time and expertise. AI can automate significant portions of this cycle, potentially launching and adapting attacks at a pace no human team could match.

A striking recent illustration came in 2025, when a tech company uncovered what is believed to be the first AI-orchestrated cyber espionage campaign [7]. In that case, an advanced AI (Anthropic's "Claude" language model) was manipulated by a threat actor to autonomously perform about 80–90 percent of a cyber operation targeting multiple organizations [7]. The AI was tasked to do everything from reconnaissance (scanning target networks) to writing exploit code and even exfiltrating data, with minimal human guidance. Over ten days, the AI system carried out thousands of hacking actions (such as testing passwords and searching for sensitive files) at machine speed, far beyond what a human hacker or even a team could do in that time [7]. Human operators intervened only at a few critical decision points; the AI handled the bulk of the labor, illustrating the concept of a "cyber agent" operating continuously and intelligently [7]. As analysts noted, this attack ran at "a speed and scale that would have been simply impossible to match" for human hackers [7].

The implications of such AI-driven cyberattacks are profound. Threat actors can use AI "agents" to automate reconnaissance, vulnerability discovery, and exploitation. For instance, an AI can simultaneously probe thousands of systems, adapt to defenses it encounters, and even write new malware on the fly, tailored to a specific target environment. In the case above, the AI identified high-value data and extracted it, even generating summaries and documentation of what it stole [7]. This level of automation lowers the barrier for conducting large-scale cyber campaigns. A less-skilled group could leverage a powerful AI to do much of the heavy lifting, potentially enabling actors who wouldn't otherwise have the resources to compete with major cyber powers [7].

On the defensive side, AI is equally a game-changer. Cyber defense has been likened to finding needles in haystacks, detecting the one malicious pattern among millions of regular events in a network. AI systems, particularly machine learning anomaly detectors, can sift through massive volumes of network traffic or system logs to flag suspicious activities that humans might miss. They can learn to identify subtle indicators of compromise or predict which vulnerabilities in a system are most likely to be targeted, allowing preemptive patching. There's an emerging concept of autonomous cyber defense, in which AI programs within a network can automatically isolate or neutralize intrusions in real time. Indeed, the future might pit AI vs. AI in cyberspace: attackers deploying AI malware and defenders using AI hunters to counter them. One cybersecurity expert noted we are entering a "new battlefield" where "AI fights AI" and humans may largely supervise or set objectives.

A significant advantage of AI in cyber is speed. Attacks that unfold in milliseconds (like an AI worm rapidly spreading) leave no time for human-in-the-loop responses. Only another automated system can react fast enough. This has driven research into things like AI-based intrusion detection systems and automated incident response, essentially, cyber firefighting bots. For example, DARPA (the U.S. Defense Advanced Research Projects Agency) ran a Cyber Grand Challenge in 2016 in which AI-driven systems competed to hack each other and patch themselves autonomously, showcasing the feasibility of machine-speed cyber operations.

However, AI in cyber warfare also raises new concerns. An AI might accidentally escalate an attack beyond intended limits (e.g., a self-spreading AI virus that goes out of control). There's also the risk of misattribution; AI attacks might cleverly mask their origin or even falsely implicate another party, potentially leading to misdirected retaliation. On the flip side, defenders worry about adversarial attacks on their own AI systems: just as one can fool a computer vision AI with specially crafted inputs (e.g., stickers on a stop sign making an AI think it's a speed limit sign), an opponent might trick an AI intrusion detector into ignoring a real threat.

One thing is clear: cyber warfare in the AI era will be more persistent and autonomous. Instead of sporadic hacks, there may be continuous AI-driven probing and dueling in networks worldwide. AI "agents" could perform espionage by quietly residing in systems and exfiltrating data in stealthy ways, or, conversely, act as active saboteurs that only trigger at key moments (such as disabling an air defense network now that a physical attack begins). The integration of AI means that cyber operations can be conducted at a scale (breadth of targets) and depth (sophistication of methods) that is orders of magnitude greater than before [7].

A real-world example illustrating both threat and defense: after the AI-led espionage incident was discovered, the defenders (Anthropic's security team) reportedly also used their AI to help analyze the attack and improve detection [7]. This shows the symmetry: the same AI that can be weaponized for attack can be a tool for defense. As the company wrote, "the very abilities that allow [our AI] to be used in attacks also make it crucial for cyber defense… our threat intel team used [AI] extensively in analyzing the enormous data" from the incident [7]. The lesson is that AI is now a necessary component of cybersecurity; one must fight fire with fire.

In sum, in Warfare 4.0's cyber domain, we will see autonomous cyber weapons, malware, and bots powered by AI and autonomous cyber defenses, AI guardians in networks. This constant low-level conflict in cyberspace can have high-level consequences, such as knocking out an adversary's critical infrastructure (power grids, financial systems) in conjunction with or even instead of kinetic military action. The silent battles between algorithms might precede or replace open battles between soldiers. This requires militaries and societies to think about security in new ways, where victory might go not to the side with more tanks or planes, but the side with better code and algorithms.

#### 6.2.4.3 AI-Augmented Command and Decision-Making

The third pillar of Warfare 4.0 is the use of AI to assist (or even partially automate) strategic and tactical decision-making, essentially serving as a brain multiplier for commanders and planners. Modern military operations generate overwhelming amounts of data: intelligence from satellites and drones, unit reports, sensor feeds, open-source information on the internet, etc. Human staff struggle to process this "fog of data" promptly. AI-enabled decision support systems (DSS) promise to parse big data, recognize patterns, and present options or predictions to human decision-makers, dramatically speeding up and improving the quality of decisions.

Military commanders are interested in AI's potential to improve situational awareness and predictive insight, especially at the operational level, where they must rapidly integrate multifaceted information to make life-and-death decisions [8]. For example, an AI system might fuse inputs from dozens of reconnaissance drones and sensors to provide a continuously updated map of enemy and friendly positions. Already, experimental systems use computer vision to scan live video for enemy units or use machine learning to predict an adversary's likely movements based on observed behavior patterns. In essence, AI serves as an extra "analyst" tirelessly sifting data that would overwhelm humans.

A 2025 report by the Center for Security and Emerging Technology highlights that AI-enabled systems can enhance situational awareness and accelerate critical decisions even in high-pressure environments [8]. For instance, if a commander has minutes to decide how to respond to a sudden missile launch, an AI could rapidly analyze sensor data to recommend which interceptor to fire and predict the impact point. These tasks might take a human longer or be prone to error under stress. In testing, some AI models have already shown superhuman performance in specific decision tasks, a well-known example being AI agents in simulated dogfights that have beaten experienced human pilots in trials by reacting faster and exploiting flight dynamics in non-intuitive ways.

Another application is AI-driven simulations and war-gaming. AI can run millions of simulated battles to explore outcomes of various tactics or strategies, far beyond what human planners with a few tabletop exercises could achieve. Such AI war games can provide decision-makers "a more nuanced understanding of the potential consequences of various actions." Militaries are incorporating AI into their training and planning simulations; for example, the US has used machine learning to simulate enemy courses of action and challenge officers in exercises with more realistic, adaptive opponents than traditional scripted scenarios.

In the realm of intelligence, AI tools can correlate disparate data to uncover patterns that human analysts might overlook. As one overview noted, AI algorithms can analyze vast datasets from various sources to identify patterns and anomalies that human analysts are unlikely to detect. This is useful for tasks such as counterterrorism (finding the needle in a terror plot in the haystack of global communications) or anti-submarine warfare (spotting subtle indicators in ocean surveillance data). Furthermore, AI can improve predictive capabilities, giving policymakers foresight, for example, forecasting political instability or upcoming enemy offensives by analyzing social media, economic indicators, and troop movements collectively.

Logistics and resource allocation also benefit from AI optimization. Modern militaries are highly complex organizations, and AI can help allocate limited resources to where they are needed most. An AI might dynamically adjust supply routes or maintenance schedules based on up-to-the-minute battlefield data, thus optimizing resource allocation in real time. This kind of AI-driven efficiency ensures that, say, fuel and ammunition reach a brigade at the right time and place, or that repair parts are prioritized for the most critical equipment, a significant advantage in fast-paced conflicts.

While AI decision aids hold great promise, their integration must be handled carefully. Commanders are advised to balance enthusiasm with understanding the limitations of these systems [8]. Several challenges have been identified: AI systems can be brittle if pushed outside the scenarios they were trained on (context shift failures) [8]. They may have hidden biases in the training data. For example, if an AI learned from historical data that the enemy consistently attacked from the north, it might underplay the possibility of a southern attack, leading to blind spots. Moreover, there are aspects of war, particularly those involving human psychology and will, that AI may find inherently difficult to predict, as human behavior is less deterministic than physics. Planners are cautioned that irreducible uncertainty will always exist in questions such as "What will the enemy do next?" and that human judgment is still required [8].

Human-AI interaction problems are also a concern. One is automation bias, the tendency of humans to over-trust suggestions from an AI, even if flawed. Military history already has cautionary tales: in 1988, a U.S. Navy ship's automated Aegis system mistakenly suggested a civilian airliner was a hostile fighter, and the crew's over-reliance on that automation contributed to a deadly error (Iran Air Flight 655 was shot down) [6]. Similar "blind trust" issues occurred with Patriot missile batteries in 2003, leading to fratricide incidents when operators assumed the automated classifications were correct [6]. These incidents underscore that no matter how advanced, AI outputs must be approached with a critical mindset. Training and doctrine need to evolve to ensure operators understand both the capabilities and limits of AI, and remain alert to when the AI might be wrong, as Stanislav Petrov did in 1983, when his intuition overrode a false computer alarm of a nuclear attack, thereby averting disaster [6].

Conversely, there's also the risk of information overload or confusion if AI systems present too much data or contradictory recommendations. Ensuring that AI decision aids are user-friendly and provide explanations for their suggestions is an active area of research (Explainable AI). A black-box recommendation ("Attack now") without a rationale is problematic for accountability and trust. Military leaders are unlikely to defer to an AI unless they understand the reasoning or have strong evidence of its reliability.

Despite these challenges, early use of AI decision support in military contexts has been encouraging. For instance, intelligence analysts using AI assistance to sort through drone surveillance footage have reported dramatic reductions in workload as the AI flags the likely frames containing objects of interest (vehicles, personnel, etc.). In wargame experiments, mixed human-AI planning teams often outperform human-only teams, as the AI can rapidly crunch outcomes of moves, while humans contribute creative thinking and strategic judgment. This synergy hints at a future model often referred to as centaur warfighting, a term borrowed from chess where human-AI teams proved superior to either pure humans or pure AIs. In warfare, centaur teams might mean a command post where officers and AI systems collaborate in real time: the AI monitors myriad feeds and suggests optimal courses of action, and the officers apply intuition and ethical judgment to select or modify those courses.

Around the world, militaries are pursuing these AI enhancements. The U.S. Project Maven (started in 2017) sought to use AI to analyze imagery more quickly. China's military, via its civil-military fusion strategy, is investing in AI for data fusion and decision support, with Chinese generals publicly emphasizing that whoever harnesses AI will gain a significant edge [6]. Other nations, from Israel to India, are also entering the fray, recognizing that AI could be the key to managing the complexity of future multi-domain operations (integrating land, sea, air, cyber, space).

In conclusion, AI-augmented strategy and command is about making better decisions faster. It leverages the machine's strengths (data processing, pattern recognition, speed) to support human strengths (strategic insight, value-based judgment, creative problem-solving). If Warfare 3.0 NCW gave commanders unprecedented information, Warfare 4.0 risks drowning them in data, so AI is both the cause (more sensors = more data) and the cure (helping digest data) for this dilemma. The end goal is a sort of hyper-enabled commander, not replaced by AI but empowered by it, able to direct forces with a clarity and swiftness that outmatches any adversary stuck with legacy decision speeds. Whether this translates into battlefield success will depend on effective human-machine teaming and trust, which are challenges actively being explored through military exercises and trials.

Having outlined what Warfare 4.0 entails, we now turn to the broader opportunities this transformation offers and the risks and ethical quandaries it presents.

## 6.3 OPPORTUNITIES AND TRANSFORMATIONAL POTENTIAL

The advent of AI and autonomous systems in warfare offers a range of potential benefits and transformative opportunities. Proponents argue that if harnessed responsibly, these technologies can make military operations more efficient, precise, and even more humane (by sparing humans from danger and reducing accidental harm). Here, we outline some key opportunities presented by Warfare 4.0:

### 6.3.1 Enhanced Decision-Making and Speed

One of AI's most lauded advantages is its ability to process information and execute tasks at machine speed. In military OODA loops (Observe–Orient–Decide–Act), shaving off seconds or minutes can be decisive. AI-enabled DSSs can analyze complex, dynamic battlespaces faster than any human staff, providing commanders with timely insights. They enhance situational awareness and accelerate critical operational decisions, even in high-pressure environments [8]. For example, AI can rapidly collate satellite images, signals intelligence, and field reports to present a near-real-time map of threats. With the aid of AI, a headquarters might compress planning cycles from days to hours, enabling forces to outmaneuver opponents. Faster defensive reactions (such as automated cyber defenses or air defense responses) can also blunt enemy attacks that are too fast for humans. In essence, AI offers to partly overcome the "fog of war" by digesting big data into actionable intelligence with unprecedented speed.

### 6.3.2 Improved Precision and Reduced Collateral Damage

Precision in warfare saves lives and resources. AI and autonomous systems can contribute to greater accuracy in several ways. First, AI-powered target recognition can help identify military targets more accurately and avoid civilian objects. If an autonomous drone or a decision aid can tell a civilian truck from an army transport with high confidence, it reduces the chance of mistaken strikes. Moreover, AI can guide weapons or maneuver units with fine-grained control. An autonomous munition might adjust its course mid-flight to avoid a school next to a valid target. Proponents of autonomous weapons even claim that such systems, not prone to human panic or revenge impulses, could make combat more ethical by adhering strictly to their programming to minimize unnecessary harm (though this is debated). We already see elements of this: loitering munitions can wait and attack only when a target is confirmed, rather than a pilot under pressure releasing ordnance with partial info. The result is potentially fewer errant munitions and fewer collateral damage and friendly fire incidents.

### 6.3.3 Force Multiplication and Efficiency

AI allows militaries to do more with less. Autonomous systems can act as force multipliers, meaning a smaller force equipped with advanced AI can achieve results comparable to a much larger traditional force. We see this with drone swarms: dozens of cheap drones guided by AI might overwhelm a far costlier defense system [5]. This inverts the old paradigm of warfare, where victory often required numerical or industrial superiority. For nations with smaller populations or budgets, AI is an attractive equalizer. Even for large militaries, AI can increase efficiency: predictive maintenance algorithms can analyze equipment sensor data to predict failures and schedule repairs preemptively, significantly improving vehicle and aircraft readiness rates. Logistics can be optimized with AI to find the best supply routes and methods, reduce waste, and ensure troops have what they need when they need it. In training, AI-driven simulators can cheaply train pilots or soldiers in virtual environments, preserving expensive fuel and equipment life. Collectively, these efficiencies mean a military can maintain a high operational tempo and capability with potentially lower cost and manpower footprints.

### 6.3.4 Risk Reduction to Human Soldiers

A very tangible benefit of autonomous systems is removing humans from immediate harm's way. Dull, dirty, and dangerous missions can be offloaded to robots. Reconnaissance in a sniper-filled city, logistics convoys through IED-laden roads, or initial breaching of a mined beachhead, these are scenarios where previously many lives might be lost. With robots or drones, militaries can accomplish the task with reduced risk to personnel. The U.S. military, for example, uses unmanned systems to clear mines and IEDs, a job that in the past cost many sappers' lives. In the future, entire forward operating bases might be run with largely robotic perimeter security, so fewer troops are exposed. As noted earlier, AI-powered autonomous weapons can operate without human intervention, potentially reducing the danger to human soldiers [9]. This could also have strategic effects: nations may be less hesitant to undertake specific operations if they expect minimal casualties, possibly acting as a deterrent advantage (the enemy knows you can strike them without suffering losses, which might dissuade the enemy from aggression in the first place).

### 6.3.5 New Capabilities and Domains

AI opens possibilities that did not exist before. For instance, swarming tactics coordinating dozens or hundreds of assets simultaneously would be impossible to manage by humans alone, but AI can handle the complex choreography. This enables qualitatively new tactics, such as swarms that saturate defenses or conduct rapid, widespread search operations over large areas. In cyberspace, AI can hunt for threats at scale, as mentioned, making networks far more defensible. Another new capability is realistic simulation and planning: generative AI can create rich simulated environments (even deepfakes or virtual populations) to test out information operations or train units in cultural contexts. AI can also enhance electronic warfare by rapidly sensing and adapting to changes in the electromagnetic spectrum (jamming frequencies, etc.) faster than adversaries. In space, autonomous satellites could maneuver and respond to threats (such as anti-satellite weapons) on their own, thereby protecting valuable assets. All these novel capabilities expand the military toolkit in ways earlier generations couldn't imagine.

### 6.3.6 Deterrence and Stability (Potentially)

While much discussion of AI in warfare focuses on the risks of instability (which we will address in the next section), there are arguments that AI could strengthen deterrence if used appropriately. For example, AI might improve early warning and attribution of attacks. One study suggests that machine learning and data fusion can more effectively detect and attribute actions such as missile launches or cyberattacks, thereby increasing the credibility of deterrence by ensuring a swift, confident response [10]. If an AI can immediately identify who is behind a cyber intrusion with high confidence, it removes the window where an aggressor might have hoped to sow doubt or avoid retaliation. Additionally, AI might enhance defensive resilience (through rapid countermeasures) such that adversaries are less tempted to attack in the first place. An analogy: if both sides know each has AI that can retaliate almost instantly to any nuclear strike (via automated early warning), that could reinforce nuclear deterrence by removal of human hesitancy in retaliation (although it introduces dangers too, as we'll note). In conventional deterrence, a nation might demonstrate AI-driven capabilities (such as drone swarms or robotic deployments) to dissuade adversaries. A recent example is how Turkey's use of drones in conflicts (Syria, Libya, Azerbaijan) showcased a new form of military power that neighboring states must factor into their calculations.

### 6.3.7 Humanitarian and Peacekeeping Applications

Beyond combat, AI can assist in operations other than war. For instance, AI can significantly aid in disaster relief missions that militaries often support by using algorithms to map damage via satellite images, coordinate logistics, or even guide autonomous rescue drones. In peacekeeping or stability operations, AI might help predict where violence could flare up by analyzing social media and local data, allowing peacekeepers to intervene early. Surveillance drones with AI could monitor vast border areas to prevent arms smuggling or protect refugee camps without requiring a massive troop presence. Some have even proposed autonomous systems for tasks like de-mining war-torn areas (robots that use AI to identify and neutralize landmines more safely). These contributions, while not as celebrated as combat roles, demonstrate that the same technology that enables more effective warfare can also save lives and reduce suffering in war's aftermath.

In aggregate, these opportunities depict a vision of warfare in which smarter, more nimble forces strike only what they intend to, defend more effectively, and expose their personnel to less hazard. It is a vision in which efficiency and control are improved to the extent that war's chaos might be somewhat tamed. To give an illustrative scenario: imagine a UN peace enforcement mission in 2035 under Warfare 4.0 doctrine. Autonomous surveillance blimps monitor ceasefire lines 24/7, AI analysis spots a buildup of hostile militia in real time and alerts peacekeepers, who deploy robotic vehicles to cordon off the area. A firefight breaks out, but autonomous gun turrets precisely target only shooters and incoming rounds, while drones broadcast warnings to civilians to evacuate. Human commanders oversee it all through augmented reality displays fed by AI. They make the final calls on the use of force, but the options they receive are generated by AI that has already calculated how to minimize collateral damage. Within minutes, the situation is contained with minimal loss of life. Such a scenario might sound optimistic, but it is the kind of outcome that AI optimists in defense circles envision: a transformation of warfare to something more intelligent, restrained, and swift, leveraging technology to reduce the worst human costs of conflict.

Of course, realizing these positive outcomes is not automatic. It depends on policy, training, and design choices that prioritize ethical use and reliability. Many of these benefits could be nullified or reversed if AI is misused or its challenges are not addressed, prompting us to examine the flip side: the risks, challenges, and ethical dilemmas of Warfare 4.0.

## 6.4 CHALLENGES, RISKS, AND ETHICAL CONSIDERATIONS

The integration of AI into warfare, while offering numerous advantages, also presents significant challenges and grave risks. Some issues are technical, others strategic or ethical. In this section, we discuss the key concerns: from the reliability and controllability of autonomous systems to the prospects of destabilizing arms races and escalation, to profound legal and moral questions about delegating life-and-death decisions to machines.

### 6.4.1 Loss of Human Control and Unintended Escalation

Perhaps the foremost fear surrounding autonomous weapons is that they could initiate or escalate conflict without explicit human intent at that moment. As AI-driven systems make more decisions, the risk grows that a mistake or unforeseen interaction could trigger a crisis. An autonomous weapon, by definition, can apply lethal force without a human in the decision loop at the time of action. This raises the nightmare scenario of a "Flash War," analogous to a flash crash in stock markets, in which opposing AI systems act and react in milliseconds, leading to a spiral of violence before humans even grasp what is happening. A RAND study on AI and deterrence warns that machine decision-making could inadvertently escalate due to the sheer speed and differences from human understanding [10]. For instance, an AI air defense might misidentify a flock of birds as incoming missiles and launch interceptors; the adversary's automated systems detect a missile launch and retaliate, and suddenly an exchange is underway that neither side's leadership intended. When decisions occur at "machine speeds," traditional buffers, such as a human double-checking an alert, may be bypassed [10]. The 1983 Petrov incident is instructive: a Soviet computer erroneously reported a U.S. missile attack, but a human (Petrov) wisely held off on retaliation [6]. If an AI had overseen that decision, it might have launched counter-strikes, accidentally triggering a nuclear war. Petrov's case is a potent reminder, as one analyst put it, of the risks of reliance on complex systems in which errors and malfunctions are not only probable but probably inevitable [6].

Relatedly, autonomous systems might not understand the subtleties of human signaling or rules of engagement. Deterrence often relies on clear communication and the ability to back down or de-escalate. Machines are likely worse at grasping human signaling an AI might interpret a training exercise as real aggression, for example [10]. The RAND study also noted that widespread AI and autonomy could introduce instability and "inadvertent escalation and crisis instability" if not properly managed [10]. Mixes of human and AI agents, each with different reaction times and logic, could create unpredictable escalatory dynamics [10]. All this suggests that a loss of human control can make conflicts harder to contain.

### 6.4.2 Reliability, Errors, and Vulnerabilities

No AI system is infallible. In fact, current machine learning models are known to be brittle; they can perform very well on training scenarios but then behave unexpectedly in novel situations. On the battlefield, this could translate to tragic mistakes. For example, a target recognition AI might be 99 percent accurate under most conditions. Still, it might misidentify a civilian vehicle as a hostile one if the lighting is unusual or the vehicle's paint is a rare color. Suppose that AI is allowed to fire, and that 1 percent error rate becomes lethal. Furthermore, adversaries can exploit AI weaknesses through adversarial attacks. These techniques involve an enemy subtly altering the environment to confuse an AI (such as camouflage patterns designed to fool a drone's vision into not seeing a tank, or, conversely, making a school bus look like a tank to the AI). Adversarial inputs can cause AI to misclassify objects completely. Unlike humans, AIs might fail in ways no human would (e.g., a sticker on a stop sign causing an AI to misclassify it as a speed limit sign). This is a serious concern for military AI – imagine an autonomous patrol drone that can be "blindfolded" by the enemy through hacking or environmental tricks.

Even without malicious interference, technical failure is a big issue. Software bugs or edge cases can cause systems to crash or behave erratically. A drone swarm might experience a communications glitch that causes drones to collide or fire wildly. Autonomous vehicles have been involved in accidents in civilian contexts; in war, the tolerance for such errors is even lower. The complexity of these systems means even their creators often do not fully predict every possible state, which complicates testing and validation.

Moreover, there is the issue of cybersecurity of AI systems. In many cases, an autonomous weapon is also an internet-connected computer and thus hackable. If an enemy hacks your lethal autonomous weapon, they could potentially turn it against you or make it target civilians. The dual-use nature of AI (civil and military) can lead militaries to adopt commercial AI components that may not meet military-grade security standards, increasing vulnerability. As noted in an earlier section, the dual-use aspect of AI also complicates regulation [9]. It's hard to control proliferation when the same AI tech is in everyone's smartphones.

### 6.4.3 Ethical and Legal Dilemmas (Accountability and Compliance with Laws of War)

From an ethical standpoint, the idea of machines making life-and-death decisions autonomously is deeply troubling to many. International humanitarian law (IHL) requires constant discrimination between combatants and non-combatants and proportionality in the use of force. Can an autonomous system uphold these requirements? Distinction is a critical principle: an autonomous weapon must be able to distinguish between civilian and combatant targets and between military targets and protected civilian objects. Current AI, especially vision-based, can struggle with this in the messy real world (imagine distinguishing a combatant who has put down his weapon and is surrendering, something humans can intuit from context, but an AI might not). The question "Can autonomous weapons, as mandated by international law, distinguish between combatants and civilians?" is a pressing one [9]. Many experts doubt that fully autonomous systems can reliably meet IHL standards, which could render their use unlawful if they cannot be adequately supervised.

Another issue is accountability. If a drone operated by an AI wrongly bombs a hospital, who is responsible? The AI can't be put on trial. Is it the commander who deployed it, the programmer who coded it, the manufacturer, or the state that chose to use such a system? This "accountability gap" is a core argument of those who advocate banning "killer robots." Without clear accountability, justice for wrongful deaths becomes elusive, a moral and legal black hole. A related concept is meaningful human control: many ethicists and representatives have argued that use of lethal force should always have meaningful human oversight and control, such that a human is morally and legally responsible for each strike. Autonomous weapons challenge this if they are allowed to engage on their own. Some countries (especially in Europe) and numerous NGOs are pushing for international regulations or treaties to ensure humans remain in the loop, or to ban fully autonomous lethal weapons outright. Meanwhile, other nations (the US, Russia, China) have been hesitant to agree to broad bans, preferring to see how the tech evolves.

Delegating moral agency to machines also feels problematic because machines lack human judgment and empathy. They cannot truly understand the value of human life, the context of a situation, or exercise mercy. There's an almost philosophical question: is it inherently wrong to permit an algorithm to decide to kill a person, even if that person is an enemy combatant? Some argue it undermines human dignity and the moral order of society to have killing done by automatons. This ties to the concept of "the moral buffer": political leaders might more readily resort to force if their own soldiers aren't directly at risk (the "bloodless war" for one side). If, say, one can send an army of robots, the internal political cost of war lowers, which could paradoxically make wars more frequent. Indeed, one analyst warned that AI could "destabilize the global security system" if used irresponsibly. The ease of conflict initiation when one's own people aren't immediately in harm's way may lower the threshold for war.

### 6.4.4 Arms Race and Proliferation

We are likely already in an AI arms race. Major powers are pouring investments into military AI. For example, as earlier mentioned, China's military-civil fusion policy and relative lack of transparency or ethical restraints in AI development have spurred U.S. fears of being overtaken [5]. The result is a classic security dilemma: each side accelerates development not to use first, but out of fear that the other will, yet that very acceleration makes conflict more likely or at least more unstable. AI capabilities are also diffusing beyond top-tier militaries. Commercial availability of drones and AI means even non-state actors (terrorists, insurgents) could deploy rudimentary autonomous weapons, for instance, a drone with simple object recognition programmed to attack anyone with a specific uniform or vehicle type. The proliferation of such technology is hard to stop because much of it is software and commercially available hardware. Unlike nuclear weapons, which require rare materials and extensive facilities, AI weapons can potentially be built in a garage with parts ordered online and code from GitHub.

An uncontrolled arms race is dangerous because it may lead to the deployment of untested, unsafe systems as rivals rush to field something first. It also raises the chance that these weapons fall into the hands of those with no regard for IHL (e.g., terror groups that might use autonomous weapons deliberately to target civilians). The international community is grappling with this; the UN Convention on Certain Conventional Weapons (CCW) has had meetings on Lethal Autonomous Weapon Systems (LAWS) for years, but achieving consensus is slow. In the meantime, technological momentum may outpace diplomacy. Some experts liken the AI arms race to the early nuclear arms race: a transformative technology introduced with little initial regulation, which then required urgent action (such as the Non-Proliferation Treaty (NPT)) once its destructive potential became clear. With AI, part of the complication is its ubiquity and dual-use nature.

### 6.4.5 Psychological and Organizational Challenges

Even aside from frontline risks, the incorporation of AI into military organizations presents challenges. There can be issues of over-reliance or under-reliance. Some operators might trust the system too much (as in the Aegis and Patriot incidents, where operators didn't question the machine's call) [6], while others might mistrust it too much and not use it when it would help. Training troops to work seamlessly with AI is non-trivial; it's a new "team member" with its own quirks. There's also risk of confusion in the chain of command: if an AI recommends a course of action that junior officers think is flawed, will they have the confidence to override what might be seen as the high-command-approved algorithm? And in coalitions, when different allies use different AI systems, interoperability issues arise (such as one force's drones not communicating with another's, or classification differences in target selection).

At the strategic level, heavy reliance on AI could create a single point of failure. If, in future wars, one can neutralize the opponent's AI (through hacking or electronic disruption) and that opponent has built its entire doctrine around AI, it could be paralyzed. A historical parallel is that countries deeply reliant on radar or communications satellites could be disproportionately crippled if those systems were taken out. So, some strategists caution against letting the human skills atrophy; keep training people in navigation, dogfighting, analysis, etc., as a fallback. This, however, is hard when AI seems to offer superior performance; over time, human proficiency might degrade if they always rely on the AI (there's evidence of this in aviation: pilots reliant on autopilot for routine flying can become less adept at manual flying, which can be dangerous in emergencies when autopilot shuts off).

### 6.4.6 Cybersecurity and Deepfakes in Info Warfare

Another ethical and security challenge is the misuse of AI in information warfare. Deepfake videos or audio generated by AI could be used to spread false information, incite hatred, or impersonate leaders (imagine a fake video of a president declaring war could sow chaos). In conflict, such techniques might be used to demoralize populations or justify a pretext for war. While not a direct battlefield issue, it's part of Warfare 4.0's broader context: the line between truth and deception blurs when AI can fabricate highly realistic false content. Societies and militaries will have to develop new methods to verify the authenticity of communications and intelligence.

### 6.4.7 Legal Arms Control Gap

Currently, there is no comprehensive international law fully addressing AI in warfare. The Geneva Conventions predate AI (though their principles still apply). As aforementioned, the CCW discussions have not yielded a binding agreement.

Without clear regulations, there's nothing to prevent a race to the bottom in terms of safety, i.e., if country A constrains itself to always have a human in control. Still, country B does not, and as a result, their systems have a speed advantage; country A may feel pressured to remove the human too to keep up. This dynamic can erode ethical standards. The absence of an agreement also means there's little transparency, unlike nuclear arms treaties that include inspections and confidence-building measures. Nations are secretive about their AI projects, which can lead to worst-case assumptions and arms buildup. As of 2025, dozens of countries have called for some ban or regulation on fully autonomous weapons, but major powers remain cautious. If this continues, the world might enter a phase where these systems are deployed with unclear rules of engagement, which is a recipe for dangerous miscalculations.

In summarizing the challenges, a United Nations University article put it succinctly: The weaponization of AI has profound implications for global security and warfare. AI can improve military capabilities by enabling faster decision-making, more accurate targeting, and more efficient resource allocation… However, these developments raise concerns about the escalation of conflicts, the potential compromise or misuse of autonomous weapons, and the possibility of an AI arms race [9]. It also asks pointedly: "Who bears responsibility if an AI-powered weapon causes inadvertent harm? Is it ethical to delegate decisions concerning life and death to machines?" [9] These questions encapsulate the crux of the ethical debate.

Ultimately, addressing these challenges requires a combination of technical safeguards, robust testing and validation regimes, clear operational doctrine (e.g., requiring a human veto over AI lethal decisions), and likely international norms or agreements. Some proposals include "keeping humans in the loop" mandates, setting technical standards for AI reliability, and even creating a global monitoring body for military AI developments to avoid surprise and build confidence. Ethics training for developers and military personnel is also advocated, so that, from design to deployment, people consider the moral implications (for instance, programming an autonomous system to refuse an illegal order, such as targeting civilians).

The coming years will be critical in determining whether the dangers of Warfare 4.0 can be mitigated or whether the risks spiral out of control. History has shown with technologies like chemical and nuclear weapons that humanity often seeks to put rules around the most destructive tools one can hope that will be the case with AI as well. In the next section, we look ahead to consider how warfare might evolve as AI matures further and what a LIFE 4.0 scenario might look like, where AI pervades conflict, hopefully under the guidance of human values.

## 6.5 THE FUTURE OUTLOOK (LIFE 4.0 SCENARIO)

Looking toward the future, one can imagine both perilous and promising scenarios for Warfare 4.0 as part of the broader LIFE 4.0 epoch, an era in which AI is ubiquitous across all aspects of society. By extrapolating current trends, we can sketch a picture of how conflict might unfold 10, 20, or even 50 years from now, and how the international community might adapt to the AI-driven security landscape. This outlook will consider technological trajectories, potential doctrines, and normative developments that could define warfare in the mid-21st century.

### 6.5.1 Hyper war and "Centaur" Command

Many analysts predict we are heading into a period of hyperwar, a term denoting warfare accelerated to unprecedented speeds by AI, automation, and possibly quantum computing. In hyper war, critical decisions might need to be made in split seconds; traditional chains of command may be supplemented by pre-delegated authorities to AI systems (e.g., automatically shooting down confirmed inbound missiles without waiting for human go-ahead, simply because humans cannot react in time). To manage this, militaries might adopt Centaur Warfighting, as mentioned earlier, with tightly integrated human-AI teams in which routine decisions and actions are handled by AI, freeing human leaders to focus on grand strategy and ethical oversight. A future command center may look like a fusion of a NASA mission control and a Silicon Valley operations room: rows of analysts (or warfighters) supervising AI feeds on giant screens, where AI handles the minutiae and flags only the most crucial anomalies to human attention. Humans would still set objectives and rules, but much of the execution could be autonomous. A positive outcome here is that humans, now unburdened from micromanaging units or sifting raw intel, might make better high-level decisions, focusing on intent and context that AI cannot grasp. This human-AI symbiosis could become the gold standard of military effectiveness, and nations that cultivate it (through training and trust-building between personnel and algorithms) will likely have an edge.

### 6.5.2 Global AI Arms Control Regimes

By the LIFE 4.0 era, we may also see the maturation of international agreements governing military AI. A possible future is one where there is a treaty analogous to the Nuclear NPT, but for AI in armed conflict. This might include bans on specific, particularly destabilizing systems, for instance, an agreement not to let AI control nuclear launch decisions, to avoid hair-trigger automated nuclear war. Indeed, the Bulletin of Atomic Scientists and others have advocated that nuclear command and control remain a human prerogative, given the existential stakes [11]. Future arms control could also establish norms, such as requiring autonomous weapons to have a mechanism for human override or self-deactivation if they lose communication (to prevent them from "going rogue"). Verification of such treaties would be challenging but not impossible; it could involve inspections of AI training data and algorithms (a novel form of arms inspection) or on-site monitoring of autonomous weapon storage (to ensure they're not activated without proper authorization protocols). Ideally, major powers will realize that an uncontrolled AI arms race is in no one's interest and will come to some detente or guardrails, much as they did (eventually) with chemical, biological, and nuclear weapons. For example, a future agreement might prohibit fully autonomous anti-personnel weapons, akin to the existing bans on antipersonnel landmines in many countries.

### 6.5.3 AI and Deterrence Paradigm Shifts

The presence of AI might change classical deterrence theory. In a LIFE 4.0 scenario, nations might possess AI-driven defense networks so effective that they create a kind of balance of power through omniscience. If every side knows that surprise attacks are unlikely to succeed because AI surveillance is so pervasive, the incentive to strike first could diminish. Alternatively, if offensive AI outpaces defensive AI, deterrence might shift to a doctrine of "mutually assured disruption" analogous to mutually assured destruction in the nuclear context, but here referring to cyber and robotic attacks. For instance, if country A's swarms can devastate country B's infrastructure in hours and vice versa, both might avoid direct war and resort more to strategic posturing and limited proxy conflicts. Deterrence in the AI age might also involve brandishing one's algorithms rather than traditional arms. We might see military exercises in which nations demonstrate AI capabilities (such as NATO war games involving drone swarms or AI logistics) as signals of strength, just as missile tests are. However, such demonstrations could be dangerously escalatory if misinterpreted.

### 6.5.4 Integration of New Domains and Technologies

By mid-century, other emerging tech will combine with AI, further transforming warfare. Quantum computing could break current encryption, leading to a race for quantum-resistant communications (critical for secure command links to AI systems). Biotechnology and AI might converge in "biocyber" warfare, for instance, with gene-edited bio-agents guided or designed by AI to target specific crops or even genotypes (a frightening prospect that would require new conventions to ban). Neurotechnology might enable direct brain-machine interfaces for soldiers or pilots, effectively turning them into faster decision-makers when linked with AI (e.g., thought-controlled drones or enhanced reaction times). If LIFE 4.0 entirely blurs the line between man and machine, the ethical line may blur as well. A cyborg soldier might be considered a lawful combatant, but what about an AI that's partly human-trained on brain data? These questions, though speculative, will challenge legal definitions.

### 6.5.5 Socio-Technical Impact and Public Perception

In a world permeated by AI, public perception of war could change. Domestic populations might become more detached from war if fewer human soldiers deploy (just robots fighting far away), potentially lowering the psychological barrier to conflict. Conversely, there might be strong public revulsion toward "killer robots" if incidents occur, leading democratic societies to pressure their governments to enter treaties or bans. The global civil society movement for banning autonomous weapons (the Campaign to Stop Killer Robots, etc.) could gain traction if, say, an autonomous weapon mishap causes civilian deaths and becomes a high-profile scandal. This suggests that political will and public opinion will be crucial in shaping the bounds of Warfare 4.0. Countries with open media might constrain AI use due to public concern, whereas authoritarian regimes might push forward unencumbered, creating an asymmetry. This asymmetry itself could become a security concern (like how democracies and authoritarian states currently clash over norms in cyberspace and information control).

### 6.5.6 Possibility of AI-Only Conflict

A striking possibility is that future conflicts might be decided by AI systems duking it out with minimal human casualties, essentially algorithmic attrition. For instance, two countries might engage in continuous cyber battles aiming to cripple each other's AI and networks, a sort of bloodless war of attrition until one's systems are dominant. Or aerial battles may occur where each side's autonomous drones fight each other in the skies (somewhat like sci-fi depictions of robot wars). The victor of these AI-on-AI engagements could force the loser to concede without significant loss of life. Some optimists have mused that perhaps wars of the future could be less deadly if machines absorb the brunt of the fighting (imagine a naval battle in which unmanned ships sink each other, with no sailors dying). However, even if direct deaths are reduced, the destruction and chaos wrought could still harm populations (loss of electricity, economic collapse, etc., from cyberattacks). And one must consider that if one side's machines lose, that side might resort to more desperate measures (including nuclear escalation) if facing defeat, so it's not so simple.

### 6.5.7 Defense Dominance or Offense Dominance

Historically, technology sometimes favors offense (e.g., tanks and blitzkrieg in World War II) and sometimes defense (machine guns in World War I trenches). With AI, it's debated. Will AI make attacks so swift that "the best defense is a good offense" (strike first before you're hit)? Or will AI provide such strong early warning and precision interception that attacking is futile (any attempt is quickly neutralized)? The answer may vary by domain. In cyber, offense might reign because there are always new vulnerabilities to exploit, and it's hard to secure huge networks perfectly. In strategic military action, defense might gain ground if AI-enabled missile defenses, drone defenses, etc., become highly effective (like swarms of interceptors shooting down swarms of attackers). If defense dominates, major wars might become rarer (as in the Cold War nuclear deterrence scenario). If offense dominates, instability increases (incentive to preempt).

### 6.5.8 Small States and Non-State Actors

AI proliferation means even small states or wealthy militias could wield disproportionate power. We might see private military corporations or terror groups deploying AI-driven capabilities. Imagine a terrorist cell using a swarm of explosive drones to attack a city, a very real concern as components get cheaper. Counter-AI security for VIPs (like drone jammers around parliaments) may become routine. There's also the notion of "AI terrorism," in which AI is used to identify the most disruptive ways to attack society (e.g., by planning simultaneous attacks for maximum chaos).

Governments will have to enhance domestic security using AI as well (e.g., AI surveillance to detect plots), raising yet another ethical debate about privacy and civil liberties in LIFE 4.0.

### 6.5.9 Positive Uses and Conflict Prevention

On a more hopeful note, AI might aid in conflict prevention and resolution. Predictive analytics could foresee brewing conflicts or genocides earlier, prompting diplomatic or humanitarian interventions to defuse them. AI translation and communication tools might reduce misunderstandings between nations (maybe future hotlines between leaders use AI real-time translation to allow direct communication in their own languages, avoiding misinterpretation). Autonomous peacekeepers or monitoring drones could help enforce ceasefires impartially, reducing mistrust between belligerents. In sum, while AI can fuel war, it can also be a tool for peace if applied to transparency and confidence-building.

### 6.5.10 LIFE 4.0 and Human Element

In a LIFE 4.0 world, not just warfare, but life itself is transformed by AI. That context will reflect on the military. Soldiers of the future might have grown up with AI assistants, be well-versed in interacting with algorithms, and perhaps be more comfortable taking AI advice. The general population might likewise be used to AI in daily life (self-driving cars, AI doctors). This could mean less skepticism about using AI in national defense, or it could mean higher expectations that AI be used ethically (as people will see AI doing good in other fields and demand the military adhere to similar standards).

In the end, the LIFE 4.0 scenario for warfare can be shaped by conscious choices we make today. The tools are in development; how we choose to deploy or constrain them will decide if the future battlefield is an AI-governed bloodbath, a sterile robot duel, or perhaps avoided entirely through more innovative strategy and diplomacy. Ideally, global society will navigate toward a framework in which AI enhances security rather than undermines it, where, for example, AI aids rapid de-escalation (maybe by negotiating with each other under preset human parameters to stand down forces when crises begin, a fanciful but intriguing idea).

Some visionaries even suggest a future where war itself becomes obsolete under LIFE 4.0. If AI and automation address many root causes of conflict (resource scarcity via better resource management, or ideological differences via better communication and education), perhaps nations will find few reasons to fight. That is admittedly optimistic. History suggests new technology doesn't erase old human grievances. However, in the very long term, one could imagine a "strong AI" or artificial general intelligence providing counsel to humanity to avoid self-destruction, or advanced autonomous systems handling defense so effectively that no adversary would bother to attack.

In summary, the future of Warfare 4.0 holds both great uncertainty and potential. It will likely be characterized by ultra-fast engagements, ubiquitous sensing and information warfare, and a premium on integrating human judgment with machine efficiency. It could usher in an era of more surgical, less costly conflicts, or it could magnify destruction to frightening levels. The difference lies in governance: technological evolution is inevitable, but whether our norms, laws, and institutions keep pace is the crucial question. As we move from LIFE 3.0 to LIFE 4.0, the hope is that humanity can apply wisdom to match its technological prowess, ensuring that AI becomes a tool that safeguards peace and justice rather than a Pandora's box unleashing new horrors.

## 6.6 CONCLUSION

The march from Warfare 1.0 to Warfare 4.0 reflects an extraordinary journey of human conflict alongside human innovation. We began with battles determined largely by valor, strategy, and muscle from the phalanxes and legions of antiquity to the knights and archers of feudal fields. The industrial age then brought machines, mass, and the harrowing concept of total war, culminating in world wars that pushed humanity to the brink. The late 20th century's digital revolution heralded smarter, more networked warfare, enabling precision and coordination that past generals could only dream of. Now, as we fully enter the age of AI, the dawn of Warfare 4.0, we are poised at another inflection point.

AI is reshaping conflict in real time. Autonomous drones and robots are gradually moving from labs to battlefields, taking on roles ranging from reconnaissance to strike missions. Cyber warfare driven by AI threatens critical infrastructure and blurs the lines between war and peace, front line and home front. And in command centers, AI decision support tools are beginning to inform planning and targeting, potentially revolutionizing how wars are directed. These developments carry profound implications: tactically, they promise faster ops tempos and new capabilities; strategically, they could alter the balance of power and the calculus of deterrence; ethically, they force us to reexamine fundamental questions about accountability, the value of human judgment, and the laws of war.

Throughout the chapter, we've underscored a dual theme: transformation and continuity. Technology transforms the face of war from swords to muskets to drones, yet continuity persists in war's nature and in the importance of human factors. Even the most advanced AI today lacks accurate understanding, intent, or empathy; those remain uniquely human (at least for now). Thus, while warfare in the AI era may look radically different from that of ancient times, the core issues of war and peace are still rooted in human decisions and values. In essence, war is a human endeavor, even when mediated by machines. This means that the emergence of AI doesn't remove human responsibility; it amplifies it by giving us more powerful tools to control or to kill.

One major takeaway is that in LIFE 4.0, security will increasingly depend on how healthy societies manage the integration of AI into military affairs. This includes technical management (building safe, secure systems), policy management (creating doctrines and perhaps treaties to govern use), and ethical management (instilling appropriate use and constraints). The chapter highlighted many positive potentials of Warfare 4.0: fewer friendly casualties, more precise engagements, and possibly shorter conflicts if AI helps achieve quick dominance or deterrence. But we also highlighted the myriad risks: accidental escalation, loss of control, an AI arms race, and moral hazards. Which side of the ledger becomes dominant is not foreordained; it will depend on choices made by soldiers, engineers, policymakers, and citizens in the coming years and decades.

From a global perspective, AI in warfare is a great equalizer but also a possible destabilizer. Small states and even non-state actors can harness AI to punch above their weight, potentially deterring aggression (a small state might ward off a bigger foe with smart drones) but also potentially empowering terrorist violence or regional arms races. Meanwhile, great power competition, notably the U.S.-China rivalry, is driving rapid AI militarization, as each fears falling behind the other. This competition could yield breakthroughs that benefit humanity (like AI systems that improve disaster response or global surveillance to enforce arms control) or it could entrench distrust and hair-trigger postures. International cooperation, or at least communication, will be vital to maximize the former and minimize the latter.

In conclusion, it's worth reflecting on an enduring truth: technology shapes warfare, but warfare also shapes technology's development and use. The quest for advantage in conflict has historically accelerated innovations (the internet itself has roots in defense networks, many advances in computing and robotics are defense-funded). So, AI's trajectory will in part be driven by military investment and requirements. This makes it more important to instill ethical considerations at the ground level – to ensure the push for military AI prioritizes control, compliance with international law, and alignment with human values. Military organizations are beginning to recognize this; concepts like "Responsible AI" are being adopted by some defense departments, aiming to create AI systems that are testable, transparent, and governable.

Finally, as academics, professionals, and citizens reading this in the context of LIFE 4.0, we should remember that the story of Warfare 4.0 is still being written. Unlike past revolutions that we can only analyze in hindsight, this one is unfolding in our own time. There is an opportunity and indeed a responsibility for the current generation to guide this transformation conscientiously. Multidisciplinary engagement is needed: engineers must work with ethicists, military officers with legal scholars, diplomats with AI experts. By doing so, we can strive to ensure that, as AI augments the arsenal and the command post, it ultimately upholds security and justice rather than undermines them.

In the end, the goal should be to leverage AI to prevent and shorten wars, protect civilians and soldiers, and, when necessary, enforce peace, effectively bending the arc of Warfare 4.0 toward a more humane horizon. Achieving that will be challenging, but it is a worthy quest. The alternative, letting the technology drive itself without ethical brakes or human oversight, could lead to a future as dark as any dystopian fiction. The choices we make today will determine which future materializes. Warfare 4.0 is not just about new battlefronts in the age of AI; it's about who we are and what we stand for as we navigate the most recent chapter in the age-old story of conflict and cooperation.

## REFERENCES

1. Atoui, Y., Startups, AI and Connected Insects: Germany Charts Its Path Toward War 4.0, in *Tunisia, Numerous*, July 24, 2025.

2. Colwell, B.D., Innovation and Invention in Warfare Techniques and Military Technologies: The Middle Ages, in *Blog Post*, June 26, 2025.
3. Ibrahim, A., The Industrial Revolution and Its Transformative Impact on Warfare, in *Global Defense Insight*, 2023.
4. Collmer, S., *Information as a Key Resource: The Influence of RMA and Network-Centric Operations on the Transformation of the German Armed Forces*, 2007.
5. Kass, H., China's AI Drone Swarms Should Terrify the United States, in *National Interest*, November 16, 2025.
6. Kania, E.B., The Critical Human Element in the Machine Age of Warfare. *Bulletin of the Atomic Scientists*, 2018. 74(6): p. 358–364.
7. Anthropic. Disrupting the First Reported AI-Orchestrated Cyber Espionage Campaign, in *Anthropic Blog*, November 13, 2025.
8. Probasco, E.S., et al., AI for Military Decision-Making: Harnessing the Advantages and Avoiding the Risks, in *Center for Security and Emerging Technology*, April 2025.
9. Marwala, T., Militarization of AI Has Severe Implications for Global Security and Warfare, in *United Nations University*, 2023.
10. Laird, B., et al., *Deterrence in the Age of Thinking Machines*. London: Dolphine, 2020.
11. Kallenborn, Z., Giving AI Control of Nuclear Weapons: What Could Go Wrong? 2022.

# 7 Governance 4.0

**Abstract**

Governance has continually evolved through epochs alongside social and technological transformations. This chapter surveys four broad stages from the era of monarchies and feudal rule (Governance 1.0), through the rise of constitutional democracy and nation-states (Governance 2.0), to the digital revolution of e-governance and globalization (Governance 3.0), and finally into the current frontier of Governance 4.0, defined by artificial intelligence (AI) integration into policymaking and public administration. We examine how each stage reshaped institutions and citizen engagement, laying the groundwork for today's AI-driven governance. In the AI epoch, governments leverage big data analytics, machine learning, and automation to inform decisions, deliver services, and regulate proactively, marking a paradigm shift toward innovative governance. The chapter explores the transformative potential of AI to enhance civic engagement, transparency, and the effectiveness of public institutions, while also highlighting the profound challenges and ethical considerations it presents. Issues such as algorithmic bias, privacy, accountability, and the digital divide are analyzed, recognizing that while AI offers powerful tools to improve governance (from predictive policy simulations to real-time service delivery), it can also magnify risks if unchecked. A global perspective is maintained throughout, comparing how different societies and governments adapt to AI, from local municipalities employing AI chatbots for citizen services to national and international bodies crafting AI governance frameworks. In envisioning a LIFE 4.0 future, where AI is ubiquitous, the chapter discusses scenarios of governance that could emerge, including more participatory and data-driven democracies as well as potential techno-authoritarian pitfalls. The chapter concludes with a forward-looking outlook on steering society in this new epoch: the imperative of balancing innovation with ethical guardrails, ensuring that Governance 4.0 augments human well-being, upholds democratic values, and equitably serves all citizens on a global scale.

## 7.1 INTRODUCTION

Human governance systems have always been shaped by the prevailing technologies and social paradigms of their era. From the divine-right monarchies of pre-modern history to the modern constitutional states and the digital government initiatives of recent decades, each phase of governance reflects an adaptation to new challenges and capabilities. We now stand at the cusp of an AI epoch, where advanced information technologies, especially artificial intelligence (AI), are poised to redefine how we make collective decisions and deliver public services. The concept of Governance 4.0 encapsulates this emerging paradigm: governance powered by AI-driven policy tools, predictive analytics, and "smart" adaptive institutions.


DOI: 10.1201/9781003455219-7

In this chapter, we navigate through the historical progression of governance models, labeled 1.0 through 4.0, to contextualize the revolutionary changes introduced by AI. We begin by examining Governance 1.0, characterized by traditional monarchical control and centralized rule, which dominated societies for centuries. We then move to Governance 2.0, marked by the advent of constitutional democracy and nation-state institutions that introduced the rule of law, popular representation, and bureaucratic administration. Building on that, we discuss Governance 3.0, the era of electronic governance (e-governance) and globalization, when digital communication and international interdependence started transforming government operations and enabling new forms of civic engagement across borders. This evolution sets the stage for Governance 4.0, the focus of our analysis, where we explore how AI and related technologies (such as the Internet of Things (IoT) and big data) drive policy innovation, enable predictive regulation, and facilitate more innovative governance mechanisms.

Throughout our discussion, a global perspective is maintained. Governance 4.0 is not unfolding uniformly around the world; different countries and cultures approach AI in governance in varied ways. We will compare examples from across the globe, from data-driven governance experiments in smart cities to national AI policy frameworks and international cooperative efforts on AI ethics. This diversity highlights how context, values, and political systems influence the adoption of AI in governance. It also underscores a central theme of the chapter: while AI has enormous potential to improve governance, making it more efficient, responsive, and transparent, realizing these benefits in practice requires careful attention to challenges and ethical considerations. Issues of bias, accountability, privacy, and the risk of authoritarian abuse of AI tools are critical to address if society is to steer effectively in the AI epoch [1].

Finally, the chapter looks forward to the outlook in a "LIFE 4.0" scenario. Here, we imagine how governance might function in a world where AI is deeply integrated into every facet of life, a future where, for instance, policymaking could be aided by AI simulations of societal outcomes, public opinion might be gauged through AI analysis of digital forums, and day-to-day administration could be largely automated. This scenario analysis will consider both optimistic possibilities (such as more inclusive, evidence-based decision-making and enhanced global cooperation through AI) and cautionary scenarios (such as "Cyberocracy" ruled by algorithms that could undermine human agency or democratic accountability). The goal is to identify guiding principles and strategies to steer society with foresight, ensuring that Governance 4.0 enhances human flourishing and democratic governance rather than constraining them.

In sum, this introduction frames the journey we will take through the past, present, and future of governance to illuminate the stakes of the current AI-driven transformation. Each section of the chapter corresponds to a stage of governance evolution or a critical thematic area (opportunities, risks, outlook), as outlined in the structure. By the end, readers should have a comprehensive understanding of how AI is reshaping policymaking, civic engagement, transparency, regulation, and institutional design – and what this means for the future of governance and society at large (Table 7.1).

**TABLE 7.1**
**Comparative Summary**

| Governance Era | Structure | Decision Style | Key Driver |
|---|---|---|---|
| 1.0 | Centralized hierarchy | Authority-based | Tradition |
| 2.0 | Institutional bureaucracy | Rule-based | Industrialization |
| 3.0 | Networked systems | Participatory | Digitalization |
| 4.0 | Adaptive ecosystems | AI-assisted | Intelligence and data |

## 7.2 GOVERNANCE 1.0–4.0

### 7.2.1 Governance 1.0: Traditional Rule and Monarchical Control

Governance 1.0 refers to the earliest structured forms of governance in complex societies, epitomized by traditional monarchies and other autocratic systems of rule. In this era, political authority was typically centralized in a single ruler or a small elite, with legitimacy often derived from heritage, conquest, or divine sanction. A monarchy is a classic example: it is a political system in which supreme authority is vested in one individual, the monarch, who functions as the head of state [2]. Under absolute monarchies, the king or queen wielded broad or even unchecked powers over lawmaking, justice, and administration, based on the belief that their right to govern was ordained by higher forces or tradition. Many medieval and early modern monarchs, for instance, governed under the doctrine of the "divine right of kings," claiming to be God's appointed rulers on earth. This doctrine not only provided theological justification for absolute rule but also often discouraged dissent; to oppose the monarch could be framed as opposing the divine will.

In practice, however, even the most autocratic monarchs did not rule alone. Monarchies typically rested on a hierarchical social structure, a "court society" of nobles, courtiers, and advisors who helped the sovereign govern [2]. The monarch's court and council would manage various realms of administration (treasury, military, justice, etc.), albeit at the monarch's discretion. Governance 1.0 was thus characterized by personalized rule, where state affairs were deeply intertwined with the ruler's household and persona. Laws and policies could change at the whim or for the sake of the ruler, and institutions as we know them today (such as parliaments or independent judiciaries) were either absent or subordinate to royal authority.

Despite its concentration of power, monarchical governance performed the basic functions of providing order, defense, and some public goods within the polity, albeit with significant variation in effectiveness and fairness. Hereditary succession was a standard mechanism for continuity. However, it could lead to instability when succession was disputed, or rulers were incompetent (a noted risk when leadership depended on birth rather than merit). Over time, many monarchies developed customary laws and councils of nobles or clergy that placed some checks on rulers, especially as societies grew more complex. For instance, even absolute monarchs often had to negotiate with aristocrats or acknowledge established customs to maintain legitimacy and prevent rebellion. Early forms of consultative bodies emerged (such as the English Parliament's medieval forerunners or advisory estates in various kingdoms), setting precedents for more inclusive governance later.

From a global perspective, Governance 1.0 was not uniform across all civilizations; different regions had distinct forms of centralized rule. In Europe and parts of the Middle East, feudal monarchies and empires prevailed; in East Asia, imperial bureaucracies (such as China's dynasties under the "Mandate of Heaven" concept, a parallel to divine right) dominated; in other regions, tribal kingships or sultanates existed. Despite cultural differences, these systems shared standard features: hierarchical authority, limited formal input from the general populace, and governance justified by tradition or religion rather than popular consent. Communication and administration in this era were relatively slow and localized by today's standards, relying on messengers, decrees, and physical travel, a stark contrast to the instant information flows of the digital age. It's important to note that technology's role in Governance 1.0 was minimal beyond basic administrative tools (writing, record-keeping, and later innovations like the printing press). Rulers made decisions based on counsel and personal judgment, without the aid of data analytics or complex information systems. This meant governance was often reactive and based on limited information, and it heavily depended on the ruler's personal capacity and the loyalty of subordinates [3].

By the late medieval period, cracks in the edifice of absolute monarchical governance began to appear. The growth of commerce, the spread of knowledge (helped by the printing press), and philosophical challenges during the Enlightenment questioned the legitimacy of rule by divine right or birth. Thinkers began advocating that sovereignty should reside with the people (popular sovereignty) or at least be constrained by laws, ideas that set the stage for a dramatic shift in governance. In summary, Governance 1.0 established foundational concepts of statehood and authority but also revealed the drawbacks of concentrated power and unaccountable rule, thereby paving the way for new governance models that sought to distribute authority and embed it in laws and institutions rather than individuals.

### 7.2.2 Governance 2.0: Constitutional Democracy and Nation-State Institutions

Governance 2.0 represents the era in which the principles of constitutionalism, democracy, and the nation-state became the dominant framework for organizing political life. This shift, which gained momentum from the 18th century onward (through revolutions, enlightenment thought, and gradual reforms), fundamentally transformed the relationship between rulers and the governed. In place of personal monarchic rule, Governance 2.0 introduced the idea that government authority derives from the consent of the governed and is limited by a constitution or set of laws. Power was increasingly vested not in a single sovereign but in institutions, parliaments, elected executives, and independent courts that operate under a legal framework defining their roles and citizens' rights [4].

A hallmark of this era is the rise of the nation-state as the primary unit of governance. The Peace of Westphalia (1648) is often cited as a starting point for the modern state system, entrenching the concept of territorial sovereignty and non-interference in other states' affairs. Over the following centuries, states formalized their governance through written constitutions (the United States Constitution of 1787 and the French Declaration of the Rights of Man in 1789 being early influential examples). They developed bureaucratic institutions to administer law, taxation, defense, and public services. These constitutions typically enshrined separation of powers, dividing government into legislative, executive, and judicial branches to provide checks and balances (a concept articulated by Montesquieu), and they recognized, at least in principle, that citizens had certain inalienable rights against the state.

Democracy in this context evolved as a system of decision-making where citizens have a role (directly or through representatives) in determining laws and policies. Early constitutional democracies were often limited (e.g., only property-owning men could vote), but over time, suffrage expanded, and democratic institutions deepened. By the late 20th century, democratic governance had spread globally at an unprecedented rate. Indeed, at the beginning of the 21st century, observers noted that over one-third of the world's independent countries had robust representative democratic institutions comparable to those in long-established Western democracies, and many others had partially democratic systems collectively covering nearly half of the world's population. This dramatic expansion (sometimes called the "third wave of democracy" in the late 20th century) reflected the failure or loss of legitimacy of alternative systems. After World War I, hereditary monarchies and aristocratic empires largely fell out of favor; after World War II and the Cold War, fascist and communist dictatorships gave way in many places to market-based democracies. The appeal of constitutional democracy lies in its promise of accountable governance, protection of individual freedoms, and the ability to correct course through peaceful elections [4].

Nation-state institutions in Governance 2.0 are characterized by structured bureaucracy and the rule of law. Government became an impersonal entity distinct from rulers; civil services were established to implement policies consistently, taxation and budgets followed legal procedures, and standing professional militaries and police forces maintained order under civilian control. The idea that no one is above the law (even leaders) took root. For example, in constitutional monarchies that persisted (such as the UK or later Japan), the monarch's powers were circumscribed by law and largely ceremonial, with real governance carried out by elected bodies and officials under a constitution or established legal framework. This was a profound change from Governance 1.0, whereas a monarch might say, "I am the state," a constitutional system says, "the law is king."

From a global perspective, Governance 2.0 took various forms. Some countries became republics (with no monarch at all, e.g., France or the US). In contrast, others retained monarchs in constitutional monarchies where the sovereign's role is defined and limited by law (e.g., the UK, Sweden, and Japan). Many newly independent states in the 19th and 20th centuries (from Latin American republics to post-colonial nations in Asia and Africa after World War II) adopted written constitutions and some form of electoral politics, at least nominally. The strength and quality of these democracies varied widely, but the aspiration to constitutional governance became nearly universal. By the late 1900s, international norms (e.g., the Universal Declaration of Human Rights in 1948) further reinforced the notion that governments should be accountable and respect human rights, reflecting the Governance 2.0 values on a global stage.

Notably, this era also coincided with the rapid development of technology and communication that, while modest by today's standards, greatly aided governance. The spread of newspapers, telegraphy, and later radio and television allowed governments in the 19th and 20th centuries to communicate with citizens and administer large territories more effectively than before. It also enabled the public to be more informed (and sometimes more critical) about government actions. For instance, a literate electorate reading pamphlets and news could debate policies and hold elected officials to account in ways unimaginable in the age of absolute kings.

Institutional design innovations also emerged in Governance 2.0. Federations and unions formed (e.g., the US, the Swiss confederation, eventually the European Union) to balance governance across large or diverse populations by distributing power vertically (federal, state, local) as well as horizontally. Independent judiciaries became key arbiters protecting constitutional rules and rights. Electoral systems were experimented with from first-past-the-post to proportional representation, affecting how representation translated into governance. While challenges such as corruption, patronage, and the exclusion of certain groups persisted (and indeed continue in many places), the overarching trend was toward more inclusive and rules-bound governance compared to the arbitrariness of Governance 1.0.

By establishing that legitimacy stems from laws and the people's will (rather than divine right), Governance 2.0 laid the foundation for the following transformations. It created the stable institutions and educated citizenry that would later demand even greater transparency and efficiency – pressures that helped usher in the era of e-governance (Governance 3.0). Additionally, the nation-state framework meant governance was compartmentalized within borders, even as global problems mounted. This tension would become more evident with globalization and the need for international governance in the late 20th century. In summary, Governance 2.0 dramatically shifted the paradigm: rulers became public servants, subjects became citizens, and governance became a matter of laws, rights, and institutional processes. This model, though continuously evolving and facing backlash in some regions, remains the baseline of modern governance onto which the technologies of the 21st century, notably AI, are now being layered.

### 7.2.3 Governance 3.0: E-Governance and Globalization

Governance 3.0 describes the period roughly from the late 20th century into the early 21st century, when the advent of digital technologies and the intensification of globalization profoundly impacted governmental operations and the scope of governance. Two key intertwined themes define this era: e-governance (electronic governance), the use of information and communication technologies (ICTs) to improve government service delivery and public administration, and globalization, the increasing interconnectedness of economies, societies, and institutions across national borders, which has given rise to new forms of global governance and transnational challenges [5].

E-Governance: With the proliferation of personal computers, the internet, and mobile communications, governments began adopting digital tools to modernize bureaucracies and engage with citizens. Electronic governance refers to leveraging ICTs to make government more efficient, transparent, and responsive. This includes a wide array of innovations: online portals for public services (e.g., filing taxes and renewing licenses), electronic records and databases replacing paper archives, email and social media for communication, and, more recently, mobile apps and open data platforms. The goal of e-governance is often summarized as bringing government closer to citizens, making interactions convenient, and reducing administrative bottlenecks. For example, a citizen can apply for a passport or report a civic issue online without needing to travel to a government office and wait in line, which increases efficiency and public satisfaction. At its core, e-governance integrates digital solutions into government operations, fostering a more accessible and responsive public administration [5].

E-governance also emphasizes transparency and accountability. By publishing information on government websites (budgets, project statuses, performance metrics) and enabling two-way digital communication, it becomes easier for citizens and media to scrutinize government activity. Many countries implemented Right-to-Information laws and open data initiatives in this period, reflecting a cultural shift that public information should be accessible by default. Additionally, digital record-keeping reduces opportunities for certain forms of corruption by leaving audit trails. Some nations pioneered "e-procurement" systems for government contracts to minimize human discretion in awarding contracts, thereby curbing favoritism [5].

A related aspect is e-participation or digital civic engagement. Governance 3.0 saw the rise of online public consultation, e-petitions, and even experiments with electronic voting. Governments started using the internet to gather citizen feedback on policies and to allow the public to participate in decision-making forums from the convenience of their home computers or phones. For instance, two-way communication platforms have enabled citizens to directly message administrators or vote in polls on local issues, enhancing participation. Though not a replacement for formal elections or representative processes, these tools provided supplementary channels for gauging public opinion and involving stakeholders, often coined as elements of "digital democracy."

By the early 21st century, many governments had embraced e-governance to varying degrees. International indices, like the United Nations' E-Government Development Index, tracked progress as countries digitized more services. Advanced examples included Estonia, which became known for its ambitious e-governance ecosystem offering almost all services online (including e-residency for foreigners) and even nationwide internet voting. Other countries focused on specific improvements, such as India's massive digitization of government services in regional languages or Singapore and South Korea's sophisticated one-stop citizen portals. These innovations showed that digital infrastructure has become as important as physical infrastructure for modern governance.

However, e-governance also faced challenges: not everyone had internet access or digital literacy (the digital divide left rural or poor populations behind), and there were concerns about privacy and Cyberocracy. Massive data breaches or the hacking of government systems have become a new security threat. Additionally, early automated decision systems (such as simple algorithms to determine benefit eligibility) sometimes lacked transparency or recourse, foreshadowing debates that have only grown in the AI age [5].

Globalization: While governments were going digital, they also found themselves operating in an increasingly global context. Issues such as climate change, trade, financial regulation, public health (e.g., pandemics), and migration transcended national borders and required coordination beyond the nation-state. Thus, Governance 3.0 is also characterized by the rise of global governance mechanisms, international treaties, organizations, and multi-stakeholder initiatives aimed at addressing cross-border challenges. Effective global governance cannot be achieved by any single country acting alone; it demands international cooperation and institutions that facilitate collective decision-making. For example, the United Nations and its agencies, the World Trade Organization, the Bretton Woods financial institutions (World Bank, IMF), and newer forums like the G20 are all part of the global governance landscape that expanded and gained prominence in the late 20th century.

In this era, the concept of sovereignty began to be balanced with the idea of interdependence. States remain sovereign within their territories, but they increasingly acknowledge that many problems require shared sovereignty or at least shared commitments (like reducing carbon emissions or coordinating on internet governance standards). We also see the emergence of international norms around human rights, democracy, and, relevant to our focus, technology governance. For instance, agreements on data privacy (e.g., the EU's General Data Protection Regulation, which has global implications for how companies and governments handle data) and cyber norms are nascent forms of international governance of the digital realm.

Globalization also meant that policies in one country could have immediate effects elsewhere, and that best governance practices were rapidly disseminated. Through global networks and organizations, countries learned from each other's e-governance initiatives or public sector reforms. A notable example is how many countries in the 2000s set up "one-stop" government service portals or introduced digital national IDs after seeing successes elsewhere. Global indices and rankings (on governance, competitiveness, and e-readiness) spurred a healthy competition and knowledge exchange.

Additionally, non-state actors rose in influence on the global stage. Transnational NGOs, multinational corporations, and even loose networks of activists (amplified by social media) could pressure governments or provide governance functions. For example, international NGOs started playing a monitoring role (e.g., Transparency International scoring countries on corruption, influencing governance reforms). Tech companies in Silicon Valley, for better or worse, developed platforms that affected political communication worldwide (think of social networks influencing election discourse in multiple countries). Thus, Governance 3.0 saw a diffusion of power both downward (to citizens via digital tools) and upward/laterally (to international and non-state actors).

In summary, Governance 3.0 can be seen as the bridge between the traditional bureaucratic nation-state model and the AI-driven future. It brought speed, connectivity, and a broader scope to governance. By implementing e-governance, states made their operations more data-centric and transparent, planting the seeds for the data-driven decision-making that AI would later supercharge. Through globalization, it became clear that governance is not just about managing a country's internal affairs but also about contributing to and navigating an international order with shared rules and challenges. The experiences of this era, the successes and pitfalls of digitization, and the necessity of global thinking have directly informed how we approach Governance 4.0, which inherits this digitally networked world and pushes it into the realm of autonomous systems and intelligent algorithms.

### 7.2.4 Governance 4.0: AI-Driven Policy, Predictive Regulation, and Smart Governance

Governance 4.0 refers to the cutting-edge transformation of governance facilitated by AI, advanced data analytics, and interconnected "smart" systems. In this phase, governments are not merely using computers as tools (as in e-governance). Still, they are increasingly relying on AI algorithms to inform decisions, automate processes, and even predict and shape future outcomes. It represents a paradigm shift toward governance that is proactive, data-driven, and highly responsive, often termed "smart governance." In innovative governance frameworks, the integration of AI, along with technologies like the IoT and blockchain, aims to create governance systems that are intelligent, adaptive, and anticipatory [6].

At the heart of Governance 4.0 is the use of AI-driven policymaking tools. This entails employing machine learning models and other AI techniques to analyze vast amounts of data, including economic indicators, mobility patterns, and social media sentiment, to provide insights for public policy. For example, AI can help identify trends or problems that might not be apparent to human decision-makers. A city government might use AI to analyze real-time traffic data and automatically adjust traffic light patterns, or a national tax agency might deploy algorithms to detect fraudulent transactions and flag them for human review. By leveraging such capabilities, AI-driven governance can make public services more efficient and decisions more evidence-based. Indeed, a recent OECD analysis found that government AI use cases often target automation of routine tasks, personalization of services, and improved decision-support, rather than purely replacing human judgment. In fact, 57 percent of documented government AI initiatives focused on streamlining or tailoring processes (such as automating form processing or customizing service delivery to individual needs), and about 45 percent aimed to enhance decision-making, sense-making, or forecasting for policymakers [7].

A key concept within Governance 4.0 is predictive regulation. Traditional regulation often involves reactive rules that are made in response to issues and enforced after violations occur. Predictive regulation uses data and AI models to foresee potential problems and intervene early. For instance, regulators can use AI to simulate the impact of a proposed policy on different social groups or economic sectors before implementing it, allowing them to refine rules proactively. Environmental agencies are employing machine learning to predict which industrial facilities are most likely to violate pollution limits, enabling them to target inspections more effectively. In law enforcement, some jurisdictions have experimented (controversially) with predictive policing algorithms that analyze historical crime data to forecast where crimes are likely to occur, thereby allocating police patrols accordingly. While predictive policing raises ethical concerns (notably bias and over policing of specific communities, as discussed later), it exemplifies the broader trend of using predictive analytics in governance. Even in social services, algorithms might predict which citizens are at risk of facing homelessness or unemployment, enabling early intervention programs. The idea is that, with AI's pattern recognition, governments can move from a "firefighting" stance to a preventive, strategic approach across many domains of public policy [8].

Another hallmark of Governance 4.0 is the emergence of innovative governance ecosystems, often first visible at the city level ("smart cities") but increasingly at larger scales. These ecosystems integrate IoT sensors, AI analytics, and networked platforms to manage resources and services in real time. Consider a smart city governance scenario: sensors across a city feed data on traffic, air quality, energy usage, and public safety into central AI systems. City administrators then use a unified dashboard (sometimes aided by AI recommendations) to make decisions like rerouting public transit, adjusting electricity grids, dispatching repair crews, or issuing public health advisories, all in a highly responsive manner. Some cities, for example, have implemented AI-managed traffic systems that predict congestion and dynamically adjust traffic signals or suggest alternate routes, significantly reducing commute times. Others use AI to optimize waste collection routes, detect water leaks, or control lighting and heating in public buildings for energy efficiency. At the national level, "smart governance" might include AI chatbots handling millions of citizen queries (improving customer service), algorithms helping to draft or summarize policy documents, and data-driven performance monitoring of government programs [9].

Crucially, Governance 4.0 also transforms civic engagement and transparency in new ways. Governments can deploy AI tools to process and summarize citizen feedback from large-scale consultations or social media, enabling leaders to grasp public sentiment quickly and incorporate it into policy design. There are experimental platforms where AI moderates online deliberations among citizens to highlight consensus points or innovative ideas for government consideration. On the transparency front, some governments are beginning to use AI to sift through their own data for anomalies or patterns of waste and fraud, essentially self-auditing mechanisms that bolster accountability. For instance, machine learning can detect unusual spending patterns in procurement data that might indicate corruption or collusion, flagging them for investigators. The Spanish city of Valencia's SALER system, cited as an example, allows officials and even citizens to use an AI-backed system to identify conflicts of interest in public procurement, enhancing integrity.

Global adoption of Governance 4.0 practices is underway, though at varying speeds. Dozens of countries have launched national AI strategies to integrate AI into public sector innovation and governance frameworks. By 2024, for instance, Oxford Insights reported that more than 60 countries had published or announced national AI strategic plans, including not only advanced economies but also a growing number of lower-income countries that see AI as a leapfrogging opportunity [10]. This demonstrates a worldwide recognition that AI is a cornerstone of future governance. Countries like Singapore and the United Arab Emirates have been early adopters, appointing chief AI officers and implementing AI in domains such as transportation planning, healthcare administration, and judicial analytics (e.g., AI-assisted legal research for judges). In Europe, Estonia (already a digital government leader) has piloted AI judges for small-claims courts and AI advisory systems for policy drafting. China has been aggressively implementing AI in governance, from innovative city projects to the controversial Social Credit System, which uses data analytics to monitor and influence citizens' and corporate behavior (an example of state-driven predictive regulation, albeit with significant concerns about surveillance and rights). These examples underscore that Governance 4.0 is not a theoretical future idea; it is already being implemented in parts and pieces around the world [9].

However, the deployment of AI in governance also varies according to political and cultural contexts. Democracies tend to emphasize using AI to assist human decision-makers and improve service delivery, often under principles of transparency and human oversight (e.g., the European approach, guided by ethics frameworks such as the EU's AI Act, calls for human-centric AI and risk-based regulation). Authoritarian regimes might lean more toward using AI for surveillance and control (e.g., facial recognition in public spaces for security purposes, without the same level of oversight or privacy constraint). Most countries fall somewhere in between, experimenting with AI's potential while grappling with its pitfalls.

As Governance 4.0 unfolds, new roles and institutional capacities are emerging. Governments are hiring data scientists and establishing "digital government" or "AI governance" units to drive these initiatives. Some have created algorithmic oversight committees or transparency registers for AI systems used in the public sector [1]. This is a recognition that implementing AI in governance is not just a technical matter but also an issue of public trust and ethics. The concept of algorithmic governance has dual meanings: it refers to governing by algorithms (the focus of this section), but it also calls attention to the need to govern the algorithms, i.e., to establish rules and norms for how AI is used. We will delve into the challenges and ethics in the next section. Still, it's worth noting that leading practice recommends that AI systems in government be transparent, accountable, and free from unfair bias to maintain citizens' trust in this new mode of governance [1].

In conclusion, Governance 4.0 represents a frontier where the science of algorithms enhances the age-old art of governance. It holds the promise of governments that can "think" faster, analyzing more information than a human bureaucracy ever could and acting more nimbly in response to complex, fast-changing societal needs. Imagine real-time policy adjustments as new data streams in, or highly personalized public services tailored by AI to everyone's circumstances (e.g., automatic enrollment in benefits one is eligible for, proactively delivered without bureaucratic hurdles). These are the kinds of innovations on the horizon. Yet, as the following sections will address, these advances are not without profound challenges. As we embrace the tools of Governance 4.0, we must concurrently evolve our institutional checks and ethical frameworks, essentially, our governance of AI, to ensure these powerful tools are used in the service of society's best interests.

## 7.3 OPPORTUNITIES AND TRANSFORMATIONAL POTENTIAL

The integration of AI and advanced technologies into governance (Governance 4.0) brings a host of opportunities that could fundamentally transform government performance and the citizen experience for the better. In this section, we outline the key opportunities and transformational potentials, providing examples and evidence of how AI-driven governance can lead to improvements in efficiency, effectiveness, and inclusiveness. Many of these opportunities build upon trends that began in the e-governance era but are accelerated and enhanced by AI's capabilities.

### 7.3.1 Enhanced Operational Efficiency and Cost Savings

One of the most immediate benefits of AI in government is the ability to automate routine tasks and optimize internal workflows, leading to faster service delivery and reduced costs. AI-driven automation can handle repetitive processes, such as processing applications, issuing licenses, and sorting and routing inquiries, much more quickly than human clerks and on a much larger scale [11]. By streamlining administrative processes, governments can reallocate human workers to more complex tasks that require judgment or empathy. According to analysis by experts, AI can process large datasets far more efficiently than traditional methods, which "leads to improved cost savings" and better resource allocation in the public sector. For instance, chatbots powered by natural language processing (NLP) now handle a significant portion of citizen inquiries in some countries (answering FAQs, guiding users through forms, scheduling appointments). This reduces wait times for the public and frees up staff time. In document processing, machine learning algorithms can scan and categorize millions of pages of records (such as health records or legal documents) in seconds, greatly accelerating tasks like benefits enrollment or evidence review for cases. These efficiency gains not only cut costs but also improve citizens' experiences by minimizing bureaucratic delays.

### 7.3.2 Data-Driven Decision-Making and Predictive Insights

AI provides governments with powerful tools to harness the massive amounts of data they collect (or could collect) to support more informed policymaking. Traditionally, policy decisions relied on limited-sample studies or experience; AI enables the analysis of real-time data streams and the simulation of complex systems to support evidence-based decisions. Public agencies using machine learning can identify patterns and correlations that humans might miss. For example, economic planners might use AI to analyze employment, education, and industry data to predict which regions are at risk of job losses and proactively design retraining programs. Likewise, AI can simulate the impact of various policy scenarios, such as changes in tax rates or the introduction of a new welfare program, on different demographics, helping policymakers understand potential outcomes before implementation. Such simulation and scenario-testing capabilities mark a transformation from reactive governance to anticipatory governance. Governments are beginning to use digital twins (virtual models of cities or systems) augmented by AI to test urban planning decisions (like where to build infrastructure or how an evacuation would work in a disaster). Overall, AI's ability to parse tens of thousands of data points daily and produce forecasts enables more nuanced, timely decisions. By leveraging predictive analytics, governments can also better manage risks, for instance, predicting spikes in hospital emergency visits during certain seasons and allocating medical resources accordingly, or forecasting which areas might suffer severe droughts and adjusting water management policies in advance.

### 7.3.3 Improved Public Services and Personalization

AI has the potential to vastly improve the quality and accessibility of public services, making them more citizen-centric. One dimension is the personalization of services, tailoring government interactions to individual needs. For example, employment agencies using AI can match job seekers to vacancies more effectively by analyzing individual skills and employer requirements (akin to how private job platforms do). Education systems can implement AI tutors that personalize learning for students based on their progress and learning style. In healthcare, AI can help triage patients (via symptom-checker apps) and even assist doctors in diagnosis by analyzing medical images or patterns in patient data, enabling more timely and accurate public health services. Another dimension is the creation of new services that weren't feasible before. For instance, some cities have introduced predictive maintenance services. AI analyzes data from sensors on infrastructure (bridges, pipelines, roads) to predict when maintenance is needed, preventing service outages or accidents. Autonomous vehicles and drones managed by AI could provide innovative public services, such as delivering medical supplies to remote areas or automating garbage collection. Notably, AI is helping government agencies become not just service providers but problem solvers. In social services, predictive models can identify families likely to face hardships (like a child at risk of neglect or individuals likely to become chronically homeless) so that agencies can intervene early with supportive measures, effectively preventing worse outcomes. These improvements align with what analysts have noted: AI enables new and improved services that enhance the customer (citizen) experience, including reaching underserved communities. For example, AI-driven language translation can allow citizens to interact with the government seamlessly in their native language, breaking down language barriers in multicultural societies.

### 7.3.4 Increased Transparency and Accountability

Paradoxically, the same technologies often criticized for opacity can, if used correctly, significantly enhance transparency in governance. AI can assist in monitoring government performance and ensuring accountability by detecting irregularities and encouraging the release of open data. As mentioned earlier, machine learning algorithms can scan financial transactions, contracting data, and budget expenditures to flag anomalies that indicate possible corruption or misuse of funds. This augments the work of audit institutions and anti-corruption agencies. Some governments have started public-facing AI tools that allow citizens to explore public data, for instance, chatbots that answer questions about the budget ("How much was spent on education in my city this year and how does that compare to last year?") by querying financial databases. By simplifying complex data, AI can empower citizens and watchdog groups to hold officials accountable with facts. Moreover, AI can generate explainable summaries of complicated policies or legal texts, helping the public understand government initiatives without needing to parse hundreds of pages of legalese. Such capabilities foster a more informed citizenry and therefore a more accountable government. In addition, predictive analytics can help enforce regulations fairly and consistently, which is another aspect of accountability. If algorithms are used to guide inspections (say, tax audits or environmental enforcement) based on risk models, it can reduce human bias or undue targeting in selecting who gets inspected, thereby treating entities more equally under the law. However, the algorithms themselves must be transparent (the concept of algorithmic transparency is increasingly emphasized) [1], something we will revisit in the ethics section.

### 7.3.5 Proactive and Preventive Governance

One of the most transformative potentials of AI in governance is moving from solving problems after they occur to preventing problems before they happen. We touched on predictive regulation above; here, we emphasize that it can be a general approach across many fields. In public health, for example, AI analysis of epidemiological data (including novel sources such as Google search trends or social media mentions of symptoms) can provide early warnings of disease outbreaks, enabling faster public health responses, an advantage that could save lives in epidemic situations. In disaster management, AI models that use satellite imagery and climate data can forecast areas at high risk of floods or wildfires with greater accuracy and lead time than traditional methods. This allows authorities to evacuate, strengthen infrastructure, or proactively stockpile relief supplies. In economic policy, real-time analytics might detect early signs of a recession or sectoral downturn, prompting preemptive stimulus or support for affected industries or communities. Even within governance processes, some have envisioned AI assisting with conflict detection, scanning political discourse or signals of social tension to alert leaders to brewing unrest, so they can address grievances or misinformation before they escalate. While these uses are in their early stages, the overarching opportunity is anticipatory governance: identifying trends, simulating impacts, and acting in advance. This is a significant shift from the often reactive nature of policymaking and could lead to more resilient societies.

### 7.3.6 Greater Civic Engagement and Inclusion

AI and digital platforms can also deepen civic engagement by efficiently handling large-scale citizen input. Governments can crowdsource ideas or feedback and use AI to process the responses [12]. For example, if 100,000 citizens submit suggestions for a new city development plan, NLP algorithms can categorize and summarize the main themes, ensuring every voice is, in a sense, "heard" and considered, which would be impossible to do manually in a reasonable time. Such processes can encourage more people to participate because they know their input will not be lost in a black hole. Additionally, AI-powered interfaces (such as voice assistants and simple chatbots) can make it easier for individuals with disabilities, low literacy, or limited tech skills to interact with government services, thereby broadening inclusion. For instance, someone who cannot read forms could speak to an AI assistant that fills out the form for them. There are also projects exploring AI tools to help translate civic documents into plain language for those not versed in bureaucratic jargon. By lowering barriers to engagement, AI can help ensure a more diverse range of citizens actively engage with governance. Some municipalities have tried AI-facilitated "participatory budgeting," in which residents propose and vote on community projects via online platforms. Here, AI helps verify users, prevent ballot stuffing, and maybe even propose compromises based on project data and budget constraints.

In summary, the opportunities of Governance 4.0 span efficiency gains, more intelligent decision-making, improved service quality, enhanced transparency, proactive governance, and inclusive engagement. These translate into real-world impacts, such as faster response times (e.g., immediate disaster warnings), cost savings (automating mundane tasks, reducing waste or fraud), and potentially better outcomes in areas like health, safety, and education through data-informed interventions. A concrete example tying many of these threads together is the concept of an AI-assisted "Policy Lab." Imagine a government policy lab where officials pose questions ("How can we reduce traffic fatalities?"). An AI system analyzes a plethora of data, road sensors, accident records, hospital data, even weather and smartphone mobility data to identify key risk factors, simulate the effect of different interventions (like changing speed limits, improving lighting, or launching a safety campaign), and then recommends the mix of policies likely to save the most lives. The officials can then test those recommendations in a pilot, monitor outcomes in real time with AI analytics, and adjust dynamically. This continuous improvement cycle, accelerated by AI, is a far cry from the slow, one-shot policy process of the past. AI holds tremendous promise to revolutionize governance, much as it is revolutionizing industries. However, realizing these opportunities depends on thoughtful implementation. Technology itself can deliver capabilities, but governments must reform processes, build technical capacity, and maintain public trust to fully harness its benefits. The following section on challenges will address why these opportunities are not automatically realized and what pitfalls must be navigated. Nonetheless, the vision of Governance 4.0's positive potential is compelling: more agile, responsive governments that can handle complexity and scale in serving citizens, ushering in a new era of public sector innovation and effectiveness.

## 7.4 CHALLENGES, RISKS, AND ETHICAL CONSIDERATIONS

While the promises of Governance 4.0 are significant, incorporating AI and algorithmic systems into governance also introduces a range of challenges, risks, and ethical dilemmas. It is crucial to acknowledge and address these issues, as failure to do so can harm citizens, erode trust in public institutions, or even undermine democratic principles. In this section, we dissect the major concerns associated with AI-driven governance, supported by examples and expert findings, and discuss emerging strategies to mitigate these risks.

### 7.4.1 Algorithmic Bias and Fairness

One of the most documented risks of using AI in decision-making is that algorithms can inadvertently perpetuate or even amplify existing biases present in data. If an AI system is trained on historical data that contains biases (reflecting societal or institutional biases), it may learn and reinforce those biases in its outputs. In a governance context, this can have serious consequences. For instance, an AI used to determine eligibility for government benefits might deny assistance to specific groups if the training data encoded past prejudices or underserved patterns [1]. Likewise, predictive policing tools have been shown to unfairly target minority neighborhoods, largely because historical crime data (and enforcement patterns) are skewed by over-policing in those areas, creating a feedback loop. Studies have found that predictive algorithms in criminal justice can misidentify risks – a well-known example is bias in recidivism risk software that was more likely to label black defendants as high-risk falsely [1]. If governments rely on such tools for bail or parole decisions without correction, it can institutionalize discrimination. This raises the ethical issue of fairness: governance must serve all citizens equitably, and deploying AI that disadvantages certain groups violates that principle. Furthermore, biased AI can erode the legitimacy of governance; people will not trust or accept decisions they perceive as unjust or biased. As a result, numerous voices (from the tech community, civil rights groups, and international organizations) are calling for rigorous bias audits and fairness checks for any algorithm used by the government [1]. Governments need to implement standards and, importantly, certification processes to ensure algorithms meet fairness criteria, and to allow those affected by algorithmic decisions to have recourse or a human review process (the "human-in-the-loop" concept for critical decisions).

### 7.4.2 Lack of Transparency ("Black Box" Decisions) and Accountability

Many AI systems, especially complex machine learning models like deep neural networks, operate as "black boxes"; their internal logic is not easily interpretable by humans. In a governance setting, this opacity conflicts with the requirement of transparency and reason-giving in public decisions. Citizens have a right to understand why a government made a confident decision about them [1]. If, say, an AI system denies someone a building permit or flags them as a fraud risk, it may be challenging to explain the specific rationale if the model is not interpretable. A lack of algorithmic transparency can thus undermine accountability: officials might deflect responsibility by saying "the computer decided it." This is problematic because public accountability means that decisions should be explainable and traceable to accountable entities [1]. In some jurisdictions, this principle is being codified. For example, the European Union's forthcoming AI regulations emphasize a right to explanation for automated decisions, and some countries have instituted algorithmic transparency registers that publish information about how government algorithms work [1]. The ethical stance is that AI should augment, not obfuscate, accountability. To address this, there is a push for explainable AI in the public sector, techniques that make AI's decision process more interpretable. Additionally, governance frameworks may need to assign clear responsibility: e.g., an agency deploying an algorithm must take responsibility for its outcomes just as if a human official made the decision. Another approach is to require meaningful human oversight of automated decisions, especially those with significant impact on individuals' rights (such as liberty and access to services), to ensure that someone can be held accountable. The principle of transparency also ties into public trust: an open algorithmic policy (where code or logic is published for public scrutiny when feasible) can build confidence that the AI is doing what it is intended to do and not something nefarious or erroneous.

### 7.4.3 Privacy and Surveillance Concerns

AI's hunger for data raises serious privacy issues. Effective AI systems often require large datasets, including personal data about citizens. There is a risk that in trying to harness AI, governments may over-collect data or repurpose it in ways that infringe on individuals' privacy rights. For example, smart city sensors and facial recognition cameras can create a pervasive surveillance environment, constantly tracking people's movements and activities. Authoritarian applications of AI have demonstrated how this can slide into a tool for oppression: the use of facial recognition and AI to identify protestors or monitor ethnic minorities (as reported in the case of China's Xinjiang region) violates fundamental rights to privacy and freedom of expression. Even in democratic contexts, the deployment of AI for surveillance (e.g., law enforcement) is controversial. A notable incident is that major tech companies like IBM have pulled back on offering facial recognition technology to governments due to concerns about its misuse and documented bias (such as misidentifying people of color) [13]. The ethical imperative is striking a balance between public safety or administrative efficiency and citizens' right to privacy. Laws such as GDPR in Europe enforce data minimization and purpose limitation. AI in governance must adhere to such principles, collecting only what is necessary and using it only for legitimate, consented purposes. Technical methods like data anonymization, federated learning (where AI models learn from data without centralizing raw data), and robust Cyberocracy measures are needed to mitigate privacy risks. Moreover, oversight institutions (data protection authorities, ethics councils) should be empowered to audit government AI projects for compliance with privacy regulations. The worst-case scenario to guard against is the emergence of a digital Big Brother, a state that knows everything about everyone and uses AI to micromanage or control society. Preventing that dystopia means instituting legal and technical checks now, as AI capabilities in surveillance continue to advance rapidly.

### 7.4.4 Security and Reliability Risks

As governments become more reliant on AI and digital systems, they also become more vulnerable to system failures, errors, or cyber-attacks. An AI system, like any software, can have bugs or vulnerabilities that lead to incorrect decisions or exploitation by malicious actors. For example, if an AI is controlling critical infrastructure (power grids, traffic management), a glitch or hack could cause widespread disruption. There have been instances of algorithms behaving unpredictably or being "gamed." In one case, researchers manipulated a machine-learning model used in self-driving cars by subtly altering street signs. The AI misread a stop sign as a speed limit sign, illustrating how adversarial attacks could trick AI. In governance, one could imagine adversaries trying to feed disinformation to government AI systems (for instance, injecting fake data into health surveillance systems to mislead public health responses). Furthermore, heavy reliance on AI might lead to automation bias among officials, who over-trust AI outputs even when they are wrong, thereby compounding errors. The challenge is ensuring the robustness and security of AI systems. Governments need to invest in rigorous testing, validation, and "red-teaming" of AI (actively searching for how things could go wrong). Cyberocracy protocols must extend to AI models and data streams, not just traditional IT networks. There is also a case for keeping a human fallback option: if an AI system fails, human officials should be able to step in. The notion of reliability extends to day-to-day consistency as well. Many AI systems can be "brittle," meaning they perform well in the conditions they were trained on but poorly if the situation changes even slightly beyond their training data. If a welfare benefits AI was trained on data from normal economic conditions, its predictions might break down in an unprecedented pandemic scenario, for example. Thus, continuous monitoring and periodic retraining with up-to-date data are necessary to maintain reliability.

### 7.4.5 Loss of Human Judgment and the Human Touch

Governance is not only about cold data processing; it involves values, empathy, and deliberation. There is a concern that over-automation could remove human judgment from essential decisions that have ethical nuances. For instance, consider an AI that screens candidates for a public service job: it might efficiently shortlist based on quantified criteria, but it might not capture holistic qualities like leadership potential or community engagement that a human panel would pick up. Similarly, in areas such as justice or social work, professionals often rely on intuition and personal interaction, which are hard to replicate in algorithms. The human touch in public service (like a social worker understanding a family's situation or a judge tempering justice with mercy in sentencing) is a value that might be diminished if we lean too heavily on algorithmic decisions. Ethically, we must ask whether confident choices should not be delegated to AI at all. Many argue yes, for example, fully autonomous lethal decision-making in policing or military (often phrased as banning "killer robots") is widely opposed on ethical grounds, insisting a human must always make life-and-death decisions. In daily governance, maintaining human oversight and intervention possibilities helps ensure that compassion and context-specific wisdom guide outcomes alongside data. Also, completely automated processes can alienate citizens – think of trying to appeal a decision to a machine that cannot understand your unique predicament. Governance must remain citizen-centric, and sometimes that means a person-to-person interaction, or at least a person reviewing an algorithm's recommendation. The principle of augmenting rather than replacing human decision-makers is a common refrain in discussions on public sector AI.

### 7.4.6 Democratic Legitimacy and Consent

Another macro-level concern is how AI might shift power dynamics and decision-making processes in ways that challenge democratic norms. Algorithms created by technical experts (or private vendors) could exert significant influence over public policy outcomes without being visible to or understood by the public or their representatives. If, for example, a city uses a privately developed AI to plan urban zoning, to what extent can citizens debate or challenge those plans if they come essentially as algorithmic outputs? This touches on legitimacy in a democracy, policies gain legitimacy through transparent processes and the consent (direct or via representation) of the governed. A risk is the emergence of a technocratic governance in which vital choices are made by AI or by those who control it, rather than through open political debate. This could undermine the role of elected officials, and officials might defer to AI analyses rather than engage with their constituents' opinions. We must ensure that AI is used as a tool for elected officials and citizens, not as a replacement for political decision-making. One way to think of it: AI can supply evidence and options, but the ultimate value judgments (like how to trade off economic growth versus environmental protection, or liberty versus security in policy) must be made via the political process. Additionally, deploying AI systems that affect the public should involve some form of public consultation or at least transparent communication. For example, if a city installs an AI traffic management system, informing the public about how it works and how it benefits them (and addressing concerns like data use) is essential for maintaining trust. We see initial moves in this direction: some cities have citizen advisory boards for technology and AI deployments, and participatory frameworks to determine what is acceptable (such as some police departments seeking community input on the use of face recognition).

### 7.4.7 Institutional Readiness and Skill Gaps

A practical challenge is whether governments have the expertise and capacity to implement and manage AI responsibly. Many public sector organizations struggle with outdated IT infrastructure, limited budgets for innovation, and difficulty attracting AI talent (who are often lured by higher salaries in the private sector). As a result, governments might rely heavily on private vendors for AI solutions, creating dependencies and potentially conflicts with public values if not managed by knowledgeable public servants. The skill gap is fundamental; understanding and governing AI requires data scientists, ethicists, and tech-savvy administrators, who many agencies currently lack. Building this capacity is an ethical imperative as well, because without it, governments may either misuse AI out of ignorance or miss out on its benefits for those who need them. Addressing this involves investing in training, adopting new hiring approaches, and possibly forming public-private partnerships that include knowledge transfer. International bodies like the OECD have noted that government adoption of AI lags partly due to skill gaps and legacy systems, and have called for more concrete guidance and capacity-building in the public sector [7]. An associated risk is that not all countries can participate equally in the AI revolution; poorer nations could be left behind or become reliant on AI systems from more technologically advanced countries, raising issues of digital sovereignty and inequality between states.

### 7.4.8 Mitigating Measures and Ethical Frameworks

Considering these challenges, numerous efforts are underway to establish ethical guidelines and regulatory frameworks for AI in governance. For instance, the UNESCO 2021 Recommendation on AI Ethics, adopted by 193 countries, provides a global framework that emphasizes principles such as fairness, transparency, and human oversight in AI [14]. The OECD's AI Principles (backed by many governments) similarly stress inclusive growth, human-centered values, robustness, and accountability. Some countries have created AI ethics councils to advise on government AI projects. Concrete measures include algorithmic impact assessments (analogous to environmental impact assessments, but for algorithms' societal effects), which some jurisdictions now mandate for high-risk AI use in the public sector. These assessments force agencies to think through and document how they mitigate biases, protect data, ensure explainability, etc., before deploying the system. Another crucial measure is public engagement and education: involving citizens in dialogue about AI plans (what comfort levels are, what trade-offs they accept) and educating them about AI so that discourse is informed. An engaged citizenry can better hold government accountable on these issues.

Finally, a cultural point: Governance 4.0 requires a continuous ethical vigilance. Technology evolves rapidly, and with it, new unforeseen challenges will emerge. Governments will need to adopt a learning mindset, monitoring outcomes, welcoming independent audits, and be ready to adjust policies (or even roll back an AI system) if it produces unjust or harmful effects. This adaptive governance is itself a challenge given bureaucratic inertia, but it is necessary in such a fast-moving area.

In conclusion, the challenges of Governance 4.0 are substantial but not insurmountable. Each risk, whether it's bias, opacity, privacy invasion, security vulnerability, loss of human judgment, or democratic strain, can be mitigated by deliberate strategies: rigorous oversight, clear ethical guidelines, technical fixes, legal constraints, and inclusive governance processes that keep humans in control of AI, not the other way around. The following section, looking to the future, will build on this by considering how, in a LIFE 4.0 scenario, we might institutionalize these safeguards while still capitalizing on AI's benefits.

## 7.5 THE FUTURE OUTLOOK (LIFE 4.0 SCENARIO)

Envisioning the future of governance in the context of LIFE 4.0 involves imagining how society might function when AI and related technologies reach a level of maturity and ubiquity that they become ingrained in nearly every aspect of daily life. "LIFE 4.0" implies a stage beyond the current paradigm – a world in which AI is not just an external tool, but a deeply integrated partner in human existence, perhaps analogous to how electricity or the internet became pervasive in previous eras. In such a scenario, Governance 4.0 would itself evolve into new forms, potentially warranting the term Governance 5.0 or similar, but for our purposes, we'll discuss it as the culmination of Governance 4.0 trends. This outlook is necessarily speculative, but grounded in present trajectories and ethical foresight. We will consider how policymaking, civic engagement, institutional structures, and global governance might look, highlighting both the bright possibilities and the critical pitfalls to avoid.

In a fully realized Governance 4.0 world, governance could become highly data-driven, personalized, and perhaps even partially autonomous. Routine administrative decisions might be handled almost entirely by AI with minimal human intervention. For instance, consider the notion of an "autonomous bureaucracy." Basic government services, licensing, certifications, and benefit disbursements could be managed by AI systems that verify eligibility, detect changes in life circumstances, and process claims instantly. Errors and fraud might be minimized through constant AI auditing and cross-checking across interconnected databases. Citizens might interact with a unified AI assistant for all government needs, a sort of always-available civil servant that can answer questions (in any language), guide one through complex procedures, and proactively do things like fill your tax return or enroll you in programs you qualify for without you even asking. Such an assistant, drawing on central government intelligence, could democratize access to information; no longer would savvy insiders or those who can hire lawyers have an advantage in dealing with the government; the AI assistant would ensure everyone gets expert guidance.

This future also holds the promise of radically enhanced civic participation. Imagine a continuously operating digital forum where issues are debated with AI moderators that ensure conversations remain civil, factual (flagging misinformation in real-time), and inclusive of diverse viewpoints. Through such platforms, governments might run ongoing referendums or consultations on various issues, with AI aggregating the input in meaningful ways for policymakers. For example, city residents could weigh in on budget priorities via an app; AI analyzes the input to help elected officials see what the community values most, segmented by neighborhood or demographic group, enabling more responsive budgeting. Some futurists even envision a direct democracy augmented by AI, in which citizens could vote on many issues. At the same time, AI provides personalized briefings on the implications of each issue and helps ensure that the scale of participation doesn't lead to chaos. Essentially, AI could handle the complexity that historically made direct democracy impractical in large societies by digesting proposals, checking consistency with laws, preventing manipulation (through identity verification on blockchain, for instance), and summarizing results.

From an institutional perspective, we might see new bodies like a Ministry of AI or an Office of Algorithmic Accountability become as standard as the Ministry of Finance is today. These institutions would continuously oversee AI integration, ethics, and performance across the government. The civil service of the future could be a tight symbiosis of humans and AI: each human official working with an AI counterpart that provides advice, much as a competent aide would. In such a model, hiring might emphasize skills like ethical reasoning, empathy, and leadership, which humans excel at, since analytical and procedural tasks can be delegated to machines. The result could be a more human-centric service, ironically, because humans would be freed from drudgery to focus on creative problem-solving and direct engagement with citizens. A social worker, for example, in LIFE 4.0 might let AI handle case file paperwork and pattern analysis of a family's needs. At the same time, they spend more time in the field, personally interacting with that family and building trust.

One fascinating possibility is the emergence of AI-mediated policymaking. In a complex world where decisions have many ripple effects, future governments might use advanced simulations (sometimes called policy sandboxes or digital twins) to test out policies in a virtual environment. Before passing a new healthcare law, for instance, the government simulates with AI agents representing individual citizens, hospitals, insurers, etc., to see how the system would likely react and to identify unintended consequences or inequities. These simulated populations could even be partly informed by real data and partly by AI "imaginations" to fill gaps, giving a reasonably accurate preview of reality. Policymakers then iterate the law's design in the simulation until it achieves the desired outcomes with minimal downside, and only then implement it in the real world. This is a future extension of the predictive analytics we discussed, but at a far more sophisticated level, where governance decisions are stress-tested virtually, much like engineers stress-test bridge designs.

Another aspect of LIFE 4.0 might be the integration of global governance. If AI becomes a key tool for managing global challenges, such as climate change mitigation, pandemic response, and financial stability, we could see the rise of internationally networked AI systems. For example, nations might cooperate on a Global Climate AI that monitors Earth systems, advises all governments on optimal emission reduction strategies, and tracks compliance in near-real time. Such an AI could allocate carbon credits or warn of tipping points in a scientifically objective manner, potentially easing political disputes by providing a trusted neutral analysis. This assumes high levels of international trust and data sharing, which is a big if, but it's conceivable if the benefits are clear. We might also see the development of global regulatory AI bodies, akin to how Interpol facilitates police cooperation; an AI-driven agency could facilitate regulatory oversight of multinational corporations or cross-border data flows, flagging violations of agreed standards (such as an AI watchdog scanning for human rights abuses or illegal arms trade in global data).

However, this optimistic vision is tempered by potential dystopian scenarios that must be consciously averted. A key risk is if governance becomes too automated and top-down, edging into techno-authoritarianism. For instance, a government might use AI to implement a social credit scoring for citizens, combining data on individual behavior into a single "civic score" that determines one's access to services or opportunities. This has already been experimented with in some places and is widely criticized for obvious reasons: it can become a tool of coercion and social control, undermining personal freedoms and privacy. In a future with even more data (from wearables, implants, innovative environments) and more powerful AI, such control could be totalizing. A LIFE 4.0 authoritarian state could conceivably monitor not just where people go and what they say, but physiological indicators of emotion or intent, preemptively suppressing dissent. This is the dark side of predictive governance, essentially a "Minority Report" world of pre-crime, but, in a political sense, something that democratic societies must guard against vigorously. The continued existence of liberal democracy in the AI age is not guaranteed; it will require deliberate choices to use AI in ways that empower individuals and strengthen democratic processes, rather than to empower an elite or an AI overlord at the expense of individual rights.

One could imagine a tension in LIFE 4.0 between societies that choose a "cyberocracy" model ruled by AI and information management, and those that insist on "AI-augmented democracy." In a cyberocracy, as theorized by futurists, policy might be primarily determined by AI analyzing what is "best" for society, potentially sidelining messy democratic debate. For example, if an AI calculates that to maximize GDP and health, we must enforce specific diets and exercise, a cyberocratic government might be tempted to mandate those, infringing on personal choice. An AI-augmented democracy, by contrast, would use AI to provide options and evidence, but the collective human values and decisions (even if less "optimal" in some metrics) would prevail. It's plausible that different countries will experiment along this spectrum.

The future of work in governance is also notable. Many current public sector jobs might be eliminated or changed due to AI (e.g., clerical roles). But new roles will emerge, such as algorithm auditors, AI ethics officers, and data curators. The public workforce might shrink, or shift toward more front-line roles that involve human interaction (because those are the hardest to automate). The government might also adopt new decision-making structures; for instance, some propose liquid democracy (in which people can delegate their vote, issue-by-issue, to experts), facilitated by technology. AI could manage the delegation seamlessly and ensure decisions are always made by those most trusted on each topic.

On a positive note, LIFE 4.0 could see unprecedented problem-solving capacity. With AI, problems long thought intractable (like fully personalized education or eliminating bureaucratic corruption) might be addressed. Education could be tailored to each child by AI tutors, as noted earlier, possibly resulting in dramatically better outcomes and greater equality of opportunity. Corruption might be curtailed by transparent blockchain transaction records and AI anomaly detectors, thereby increasing trust in public institutions. Citizens might gain new freedoms as well, for example, if self-driving cars and AI traffic management virtually eliminate congestion, people gain time and mobility; if AI in healthcare prevents diseases, people live healthier; these quality-of-life improvements feedback into a more satisfied and engaged populace. Freed from daily struggles, people might engage more in civic life, thereby fulfilling some of the ancient ideals of democracy, in which citizens actively participate in public affairs. In that optimistic sense, AI could ironically enable a more human society: one in which drudgery is minimized and the creative, communal, and empathetic aspects of life are foregrounded.

To reach the best-case scenarios and avoid the worst, foresight and governance of AI itself are paramount. It's likely that in LIFE 4.0, we will have meta-governance structures explicitly focusing on AI. For example, an international AI Governance Council under the UN might exist, in which nations coordinate on standards and share best practices (some early inklings include the Global Partnership on AI, GPAI, and various G20 statements on AI). There might be treaties akin to arms control for specific AI applications (e.g., agreements not to automate nuclear launch decisions or to ban AI-enabled mass surveillance tools) [15]. Additionally, education systems will probably put significant emphasis on digital and civic literacy, teaching citizens from a young age how algorithms work, what their rights are, and how to critically engage with AI-powered information. This helps create an electorate that can handle the nuances of AI policy, which will likely be one of the dominant political topics, much as economics or healthcare are today.

In conclusion, the LIFE 4.0 scenario for governance is one of great promise intertwined with significant peril. We could see a flourishing of efficient, responsive, and participatory governance on a scale not previously possible, in a world where public decisions are well-informed and inclusive, and government truly acts as a catalyst for societal well-being, supported by tireless and intelligent AI assistants. But we could also stumble into a techno-authoritarian trap or a dependency on AI that dulls the human capacities for judgment and self-governance. The outcome will depend on choices made in the present and near future: setting the right policies, laws, and norms for AI. The remaining task of our generation is to steer society in the AI epoch in a direction that maximizes human dignity, freedom, and prosperity. Governance 4.0 is the steering mechanism, and if we design it well, the future of LIFE 4.0 can be one where AI serves as a powerful tool for human progress under the guidance of our collective values and democratic controls [6].

## 7.6 CONCLUSION

The evolution from Governance 1.0 to Governance 4.0 reflects a remarkable journey of societal learning, institutional development, and technological innovation. In the chapter, we examined how governance structures have transformed from the centralized authority of monarchies to the distributed, law-bound systems of democratic nation-states, through the digital turn of e-governance, and now into the emergent AI-driven paradigm. Each stage of governance is built on the previous, solving specific problems while introducing new ones. Governance 4.0, set in the AI epoch, emerges not in a vacuum but as a response to the increasing complexity and data-richness of modern societies, carrying forward the democratic values and global interconnectedness established in earlier eras and infusing them with unprecedented technological capabilities.

A key insight of our exploration is that AI is reshaping the mechanics and possibilities of governance in fundamental ways. Policymaking is becoming more evidence-based and anticipatory, powered by algorithms that can analyze trends and model scenarios with a depth and speed humans alone cannot match. Civic engagement is potentially broadened by digital platforms and AI tools that aggregate and interpret the voices of millions, offering the promise of more direct and continuous citizen participation in public decisions. Transparency could be enhanced by automated tracking and open data, even as new concerns arise about the opacity of AI systems themselves. Regulation is shifting toward a more proactive stance, predictive regulation aiming to prevent harm before it happens, whether in financial markets, environmental protection, or public safety. And institutional design is adapting, as we see the formation of new oversight bodies for digital and AI issues and the integration of data scientists and ethicists into public administration. In short, the what of governance (its goals and principles) remains rooted in enduring aims like justice, welfare, and liberty, but AI is revolutionizing the how of governance.

Throughout the chapter, we highlighted the global perspective, recognizing that there is no one-size-fits-all narrative. Different countries are at various points on the Governance 4.0 trajectory. Some smart cities and digitally advanced nations are already implementing shades of AI governance, while other governments are only beginning to digitize records. Cultural and political contexts influence the way AI is used: for instance, an algorithm used for citizen services in one country might be used for surveillance in another. Global governance in the AI age is itself an unfolding story, with international principles and cooperative efforts trying to keep pace with national developments. What's evident is that collaboration and knowledge-sharing will be vital. Many governance challenges posed by AI, such as AI ethics, cross-border data flows, or AI in warfare, are transnational in nature and will require collective action and norm-setting beyond any single country's remit.

The chapter also confronted the dual nature of AI's impact on governance, its opportunities, and its risks. We painted a picture of how AI can make governments more efficient, responsive, and inclusive. But we also underscored that these benefits will only materialize if we consciously navigate the pitfalls. Ethical considerations are not a side aspect of Governance 4.0; they are central to its design and implementation. Ensuring algorithms are fair, decisions are explainable, data is protected, and humans remain accountable is essential to aligning AI's use with democratic values and human rights. Encouragingly, there is momentum in this direction: governments and international organizations have begun to articulate AI ethical guidelines and pass relevant regulations, a trend likely to intensify. The lesson is that technology should not drive governance; instead, our collective choices about governance should drive how we design and deploy technology. In the spirit of "steering society" (as in the title of the chapter), we as citizens, policymakers, and researchers have agency in shaping AI's role; it is not an autonomous force of nature but a set of tools we mold and govern.

Looking ahead, we find ourselves at an inflection point akin to previous significant transitions (like the advent of democracy or the industrial revolution's impact on administration). The AI epoch holds both promise and peril for the nature of governance and the quality of life in society (LIFE 4.0). If we steer wisely, we could harness AI to help solve some of humanity's thorniest problems, from making healthcare and education universally accessible to managing climate change cooperatively. We could strengthen democracy by engaging citizens in new ways and by using data to make more informed, long-term decisions beyond election cycles. Conversely, poor governance of AI could exacerbate inequalities (if only tech-savvy groups benefit), erode freedoms (through surveillance and automated oppression), and even destabilize international peace (through AI arms races or misinformation campaigns undermining democracies).

In conclusion, Governance 4.0 is not an endpoint but the beginning of a new chapter in the story of how we organize society. It challenges us to revisit fundamental questions: What is the role of human judgment in a data-driven world? How do we distribute power when knowledge and decision-making become decentralized and automated? How can our institutions remain legitimate and earn public trust amid rapid change? Addressing these questions will require dialogue across disciplines and sectors; technologists, public servants, legal scholars, ethicists, and the public must all be part of co-designing the future of governance. The title of the book, *LIFE 4.0*, suggests an iteration beyond the current paradigm, much as Governance 4.0

suggests a subsequent iteration of governing. It is our generation's responsibility to ensure that iteration is a positive progression. By steering society in the AI epoch through thoughtful policy, inclusive debate, and adaptive institution-building, we can aim to create a future where technology and governance together advance the human condition. In doing so, we carry forward the torch lit by earlier governance reforms (from Magna Carta to modern constitutions, from telegraph networks to open government data) into a new era, guided by the enduring belief that governance at its best is the deliberate, collective steering of humanity toward the common good.

## REFERENCES

1. Open Government Partnership, Digital Governance: Automated Decision-Making, Algorithms, and Artificial Intelligence, in *Open Gov Guide*, 2023.
2. Britannica, Monarchy, in *Encyclopedia Britannica*. Manchester: Dolphine, 2019: p. 87–112.
3. Horning, R., The Divine Right to Rule: Past Perceptions of Monarchy, in *University of Portland, Early British Literature Survey Project*, 2019.
4. Shapiro, I., Democracy: The Spread of Democracy in the 20th Century, in *Encyclopedia Britannica*, London: Springer, 2025: p. 213–228.
5. Srinivas, H., Urban E-Governance: Defining E-Governance, in *Global Development Research Center*, Explainer Series C-081, n.d. Series C-081, n.d.
6. Schwab, K., Envisioning Governance 4.0, in *Project Syndicate*, 2022.
7. OECD Publishing Office, *Governing with Artificial Intelligence: The State of Play and Way Forward in Core Government Functions*, 2025.
8. IBM Policy Lab, *Precision Regulation for Artificial Intelligence*, 2021.
9. Yadav, M.R., and G. Ashesh, E-Governance 4.0: Integrating AI and IoT for Sustainable and Smart Management of Maha Kumbh 2025. *International Journal of Latest Technology in Engineering, Management & Applied Science*, 2025. 14(3): p. 426–433.
10. Oxford Insights, *Government AI Readiness Index 2024*, 2024.
11. AI Multiple Research, *AI in Government: Examples & Challenges*, 2025.
12. Open Government Partnership, *OGP 2023–2028 Strategy*, 2023.
13. Krishna, A., *IBM CEO's Letter to Congress on Facial Recognition. Reference to IBM's Discontinuation of General-Purpose Facial Recognition Offerings*, 2020.
14. UNESCO, *Recommendation on the Ethics of Artificial Intelligence*. Paris: UNESCO, 2021.
15. GPAI, *Global Partnership on Artificial Intelligence – Montreal Summit Report*, 2020.

# 8 Culinary 4.0

**Abstract**

Artificial intelligence (AI) is reshaping the entire food ecosystem from how we grow and process food to how we cook and consume it. This chapter traces the evolution from Culinary 1.0, defined by the origins of cooking and traditional practices, through Culinary 2.0's industrialized production and global supply chains, and Culinary 3.0's digital platforms and bright kitchens, toward Culinary 4.0, an era of AI-driven nutrition, kitchen robotics, and personalized cuisine. We examine how AI technologies are being integrated to enhance food production, improve sustainability, and personalize dietary experiences worldwide. Opportunities range from precision agriculture and reduced food waste to health optimization and new culinary creativity. At the same time, we discuss challenges and ethical considerations, including labor displacement, data privacy, cultural impacts, and global access disparities. Drawing on international perspectives, the chapter envisions a LIFE 4.0 scenario in which intelligent systems enable a more efficient, nutritious, and sustainable food future, while emphasizing the continued importance of human creativity, cultural diversity, and ethical stewardship in the culinary arts.

## 8.1 INTRODUCTION

The concept of "Culinary 4.0" emerges from the broader framework of LIFE 4.0, paralleling the Fourth Industrial Revolution in its transformative impact on everyday life. Just as Industry 4.0 introduced intelligent automation, data exchange, and artificial intelligence (AI) into manufacturing, Culinary 4.0 signifies the infusion of advanced technologies into the food domain. This revolution is driven by pressing global needs: by 2050, the world must sustainably feed nearly 10 billion people, a challenge demanding transformative changes in our food systems [1]. Traditional food production and consumption methods are straining under climate change, resource constraints, and population growth [1, 2]. In response, AI has moved from a novelty to a critical tool across the "farm-to-fork" chain, promising to reshape how we grow, distribute, and enjoy food [3].

This chapter provides a comprehensive exploration of how we arrived at Culinary 4.0 and what it entails. We begin by recounting Culinary 1.0, the origins of cooking, and the long era of traditional culinary practices that defined human diets for millennia. We then discuss Culinary 2.0, marked by the industrialization of agriculture and the emergence of global food supply chains in the 19th and 20th centuries. Next, we describe Culinary 3.0, characterized by digital technology from online food delivery platforms to smart kitchen appliances, that set the stage for today's innovations. The core of the chapter is Culinary 4.0, where AI-driven approaches in agriculture (precision farming, predictive analytics), in kitchens (robot chefs and IoT-enabled appliances), and in personalized nutrition (dietary algorithms and health


DOI: 10.1201/9781003455219-8

analytics) converge to transform the food landscape. Throughout, we include global perspectives and case studies that illustrate how different regions and communities are engaging with these changes. We also examine the profound opportunities that Culinary 4.0 presents, such as enhancing sustainability, nutrition, and culinary creativity, alongside the challenges and ethical issues it raises, including workforce disruption, data privacy, and cultural implications. Finally, we present an outlook in the context of LIFE 4.0, envision how daily life might be impacted by AI in food, and conclude with reflections on how to ensure this technological evolution benefits all of humanity (Table 8.1).

**TABLE 8.1**
**Comparative Summary**

| Culinary Era | Mode | Key Driver | Focus |
|---|---|---|---|
| 1.0 | Traditional | Culture | Survival and ritual |
| 2.0 | Industrial | Mechanization | Scale and efficiency |
| 3.0 | Digital | Platforms | Experience and identity |
| 4.0 | Intelligent | AI and data | Sustainability and health |

## 8.2 CULINARY 1.0–4.0

### 8.2.1 Culinary 1.0: Origins of Cooking and Traditional Practices

For most of human history, cooking was a purely human art and necessity, practiced with basic tools and techniques passed down through generations. The control of fire was the pivotal innovation that gave rise to cooking in the prehistoric era. Archaeological and anthropological evidence suggests that early humans began cooking with fire nearly 2 million years ago, an evolutionary milestone that not only made food more palatable but also provided the extra nutrition needed for the growth of the human brain [4]. According to Richard Wrangham's seminal work Catching Fire, the advent of cooking was a "giant evolutionary step," supplying early humans with richer caloric intake and fostering social bonding around the hearth [4]. By the Paleolithic period, people were constructing simple stone hearth circles around a fire, which became the focus of communal life, a fact reflected in the origin of the word "focus" from the Latin for "fireplace" [4]. In every culture, the act of gathering around a fire to prepare and share food laid the foundations of culinary tradition.

Early cooking practices were deeply rooted in local environments and cultures. What people ate and how they prepared it varied widely across the world, long before globalization. Traditional practices such as fermentation, drying, smoking, and salting were developed over millennia to preserve food and enhance flavors using available resources. Fermentation is an ancient technique; evidence of fermented beverages dates to at least 6000 BCE in China. Such practices were often discovered by trial and error and encoded in cultural rituals and cuisines. For example, the making of bread, cheese, yogurt, or beer through natural fermentation became staples in many societies, essentially representing early biotechnology guided by empirical knowledge.

During Culinary 1.0, kitchen technology advanced very gradually. Basic implements like mortars, pestles, stone griddles, and clay pots improved the efficiency of food preparation over time, but the methods remained manual and labor-intensive. Cooking was an intimate, hands-on craft; recipes were rarely written, instead transmitted orally or by demonstration from parents to children. Each region's cuisine evolved with its own flavors and techniques suited to local crops and tastes. Significantly, cooking was constrained by nature's rhythms: people relied on seasonal ingredients and had to develop methods to cope with scarcity (through preservation or storage) and to make tough foods edible (through soaking, pounding, or long cooking).

Despite these limitations, the Culinary 1.0 era laid the cultural groundwork for food as a cornerstone of human life. Shared meals and traditional dishes became key to social structure, religious practice, and identity. Many of these traditions persist and form the heritage that modern food innovations build upon. However, the scale of food preparation was fundamentally local, with villages and households largely self-sufficient in processing basic ingredients into meals. This began to change with the introduction of more complex tools (e.g., querns for grinding grain and ovens for baking bread) and especially with the later emergence of markets and trade in foodstuffs. But it was not until the modern age that a true revolution, Culinary 2.0, transformed how food was produced and distributed, taking it far beyond the local hearth and communal table.

### 8.2.2 Culinary 2.0: Industrial Food Production and Global Supply Chains

The era of Culinary 2.0 corresponds to the period of the Industrial Revolution and the 20th century, when food production underwent radical mechanization and scaling-up. In agriculture, industrialization introduced machines, chemicals, and new high-yield crop varieties, dramatically boosting output. For example, the invention of mechanical tools such as the seed drill, tractor, and combine harvester enabled one farmer to cultivate and harvest far more land than was previously possible by hand or with draft animals [5]. Improved agricultural techniques in Europe and elsewhere, such as crop rotation, selective breeding of livestock and plants, and later the use of synthetic fertilizers and pesticides, significantly increased food productivity. By the mid-20th century, these advances coalesced into what became known as the Green Revolution. Starting in the 1950s and 1960s, international programs promoted high-yielding varieties (HYVs) of staple crops such as wheat and rice, combined with the intensive use of fertilizers and irrigation, particularly in Asia and Latin America [6]. The Green Revolution is credited with alleviating mass starvation. It is often said to have saved up to one billion people from hunger by spreading modern farming worldwide [6]. Indeed, global food production began to outpace population growth; by the 1980s, enough calories were being produced to feed everyone, at least in theory.

However, industrial agriculture also introduced new complexities and risks. The focus on a few high-calorie crops and intensive monoculture led to unintended consequences, such as malnutrition (from diets heavy in calories but deficient in micronutrients) and environmental degradation [6]. For instance, prioritizing quantity over diversity contributed to the rise of "hidden hunger." Many communities received sufficient bulk calories (primarily carbohydrates) but lacked vitamins and protein, leading to health issues even as famine receded [6]. At the same time, heavy reliance on chemical fertilizers and pesticides, while boosting yields, caused soil depletion, water pollution, and ecosystem damage, raising concerns about long-term sustainability [6]. By the late 20th century, it became evident that industrial food systems, while highly productive, carried significant environmental footprints and vulnerabilities (like loss of crop diversity and reliance on finite resources).

In parallel with changes on the farm, food processing and distribution were revolutionized by industrial methods. The late 19th and early 20th centuries saw the advent of factories for preserving and packaging food: canned goods, pasteurization of milk, industrial bakeries, and later frozen foods all emerged, making food more portable and shelf-stable. These innovations enabled the rise of global supply chains capable of transporting food over long distances. New technologies in refrigeration, transport, and storage "liberated the food supply from 'nature' and its seasons," allowing year-round availability of many foods regardless of local climate [7]. For example, the invention of refrigerated shipping in the 1870s enabled the international trade of meat and fruit without spoiling, connecting farmers on one continent with consumers in another. The mid-20th century saw the introduction of containerized shipping and improved logistics, significantly lowering the cost of moving food globally. By the turn of the 21st century, it had become routine for a supermarket in Europe or North America to carry fresh produce and packaged goods from every corner of the world: bananas from Latin America, coffee from Africa, rice from Asia, fish from distant oceans, at any time of the year.

The formation of multinational food corporations and retail chains further integrated the global food system in Culinary 2.0. Brands like Coca-Cola, Nestlé, and McDonald's expanded worldwide, standardizing certain products and diets. Fast-food and processed-snack industries grew by applying mass-production techniques (assembly line kitchens, flavor engineering, aggressive marketing) to food, making convenience foods cheap and widely accessible. This era delivered unprecedented abundance and variety to many consumers. Still, it also raised public health concerns as highly processed, energy-dense foods were linked to rising obesity and diet-related diseases, even in developing countries [6].

Culinary 2.0 also highlighted issues of equity and resilience. While wealthy nations enjoyed surplus and choice, many poorer regions remained food-insecure, dependent on importing staples or on volatile commodity markets. The global supply chains that brought advantages also introduced fragility: a disruption in one part of the chain (due to natural disasters, conflict, or market shocks) could ripple through food availability and prices worldwide. Additionally, the homogenization of crop varieties, with an estimated 75 percent of global crop diversity having disappeared by the late 20th century due to the emphasis on monoculture, made agriculture more vulnerable to pests, diseases, and climate change impacts that could affect those few dominant varieties [6].

By the late 20th century, it was clear that while industrialization had vastly increased food supplies, new thinking was needed to address its downsides. This set the stage for the next wave of innovation. As computing and the internet era dawned, the food sector began adopting digital technologies, heralding the Culinary 3.0 transformation, which would further globalize and virtualize how we obtain and interact with food.

### 8.2.3 Culinary 3.0: Digital Platforms, Delivery Services, and Smart Kitchens

The early 21st century introduced a digital layer to the food industry, fundamentally changing consumer behavior and business models. Culinary 3.0 is defined by the rise of online platforms, data-driven services, and smart devices that have made food more convenient and connected than ever before. One of the most visible changes was the emergence of online food ordering and delivery on a global scale. In a historical milestone, the first online food purchase took place in 1994, when Pizza Hut's experimental "Pizza Net" website accepted an order for a pizza, marking the first time a physical good was ever bought over the internet [8]. What started as a novelty in the mid-1990s exploded into a massive industry by the 2010s. The spread of broadband, smartphones, and location-based apps enabled a delivery revolution, with platform companies connecting consumers, restaurants, and couriers at the tap of a screen. Global revenues for the online food delivery sector tripled between 2017 and 2021, reaching roughly $150 billion worldwide [9]. In the US, food delivery sales more than doubled during the COVID-19 pandemic, as consumers embraced the convenience of app-based ordering during lockdowns [9].

Several factors fueled this rapid growth. Appealing and user-friendly mobile apps, combined with GPS-enabled logistics and digital payment systems, unlocked ready-to-eat delivery as a mainstream option [9]. Companies like Grubhub (founded in 2004), Uber Eats (2014), DoorDash (2013), Deliveroo (2013), and many others expanded into dozens of countries, fiercely competing to get meals to customers' doorsteps. By the late 2010s, these services had transformed urban dining habits, and ordering had become so routine that by 2018, online restaurant ordering was growing 300 percent faster than dine-in traffic [10]. In China, the adoption was even more dramatic; the online food delivery market there grew from virtually zero to tens of billions of yuan within a few years [10]. Delivery platforms not only changed consumer expectations (with many people, especially younger generations, seeing on-demand food as a given) but also spurred business innovations such as cloud kitchens (restaurant facilities with no dine-in space, purely focused on delivery) and aggregator platforms that give users access to hundreds of restaurants in one app. Despite challenges (such as thin profit margins and labor issues among gig delivery workers), online delivery has become a permanent fixture in the dining landscape [9].

At the same time, digital technology seeped into home and professional kitchens, making them more intelligent and more automated. The concept of the bright kitchen took off in the 2010s, integrating the Internet of Things (IoT) and AI into everyday appliances. Refrigerators, ovens, and other appliances gained internet connectivity and sensors, enabling new functionalities. For instance, modern smart refrigerators like Samsung's Family Hub or LG's InstaView can track their own contents, alert users when groceries are running low, suggest recipes based on available ingredients, and even automatically place grocery orders online [11]. Smart ovens such as the June Oven use built-in cameras and AI to recognize food items and adjust cooking settings for optimal results. These ovens can essentially "see" if you put in a piece of salmon or a batch of cookies and set the time and temperature accordingly [11]. Even devices like coffee makers have become intelligent: AI-powered machines learn a user's brew preferences and ensure each cup is customized to taste [11].

In this Culinary 3.0 phase, AI and IoT began to assist – and in some cases, take over aspects of meal preparation. Early examples included specialty devices such as the Rotimatic, a fully automated flatbread maker that uses AI to adjust kneading, rolling, and cooking based on dough consistency and ambient conditions [11]. Such a device can produce fresh rotis or tortillas at the touch of a button, learning over time to optimize for different flour types. More general-purpose cooking assistants also appeared. Voice-activated smart speakers (Amazon Alexa, Google Assistant) were integrated with kitchen routines, allowing users to control appliances or get recipe instructions hands-free. One could say, "preheat the oven to 180°C" or "set a timer for pasta," and the connected kitchen would comply. Kitchens became fully integrated ecosystems. For example, a user could rely on a virtual assistant to generate a meal plan for the week, order the required ingredients online, and have the smart oven guide them through each step with prompts or automatically execute specific steps [11]. This interconnectivity not only added convenience but also aimed to reduce waste and improve outcomes: sensors could prevent overcooking, and inventory tracking meant fewer forgotten items spoiled in the fridge [11].

Restaurants, too, adopted digital tools as part of Culinary 3.0. Reservation platforms, digital point-of-sale systems, and inventory management software became commonplace, optimizing operations through data. Behind the scenes, the food supply chain began leveraging digital traceability (scanning and later blockchain pilots) to track food origin and freshness, and big data analytics for demand forecasting. Grocery retail saw the rise of online supermarkets and subscription meal kits (e.g., HelloFresh and Blue Apron), which used algorithms to predict customer preferences and manage supply. These trends blurred the line between food and tech sectors, giving birth to the term "Food Tech." By the mid-2010s, a significant portion of food consumption in urban centers was facilitated by digital intermediaries or smart devices.

Culinary 3.0 essentially bridged the gap between the physical and digital worlds of food. It changed consumer expectations toward immediacy and personalization. People could get almost any cuisine delivered quickly, discover new recipes on YouTube or cooking apps, and rely on devices to handle tedious cooking tasks. However, most AI applications in this era were relatively narrow, focusing on recommendations (like restaurant or recipe suggestions) or simple automation (temperature control, timers). The full potential of AI, to fundamentally optimize and reinvent food processes, was only beginning to be realized. As we move into Culinary 4.0, we see AI taking on a far more powerful role: not just assisting humans in cooking or delivering food, but reimagining how food is produced, tailored, and experienced on a holistic level.

### 8.2.4 Culinary 4.0: AI-Powered Nutrition, Robotics, and Personalized Cuisine

Culinary 4.0 represents the current frontier, where AI is deeply integrated across the food system, leading to breakthroughs in efficiency, creativity, and personalization. In this stage, often likened to the food industry's answer to the Fourth Industrial Revolution, AI and robotics are being leveraged to address the complex demands of sustainability, health, and culinary innovation. Below, we explore key dimensions of Culinary 4.0: AI in agriculture and production, robotics in kitchens and restaurants, personalized nutrition, and the global and sustainability implications of these advances.

#### 8.2.4.1 AI in Agriculture and Food Production

Perhaps the most crucial impact of AI in the Culinary 4.0 era is on agriculture, the very start of the food chain. Farming is undergoing a digital transformation into what is sometimes called "Agriculture 4.0" or "Smart Farming." By equipping farms with sensors, drones, satellite imagery, and machine learning algorithms, AI enables a shift from labor-intensive, reactive farming to data-driven, predictive agriculture [2]. For example, AI systems can continuously monitor crop conditions via computer vision and climate data, detecting early signs of plant stress, pest infestations, or nutrient deficiencies weeks before the human eye would notice them [2]. This early warning capability allows farmers to intervene proactively, applying targeted treatments or adjusting irrigation exactly when and where needed, thereby significantly increasing yields and reducing crop losses. At Fermata, an agricultural technology company, AI-based pest detection has outperformed traditional manual scouting by three to five weeks, helping growers in places like Brazil tackle emerging threats, such as whitefly infestations, much more effectively [2].

AI's predictive power also extends to weather forecasting and resource management. By analyzing historical climate patterns and real-time weather data, AI tools guide farmers on the optimal times to plant, irrigate, or harvest, thereby mitigating the risks posed by increasingly erratic weather due to climate change [2]. At the macro level, AI models can simulate the outcomes of different farming practices (for instance, how altering crop rotations or using drought-resistant seed varieties might affect yield under future climate scenarios), effectively creating "digital twins" of agricultural systems for experimentation without real-world risk [12]. This kind of modeling is invaluable for adapting agriculture to climate change and for breeding the next generation of climate-resilient crops. As Martin Bailey of the EU's AI office notes, digital twins and AI models can help identify new protein sources or crop traits that offer greater nutritional value with less environmental impact, accelerating innovation in food sources beyond the slow pace of traditional trial and error [12].

In food manufacturing and processing, AI is driving automation and quality control to new heights. Machine learning algorithms, fed with massive image datasets, are now used to sort and grade produce at speeds and with accuracies far beyond human capabilities [3]. For example, computer vision systems can inspect fruits and vegetables on a conveyor belt, instantly detecting subtle bruises, ripeness levels, or shape defects that might escape human inspectors [3]. These systems continuously improve through feedback, minimizing errors and ensuring only the best products move forward.

This reduces waste by more efficiently diverting substandard items (which might be used for juices or purees rather than being discarded). AI is also being used to predict and prevent spoilage in the supply chain: by monitoring conditions such as temperature, humidity, and transit time, AI models can anticipate how long a given batch of perishable food will last and flag at-risk shipments before spoilage occurs [3]. This allows logisticians to reroute or expedite certain lots, or adjust storage conditions, thereby extending shelf life and reducing food waste in distribution.

Another exciting domain is the formulation of novel foods and ingredients. AI is accelerating R&D for alternative proteins (such as plant-based or lab-grown meat), fortified foods, and culinary experimentation. Machine learning models can analyze the molecular properties of proteins, fats, and carbohydrates to suggest combinations that achieve desired textures and flavors, dramatically shortening the development cycle for, say, a new plant-based milk or a meat analog [3]. In the past, food product development might take years of recipe tweaking; now AI can simulate thousands of permutations in silico and highlight the most promising ones within days or weeks. For instance, in fermentation technology (used to produce everything from beer to cultured meat), AI can help optimize growth conditions and inputs by learning from each batch, thereby improving the yield of target compounds (such as proteins or vitamins) [3]. These tools not only save time and cost but often result in products that are closer to what consumers want (since AI can integrate consumer preference data into the formulation process). The ability of AI to consider multiple objectives simultaneously, taste, nutrition, cost, and sustainability, aligns well with the multifaceted goals of modern food innovation [3].

Crucially, AI is also being deployed to make food supply chains more sustainable and climate-friendly. For example, AI-powered supply chain optimization helps companies forecast demand more accurately, enabling them to produce and ship only what is needed, thereby preventing overproduction and surplus waste [3]. It can also optimize delivery routes to cut fuel usage and emissions, and design better packaging by simulating materials that balance durability with environmental impact [3]. Some supermarkets and food distributors use AI to dynamically price or promote items nearing the end of their shelf life, ensuring they are sold rather than thrown out, an approach that nudges waste reduction at the retail end. These improvements contribute to what many see as a necessary transformation of the food system: feeding more people with fewer resources and less waste, aided by intelligent systems at each step [3].

It is worth noting that even as AI automates many aspects of farming and production, it does not eliminate the need for human expertise; instead, it changes the nature of the work. Farmers, for instance, become high-tech managers, working alongside agronomist AI systems. There are challenges to adoption, especially among smallholder farmers and in developing regions (cost, digital literacy, and trust in AI solutions) [2]. Nonetheless, pilot projects have shown promise: for example, in India, AI tools for weather forecasting and pest advisories have helped farmers increase resilience, though overcoming barriers like trust and training is key. The potential of AI in agriculture is immense, leading experts to claim that AI could catalyze a new agricultural revolution that is smarter, fairer, and more sustainable than the last [2]. By bridging data from soil to stomach and linking farming practices to nutritional outcomes, AI also opens opportunities to improve not just yield but also the quality and health impact of the food produced.

#### 8.2.4.2 AI and Robotics in Kitchens and Restaurants

One of the most tangible (and sometimes theatrical) aspects of Culinary 4.0 is the introduction of robots and AI automation into kitchens and dining establishments. Long imagined in science fiction, robot chefs and servers are increasingly becoming reality, driven by advances in robotics, computer vision, and AI planning algorithms. Restaurants around the world are experimenting with robotic systems to perform cooking tasks, food assembly, and customer service. These range from sleek robotic arms that can flip burgers or fry noodles to autonomous mobile robots that ferry dishes from kitchen to table.

The motivations for kitchen robotics are both economic and practical. In many countries, the food service industry has faced labor shortages and rising wages, making automation an attractive solution to maintain output and reduce costs [13]. Robots can work around the clock, maintain consistency, and handle repetitive or dangerous tasks (such as deep-frying) with ease. At the 2024 National Restaurant Association show, numerous "kitchen robots" were showcased, signaling industry interest in scaling these solutions [13]. For example, the "PizzaBot" machine can dispense pizza sauce and toppings with precision, aiming to save labor on one of the most time-consuming steps in pizzeria kitchens [13]. In Japan, a robotic wok called I-Robo can cook 30 stir-fry meals per hour; it guides a human to add pre-portioned ingredients, then mechanically tosses and cooks the dish in minutes [13]. Another system, the Alpha Grill, can grill and assemble 200 burger patties an hour and even clean itself afterward [13], a clear demonstration of throughput and hygiene advantages. Fast-food chains and cafeterias are also trying out automated beverage mixers and fry stations (e.g., robots that automatically fry and season fries) to improve speed and consistency [13].

Robots are not limited to the back-of-house. Autonomous service robots have appeared in dining rooms, greeting guests or delivering orders to tables. Companies like Bear Robotics and Keenon have developed wheeled robots (often with a friendly design) that can carry trays of food from the kitchen to designated tables, and then return used dishes to the dishwashing area [13]. These robots navigate using AI mapping and sensors to avoid collisions in the dynamic restaurant environment. While they are not (yet) replacing human waitstaff entirely, they function as tireless runners, allowing human servers to focus on customer interaction. In some settings, such as hotels or large restaurants, autonomous robots roam the halls, delivering room service or taking dishes back, a novelty that can also reduce staff workload.

The fast-casual restaurant chain Sweetgreen made headlines by acquiring a food robotics startup (Spyce) and opening an "Infinite Kitchen" location where an automated production line with minimal human labor assembles salads [13]. Similarly, the founder of Chipotle launched a new concept called Kernel, built around a mostly robotic kitchen [13]. These pioneers suggest that fully automated restaurants may become viable, especially for simple, popular cuisines where tasks are repetitive. Indeed, some eateries globally now operate with minimal human staff: from a hotpot restaurant in China where robot arms handle slicing and cooking, to a cafe in the US with a robotic barista making dozens of coffee orders an hour.

However, the rollout of robots in kitchens has not been without hiccups. Early trials have revealed that robots have limitations in dexterity and judgment compared to skilled humans. Some highly funded startups failed to deliver on lofty promises, for instance, Zume Pizza, which attempted to automate pizza-making end-to-end (even baking pies in delivery trucks en route), ultimately shut down after struggling with technical challenges (famously, they couldn't keep the cheese from sliding off during robot handling) [13]. Other first-generation machines can be prone to malfunctions or slower than experienced cooks, especially when dealing with the unpredictability of food (varying sizes, textures, etc.) [13]. In South Korea, a program to introduce robot chefs at highway rest-stop kitchens led to "mixed reviews" while the robots significantly increased output (up to 150 meals per hour, double a human chef's rate), customers complained about taste and lost menu diversity, and some staff felt deskilled and demoralized by the change [14]. This underscores that food is not just an industrial product; consistency and efficiency must be balanced with quality, creativity, and the human touch that diners often appreciate.

Industry leaders believe it is only a matter of time and refinement before robots become more common. Michelle Korsmo, CEO of the National Restaurant Association, has stated, There's no doubt the future looks like more robotics in the kitchen. It's a question of whether robotics is in the dining room itself, which depends on the restaurant and experience, emphasizing that technology should align with the desired customer experience [13]. Many restaurateurs are excited about how technology can boost efficiency and customer insights, not just through physical robots but also through AI software [13]. Already, restaurants use AI-driven systems to optimize food inventory (predicting how much of each ingredient to prep each day), schedule staff based on demand forecasts, and even for marketing, analyzing customer data to offer personalized promotions (such as special deals on a patron's favorite dish) [13]. Fast-food drive-thrus are experimenting with AI voice assistants to take orders (replacing the crackly speaker boxes with more accurate automated agents that upsell reliably). In these ways, AI is streamlining operations behind the scenes even when customers don't explicitly see a robot.

In-home kitchens, the idea of a robot chef for consumers is also gaining attention. Companies have prototyped robotic arms that can be installed in a kitchen and programmed to execute recipes – one example is the Moley Robotic Kitchen, which uses a pair of dexterous arms to mimic human motions like chopping, stirring, and plating. While such products are still costly and niche, they hint at a future where cooking a complex meal could be as simple as selecting a program and watching a robot arm do the work. More immediately, many households benefit from simpler cooking gadgets with AI features: multicookers and thermometers that use AI to regulate cooking (preventing under- or overcooking), as well as smart meal-planning apps integrated with appliances.

Overall, Culinary 4.0 envisions kitchens (both commercial and home) as cyber-physical systems in which humans and intelligent machines collaborate. Robots handle heavy lifting, precise repetition, and timing. At the same time, human chefs and cooks can focus on creativity, presentation, and hospitality, the aspects of food that benefit most from the human touch. There are open questions about workforce impacts (chefs may need to upskill to manage and program robots, and some kitchen jobs could be eliminated). But new roles can also emerge (such as food tech engineers and AI maintenance specialists in restaurants). As in other industries, the ideal scenario is not complete replacement but rather augmentation, where AI and robotics take on drudgery and support humans in delivering better food experiences. Importantly, as advanced as robots are, cooking and eating remain sensory and cultural experiences, a theme we revisit when considering the human aspect that AI cannot fully replicate [1].

#### 8.2.4.3 Personalized Diets and AI-Driven Nutrition

One of the most profound shifts in the age of Culinary 4.0 is the move toward personalized nutrition, tailoring dietary recommendations and even food products to an individual's unique health profile, preferences, and goals. AI is the key enabler of this personalization at scale, crunching vast amounts of data from biomedical research, personal health trackers, and nutritional databases to provide customized guidance that used to be the domain of expert dietitians (and available only to a few). The premise is that individualized plans can replace one-size-fits-all nutrition guidelines: what you should eat to optimize your health, energy, and even mood on a given day.

In recent years, a proliferation of AI-powered nutrition apps and services has appeared, effectively acting as digital nutritionists. These apps allow users to log their food intake (often just by snapping a photo of a meal or scanning a barcode) and then receive analysis and feedback. Thanks to advances in computer vision, some apps can identify foods in images; for example, the app Foodvisor can recognize multiple items on a plate from a photo and instantly estimate portion sizes and nutritional content [15]. AI algorithms compare this information with the user's dietary targets (calories, macronutrients, micronutrients) and provide suggestions, such as adjusting the next meal to stay on track or healthier ingredient swaps. Unlike the calorie-counting apps of old, modern AI nutrition platforms are increasingly proactive and interactive. They learn a user's habits and preferences over time and adapt recommendations accordingly. An app might note, for instance, that a user tends to skip breakfast but over-snack at night, and then suggest a tailored meal plan or even deliver reminders to distribute food intake more evenly.

Crucially, many apps integrate with wearable devices and health data. They can incorporate data from fitness trackers (steps, exercise, heart rate), continuous glucose monitors, genetic testing, and gut microbiome analyses to refine their advice. For instance, the app HealthifyMe includes an AI nutrition coach named "Ria" that uses personal health metrics to craft diet plans and syncs with wearables to monitor progress [15]. This app boasts a database of regional foods from around the world, acknowledging that personalization must also respect cultural dietary patterns, a feature that, according to a published study, improved user adherence to nutrition plans [15]. Another platform, Levels, focuses on metabolic responses. By analyzing a user's blood glucose fluctuations (via a wearable glucose sensor) in reaction to different meals, its AI identifies which foods keep that individual's energy steady versus which cause spikes, then recommends choices optimized for their metabolic health [15]. This approach recognizes that two people might react very differently to the same food, and AI helps decode those personal responses.

Beyond apps, AI is also informing personalized meal services. Some companies use algorithms to formulate nutritionally balanced meal kits or supplements tailored to a customer's needs, for example, custom smoothie mixes based on one's dietary deficiencies, or meal kit subscriptions that adjust portion sizes and ingredient types in response to the customer's feedback and health data. AI can process a person's dietary restrictions, allergen information, and taste preferences together. IBM's famed Chef Watson was an early example of an AI exploring flavor combinations to invent new recipes [16]. While it was not personalized per user, its legacy has spurred new AI that do target individual tastes. Now, even large appliance makers are joining in: Samsung's AI-powered cooking app can generate recipes based on the ingredients you have at home and your dietary profile, then talk directly to your Samsung smart oven to execute the recipe's cooking instructions [15]. This kind of integration hints at a future where your kitchen knows exactly what's for dinner, not only based on what's in stock, but what's best for your health at that moment.

Personalized nutrition extends to medical and wellness applications. There is growing interest in "food as medicine," using diet as a primary tool for managing health conditions. AI can synthesize the vast scientific literature on nutrition and specific diseases to advise patients, for instance, helping diabetics manage blood sugar through diet or suggesting foods rich in certain compounds to reduce cholesterol or blood pressure potentially. In one debate at a major nutrition conference, researchers presented evidence that AI-tailored diets led to better health outcomes (such as improved blood sugar control) compared with generic healthy-diet advice [17]. Moreover, AI is being researched for its ability to correlate diet with the gut microbiome, an emerging frontier in health science. Early studies show that AI-based personalized diets can positively influence gut bacteria diversity, which is linked to various aspects of health [18].

Of course, personalization raises questions: Will everyone have a different meal plan, and what does that mean for communal eating or cultural food norms? There will likely be a balance that AI might personalize within the context of culinary traditions rather than completely individualistic diets. For instance, an AI might advise a person of South Asian heritage on how to adjust traditional recipes to be healthier (less ghee, more lentils, etc.) rather than suggesting entirely unfamiliar foods. In fact, one strength of AI is that it can act as a "translator" of complex nutrition science into practical choices for different populations [12]. As Dr. Giulia Menichetti noted, AI can help clarify nutritional information and nudge consumers toward healthier choices by tailoring advice in an easily understandable way [12].

Another exciting aspect is the potential for dynamic personalization. With AI and real-time data, nutrition can become an ongoing dialogue. Imagine an app that notices you didn't sleep well last night and suggests a certain breakfast for an energy boost with slow-release carbs and extra hydration. Or if you logged an intense workout, the AI might recommend a protein-rich meal sooner. If a blood test shows you are low in, say, iron or vitamin D, the AI could adjust your weekly meal plan to include more iron-rich foods or prompt you to get sunlight. This reactive capability could make dietary adjustments much more responsive than annual doctor visits or static diet programs.

The ethical and privacy dimension here is significant. For effective personalization, users must share intimate data about their health and habits. Ensuring data privacy and security is paramount, as is avoiding misuse of data for intrusive marketing. There's also a need for transparency algorithms recommending diets should ideally be able to explain why they suggest what they do, to build trust and help users learn. Experts emphasize that AI in nutrition should be inclusive and unbiased [12]; early algorithms trained mainly on Western dietary patterns, for instance, might not perform well for someone whose diet consists entirely of different foods. Companies are expanding their food databases and hiring diverse dietitians to guide the AI so that it can serve global audiences fairly.

In summary, AI-driven personalized nutrition holds promise to improve public health by making individualized healthy eating attainable for the masses. It transforms what was once a laborious process of consultation and self-tracking into a convenient, even enjoyable experience, turning one's smartphone into a nutrition coach. As this field advances, we might see deeper integration, like AI systems working with doctors to prescribe diet modifications as readily as medications. Importantly, personalized nutrition highlights a broader trend in Culinary 4.0: the convergence of food, health data, and biotechnology. It underscores the notion that food is not only fuel or pleasure but also information that AI can interpret to benefit individual and societal well-being.

#### 8.2.4.4 Global Sustainability and Access in the AI Food Era

AI's deployment in the food system comes at a critical time for global sustainability and food security. The world faces the dual challenge of producing enough food for a growing population while reducing the environmental impact of agriculture. Here, the global perspective is essential: different regions have different needs and starting points. Culinary 4.0 innovations, if applied thoughtfully, could help bridge gaps between developed and developing nations' food systems and promote a more equitable global food future. Still, if mismanaged, they risk widening those gaps. This section examines how AI is being used globally to promote sustainable practices and what steps are necessary to ensure inclusive benefits.

First, AI can significantly advance the goal of sustainable farming. Precision agriculture techniques discussed earlier (targeted input application, predictive crop maintenance) directly address issues such as the overuse of water and chemicals. For instance, European initiatives in precision farming use AI to optimize crop yields while reducing fertilizer and pesticide use, aiming to limit chemical runoff and greenhouse gas emissions [12]. AI can help develop and manage alternative farming systems, such as vertical farms and hydroponics, by controlling lighting, nutrients, and climate to maximize efficiency. These systems promise to produce more food in urban or indoor settings with minimal land and water use. One oft-cited figure is that vertical farms can produce yields hundreds of times higher per square meter than traditional fields, and AI is the "brains" that adjusts conditions in real time for each crop variety. Countries like Japan, the Netherlands, and Singapore (which are tech-forward and have land or labor constraints) are investing in such AI-controlled farms to boost self-sufficiency.

AI also aids the development of alternative proteins and food sources, which have global implications for sustainability. Livestock farming is a major contributor to deforestation, greenhouse emissions, and water use [1]. AI is being harnessed to create plant-based or cultured meat analogues that mimic the taste and texture of real meat, helping shift diets away from resource-intensive animal products. Startups use machine learning to identify plant proteins that best emulate meat characteristics or to optimize cell culture media for lab-grown meat, accelerating progress toward commercially viable products. The hope is that such innovations can reduce the environmental footprint of protein production and provide new food sources for regions lacking traditional livestock industries. Indeed, a Nature npj Science of Food paper (2025) suggests leveraging AI to discover "novel protein sources" and design foods that replicate the nutritional and sensory profile of animal-based foods, as a key strategy for a sustainable food system [1].

On the consumption side, AI can empower consumers globally to make more sustainable choices. Personalized apps might include a "carbon footprint" score for meal options, nudging users toward lower-impact foods (e.g., suggesting more plant-based meals if one's diet is very meat-heavy, or highlighting local seasonal produce). Some grocery apps use AI to educate shoppers, e.g., by scanning an item's barcode to display its sustainability rating (based on sourcing, packaging, etc.) and suggesting greener alternatives. Over time, such micro-level guidance can shift demand toward sustainability, thereby pressuring producers to adopt better practices.

Crucially, global organizations and governments are recognizing AI's potential for food security. The United Nations has highlighted that AI, along with other tech, will play a role in achieving Zero Hunger by enabling more innovative farming and distribution [19]. Public-private partnerships are emerging to ensure that AI tools reach smallholder farmers in Africa, Asia, and Latin America, who produce a large share of the world's food but often lack access to advanced technology. For example, one World Bank study in West Africa showed that even simple mobile advisories (not yet full AI) improved yields and resilience by delivering localized weather and crop tips via SMS and radio [2]. Building on such groundwork, there are efforts to develop AI systems that can work on low-end smartphones, operate offline, or use local languages so that farmers in remote areas can benefit from precision insights without needing expensive equipment [2]. Organizations like the AI Institute for Next Generation Food Systems (AIFS) in the US and similar bodies in Europe are not only doing high-tech research but also focusing on education and training, preparing a new generation of agriculturists and food scientists fluent in AI techniques. This knowledge transfer is key to democratizing Culinary 4.0.

Despite these positive trends, there is a legitimate concern about a "digital divide" in food and AI. Wealthier large-scale farms and food companies are rapidly adopting AI (due to their capital and infrastructure). In contrast, small farmers or small food businesses may lag, potentially being left at a competitive disadvantage. If AI-driven supply chains dramatically improve efficiency, traditional producers could struggle if they can't integrate. Therefore, many experts call for inclusive strategies: subsidizing AI tools for smallholders, open-sourcing specific AI models for broader use, and ensuring representation of diverse farming systems in AI development (so the tools aren't only tailored to industrial farming) [2, 12]. There are also cultural and social aspects that some communities might be skeptical of AI or find it conflicts with local knowledge. Building trust requires involving farmers and food workers in the design of AI solutions and demonstrating respect for traditional practices (for instance, using AI to complement, not override, farmers' experiential knowledge).

On the sustainability front, one irony is that AI itself consumes computational resources (training advanced models can use significant energy). Thus, there's active research on making "Green AI," ensuring that the carbon footprint of deploying AI in agriculture/food is offset by the gains it provides (e.g., if AI helps cut food waste by a significant amount, that benefit outweighs the electricity used by servers running AI algorithms). Techniques like federated learning (training AI on distributed devices without central servers) could be used to minimize data transfer and energy use, which is promising for widespread deployment in rural areas with limited connectivity [20].

One cannot talk about a global perspective without noting how AI in food intersects with policy and ethics. Some countries are incorporating AI into their national food security strategies. For example, China has invested in smart agriculture as part of its AI and modern farming initiatives, and is reportedly achieving yield improvements and cost reductions. In contrast, farmers in other regions worry that AI could bring corporate control or surveillance (if, say, large agritech companies own the AI platforms). It's a delicate balance: to leverage AI's benefits, data (such as soil health and crop performance) often needs to be gathered at scale, which raises questions about data ownership. Clear policies and possibly community data cooperatives could help ensure that farmers retain control over their data and get fair value from sharing it.

In a LIFE 4.0 scenario globally, one can imagine a more resilient and connected food system. AI could act as a planetary monitoring and coordination mechanism, identifying surpluses in one region and deficits in another and optimizing the flow of food to where it's needed most, thereby reducing both waste and hunger. Real-time data could enable adaptive trade: for example, if drought hits a major grain producer, AI might preemptively adjust global stock allocations or suggest alternative crop plantings elsewhere to compensate, cushioning the shock. Such responsiveness is part of what experts mean by building a "smarter, fairer and more resilient" food system for the world [2]. It's a vision where knowledge (such as best farming practices or pest warnings) is quickly shared across borders via AI networks, democratizing insights that were once siloed.

Yet, making this vision a reality requires addressing the challenges outlined. It demands investment in digital infrastructure in rural areas, interdisciplinary collaboration (agronomists with data scientists, chefs with technologists), and ethical governance to keep the focus on sustainability and equity. As one researcher put it, the food system transformation is "not just a tech challenge, it's a systems challenge." AI is a tool, though a powerful one; human decision-makers must guide its use. We must ensure that Culinary 4.0 doesn't inadvertently accelerate trends such as unhealthy consumption of ultra-processed food or the loss of culinary heritage; instead, it should help address the problems of Culinary 2.0 (excess, waste, environmental harm) while enhancing the positives of Culinary 1.0 (freshness, nutrition, cultural richness). The following section turns to these broader opportunities and transformational potentials, followed by the challenges to be mindful of.

## 8.3 OPPORTUNITIES AND TRANSFORMATIONAL POTENTIAL

AI's infusion into the food sector brings a host of promising opportunities that could transform how we eat, farm, and live. Here, we summarize the key potential benefits of Culinary 4.0:

### 8.3.1 Enhanced Food Security and Productivity

AI can significantly improve crop yields and reduce crop losses through precision farming and early problem detection. By making agriculture more efficient and predictive rather than reactive, AI contributes to higher and more stable food production [2]. This is critical for feeding a growing global population without necessarily expanding farmland. Higher productivity on existing land also helps protect forests and habitats from conversion to agriculture.

### 8.3.2 Sustainability and Environmental Gains

Many AI applications in food aim to do more with less, aligning with sustainability goals. Precision application of water, fertilizer, and pesticides means reduced resource waste and lower pollution (e.g., less nitrogen runoff and fewer greenhouse gas emissions) [12]. AI-optimized supply chains minimize food miles and spoilage, thereby lowering the carbon footprint of food logistics [3]. Additionally, AI is accelerating the development of sustainable alternatives, from plant-based proteins to eco-friendly packaging, by speeding up innovation cycles [3]. Collectively, these contribute to a greener food system.

### 8.3.3 Reduction of Food Waste

A staggering portion of food produced globally is never eaten, due to inefficiencies in harvesting, storage, and consumption. AI can attack waste at every level: on farms (predicting optimal harvest times and storage methods to avoid spoilage) [3], in distribution (dynamic routing and inventory management to ensure perishable foods are sold in time) [3], at retailers (AI-driven dynamic pricing or donation alerts for unsold goods), and in homes (smart fridges reminding us to use items before they expire). Cutting waste not only improves food availability but also means the resources used to produce that food were not spent in vain.

### 8.3.4 Personalized Health and Nutrition

As discussed, AI opens the door to precision nutrition, which could lead to healthier populations. By tailoring diet recommendations to individual needs and nudging people toward better choices, AI may help reduce obesity, diabetes, and other diet-related diseases in the long run [17]. This could also ease healthcare burdens by emphasizing prevention through diet. On a societal level, even if people don't use apps, insights from personalized nutrition research can inform public health guidelines that are more effective for subgroups of the population.

### 8.3.5 Culinary Innovation and Diversity

AI doesn't just optimize the old; it can create the new. We're seeing an explosion of creativity with AI generating novel recipes (sometimes pairing ingredients in unconventional ways that humans might not have thought of) [16]. Chefs are beginning to use AI as a collaborative tool to brainstorm dishes or adapt recipes for alternative ingredients (beneficial in an era of climate-induced crops or supply chain disruptions). This augments human creativity and could lead to new cuisines or fusion dishes that delight the palate. Furthermore, by analyzing cuisines globally, AI might help preserve culinary heritage. For example, by cataloguing and learning thousands of traditional recipes, it can ensure that knowledge is not lost and even suggest modern interpretations, keeping food culture alive.

### 8.3.6 Economic Efficiency and New Markets

For businesses, AI can reduce costs by automating labor-intensive tasks and reducing errors and batch failures in production. This can improve profit margins or be passed on to consumers in the form of lower prices. Entirely new markets and startups are emerging in food tech, from robotic kitchen equipment to AI-driven nutrition coaching, creating jobs in these high-tech segments. As labor is reallocated, we may see new roles (such as algorithmic farming advisors or robot maintenance specialists) that empower the workforce with tech-focused skills. In agriculture, AI can help small-scale farmers become more competitive, potentially boosting rural incomes if implemented equitably.

### 8.3.7 Consumer Convenience and Experience

On a day-to-day level, Culinary 4.0 technologies make life easier and sometimes more fun for consumers. Meal planning and cooking can be less of a chore when an intelligent assistant handles the tedious parts (generating grocery lists, setting cooking times, ensuring nothing burns). People with busy schedules, limited cooking skills, or special dietary needs stand to benefit significantly from such support. Additionally, dining experiences can become more personalized. For example, AI could power a restaurant recommendation engine that knows your taste so well that it reliably introduces you to places you love, or a digital wine sommelier that picks a wine ideally suited to your meal and preferences. These "smart" services enhance satisfaction and engagement with food.

### 8.3.8 Global Knowledge Sharing and Collaboration

AI can act as a connective tissue linking farmers, chefs, and researchers across the globe. A pest solution found in one country can be disseminated via AI platforms to farmers elsewhere encountering the same issue. Recipe and nutrition insights can be shared cross-culturally, promoting a richer global dialogue about food. This could foster greater empathy and understanding. In such instances, AI might help highlight the value of indigenous crops or cuisines, elevating their status in the global arena and encouraging biodiversity.

In sum, Culinary 4.0 holds transformative potential to solve many of the legacy problems of our food system: hunger, malnutrition, inefficiency, environmental harm, and limited access to information. It envisions a future where everyone, from rural farmers to urban home cooks, is empowered with AI-driven tools to make better decisions, yielding a food system that is more abundant, sustainable, healthy, and just. The opportunities are indeed exciting and largely unprecedented in scope.

Yet, realizing these benefits is not automatic. Each opportunity comes with challenges that require thoughtful implementation, policy, and ethical consideration. The following section delves into those challenges, risks, and moral concerns that accompany the adoption of AI in our culinary lives, to ensure that the transformation does not produce new problems or exacerbate existing inequalities.

## 8.4 CHALLENGES, RISKS, AND ETHICAL CONSIDERATIONS

While the future of food with AI is full of promise, it is equally important to recognize and navigate the challenges and risks that arise. Culinary 4.0 technologies can disrupt social norms, economic structures, and ethical boundaries. Below, we outline the key concerns and considerations:

### 8.4.1 Workforce Displacement and Labor Transformation

Automation in farming, food processing, and kitchens raises the prospect of job losses or downgrading of roles for human workers. For example, if robotic chefs become widespread, line cooks and kitchen assistants might find fewer employment opportunities. In South Korea's robot-run rest-stop kitchens, some human chefs reported that their roles were reduced to mere monitoring and cleaning, leading to resignations and a loss of morale [14]. It's essential to manage this transition by investing in retraining programs that help food industry workers acquire new skills (such as operating and maintaining AI systems) so they can transition into higher-value roles rather than being left behind. Moreover, there will likely always be demand for human chefs for creative and artisanal dining experiences; the challenge is to ensure a fair labor market and avoid a scenario in which only a few tech companies reap the benefits while many workers are displaced.

### 8.4.2 Data Privacy and Security

The personalization of nutrition and the digitization of food services rely on collecting sensitive data, such as dietary habits, health metrics, genetic information, and purchase history. This raises privacy concerns. Users must trust that their data on what they eat (and by extension, aspects of their health and lifestyle) is stored securely and not misused. There is a risk of data being exploited for commercial gain in ways consumers don't expect, for instance, could insurance companies seek access to nutrition app data to adjust premiums? Robust data protection regulations and transparent data governance are needed. Additionally, as kitchens and food systems become IoT-connected, cybersecurity becomes critical. A breach in a connected oven or a food factory's AI system could have safety implications (imagine a hacked oven that overheats, or a food supply chain system disrupted by ransomware [21]). Thus, securing food tech against cyber threats is as important as physical food safety measures.

### 8.4.3 Algorithmic Bias and Cultural Sensitivity

AI systems are only as good as the data and assumptions behind them. If not carefully developed, they might carry biases that lead to suboptimal or even harmful outcomes in the context of food. For example, an AI trained predominantly on Western dietary data might not provide appropriate advice for someone in Asia or Africa with a different staple diet, effectively exhibiting cultural bias. There's also a risk of nutritional algorithms favoring certain body types or health ideals that are not universally applicable. Ensuring diverse datasets (including recipes, ingredients, and health data from various ethnic and cultural groups) is crucial to prevent AI from becoming a one-size-fits-all voice that ignores cultural dietary practices. Another form of bias might be in how delivery platform algorithms prioritize restaurants. There have been concerns that they may disadvantage small or minority-owned restaurants in favor of those that pay higher commissions or meet specific popularity metrics. Algorithmic transparency and fairness audits should be implemented in such systems to maintain a level playing field.

### 8.4.4 Loss of Culinary Traditions and Skills

If AI and automation handle more cooking tasks, there is a subtle but real worry about the erosion of human cooking skills and food knowledge. Traditional methods and recipes passed down through generations might be neglected if people become too reliant on AI's suggestions or on ready-made automated solutions. Food is a carrier of cultural heritage. If future generations grow up with personalized AI meals and robotic cooking, will they still learn their grandparents' recipes or the nuances of regional cuisine? Some chefs argue that creativity could be stifled if everyone starts eating "optimized" diets. It's a paradox: AI can help preserve and disseminate recipes, yet it could also homogenize tastes if algorithms steer people toward similar "ideal" choices. Maintaining a balance between innovation and tradition is essential. This might mean consciously using AI to support cultural education (e.g., recommending seasonal local dishes or teaching you about the history of a recipe as it helps you cook it) rather than simply optimizing for nutrition or convenience alone.

### 8.4.5 Ethical Use of AI in Decision-Making

As AI takes on roles in food safety and allocation (e.g., deciding which region receives shipments during a shortage or which formula to feed infants in neonatal units), ethical questions arise. Who is accountable if an AI's recommendation leads to a poor outcome, such as an allergic reaction, because the AI failed to flag an allergen? Decision-making in areas like dietary advice for medical conditions can carry serious consequences; oversight by human professionals is needed. There's also the ethics of AI in consumer manipulation. If an AI knows your cravings, it could theoretically be used by a business to upsell unhealthy snacks at moments of weakness. Drawing ethical lines (perhaps even regulating AI use in marketing foods, like how we regulate ads to children) could be warranted to ensure AI nudges are always in the genuine interest of the consumer's health and well-being, not just commercial profit.

### 8.4.6 Equity and Access Disparities

A significant risk is that the advantages of Culinary 4.0 might accrue mostly to affluent individuals and regions, widening existing disparities. Healthy personalized nutrition guided by AI and access to cutting-edge food tech might be readily available in wealthy urban areas, but not in low-income communities that lack internet access or cannot afford the gadgets and services. There's a scenario to avoid where healthy, AI-curated diets become a luxury good, while poorer populations continue to face cheap, unhealthy processed foods. Efforts like open-source AI nutrition platforms, community-based initiatives (such as public health departments providing free diet apps or innovative kitchen tools in community centers) can help democratize these benefits. On the farming side, small-scale farmers need support to adopt AI so large agribusinesses with advanced tech do not outcompete them [2]. This could involve government subsidies for precision ag tools or cooperative models where farmers share AI resources.

### 8.4.7 Dependence and Resilience

Over-reliance on AI and automation can introduce fragility if not backed up by human capacity. If people lose basic cooking or farming skills and something disrupts the AI systems (be it a network outage, software bug, or geopolitical issues affecting tech supply), communities could be caught unprepared. Resilience requires maintaining human-in-the-loop approaches and fallback options. The COVID-19 pandemic, for instance, showed the value of having local food knowledge and skills when global systems falter. Thus, even as AI assists, retaining a level of manual skill and decentralization (e.g., encouraging local food growing and community cooking) is wise, so that technology complements rather than completely replaces human skills.

### 8.4.8 Transparency and Trust

AI decisions about food need to be transparent to earn user trust. If a nutrition app tells someone to cut down on red meat, is it because it genuinely benefits their health or because it partners with a plant-based product company? If AI adjusts a restaurant's recipe for cost reasons, is that disclosed to diners? Maintaining trust means clearly communicating how AI comes to its recommendations. This is especially true in cases like genetically engineered or AI-designed foods, consumers have the right to know and to choose what they consume based on informed understanding.

Engaging the public in dialogues about food technology (through education campaigns, labeling, etc.) will help prevent fear or misinformation from derailing beneficial innovations.

In summary, these challenges highlight that technology alone can't solve food system issues; how we implement and govern it matters enormously. The ethos of "not harm" that applies in medicine could well be used to food AI: innovations should ideally be introduced in ways that minimize harm to people's health, livelihoods, and cultures. Ethical frameworks and perhaps new regulations will be needed, akin to how we think about AI in healthcare or finance, because food is just as fundamental to human life.

Culinary 4.0 must therefore proceed with a human-centric approach. That means prioritizing inclusivity, ensuring human oversight for critical decisions, and preserving the values like enjoyment of food, community, and cultural meaning that make food much more than just a commodity or data point. The following section will take these considerations and imagine an outlook, painting a picture of daily life in the era of LIFE 4.0, where food and AI coalesce harmoniously (if we navigate the challenges well) or cautionary (if we don't), and what that implies for our relationship with what we eat.

## 8.5 THE FUTURE OUTLOOK (LIFE 4.0 SCENARIO)

Envisioning a LIFE 4.0 scenario means imagining a day-to-day world where AI's integration into food and life is ubiquitous and seamless, a future that is rapidly approaching. In this future, the relationship between individuals, communities, and food could be profoundly different, yet hopefully, enriched. Below, we paint a picture of how Culinary 4.0 might manifest in everyday contexts a couple of decades from now, if current trends continue to mature.

It's the year 2045, dinner time. In a suburban household, the role of "meal planner" is played by an AI that knows the family's schedule, dietary needs, and even mood. This household AI assistant (let's call it "ChefBot") starts the day by analyzing everyone's health data: it notes that one parent didn't sleep well (and thus might need an extra iron and protein boost for energy), the other parent did a heavy workout (needing more hydration and electrolytes), one child has a math exam (perhaps some brain-boosting omega-3s), and the other was coming down with a cold (requiring vitamin C and zinc). Considering these factors, plus each person's personal taste preferences and what's currently in the pantry, ChefBot designs a tailored dinner menu for the family.

Because grocery supplies are managed by AI as well, there is no need for a weekly shop; in fact, the traditional weekly grocery shopping trip has become obsolete. By 2050, as one food futurologist predicted, AI will have taken over household food management, automatically ordering what is needed and even arranging meal kit deliveries just in time. Earlier in the day, ChefBot communicated with a local vertical farm and a grocery fulfillment center. A drone delivery arrives in the afternoon with a bag of ingredients: fresh greens (harvested hours ago from a nearby AI-managed urban farm), a piece of ethically cultured fish (grown from cells in a bioreactor, pre-cut to portion), and an assortment of spices and herbs tailored to the recipe ChefBot has in mind.

In the kitchen, cooking has become a cooperative process between human and machine. The kitchen is equipped with modular robotic assistants, a smart oven, a counter-top robotic arm, and connected appliances. ChefBot guides the family through cooking with a mix of automation and instruction. For the salad, the robotic arm expertly slices vegetables with precision; for the main dish, the smart oven has already preheated and used machine vision to ensure the fish is cooked to the exact doneness. A voice prompts one of the parents to stir a sauce at a particular moment, imparting the human touch and judgment where needed. At the same time, the AI handles the more time-critical or tedious tasks (such as constant stirring or temperature adjustments). Cooking has become easier and far less time-consuming, giving the family more opportunity to bond rather than stress in the kitchen. Even cleanup is assisted: the same robotic arm can wash pans and load the dishwasher, and kitchen surfaces are made of self-cleaning materials developed through AI-driven material science.

Crucially, though automation abounds, it hasn't eliminated the social and sensory joy of cooking and eating together. In fact, it may have enhanced it: freed from drudgery, the family can focus on tasting, learning, and sharing. ChefBot often explains the cultural background of a dish or the nutrition science behind it as it cooks, turning meals into mini educational experiences. Tonight, for example, it describes how the herbs in the sauce are like those used in a traditional recipe from their grandmother's hometown, tying modern AI cuisine back to culinary heritage. In LIFE 4.0, rather than erasing traditional foods, there's a conscious effort (perhaps guided by user feedback and policy) to weave them into the system. Community recipe databases have been digitized, so ChefBot might pull a beloved family recipe from the cloud and adapt it to current health goals, ensuring continuity of culture while optimizing well-being.

The food itself in 2045 looks familiar in presentation (we still eat casseroles, stir-fries, salads), but where it comes from is radically different. Meat from slaughtered animals has become a rarity; much of the population gets its protein from plant-based or lab-cultured sources that taste and feel like meat but with far lower environmental cost. AI played a key role in perfecting these alternatives, making them indistinguishable from the originals. For those who do eat conventional meat, it often comes from regenerative farms monitored by AI for high animal welfare and low emissions. Climate change has made certain crops harder to grow in their old locales, but AI-guided plant breeding has produced new strains and recommended "crop swaps." For instance, parts of England now grow chickpeas and okra as staples (as predicted) [22], and AI helped farmers transition smoothly. Global cuisines have become even more intermingled: the family's meal might incorporate Andean super grains like kiwicha or African millet like fonio, ingredients once obscure internationally but now commonly available thanks to climate adaptability and supply chain diversity [22]. AI identified these crops as resilient and nutritious, and global trade (plus local cultivation in suitable climates) expanded as part of a push for food system resilience.

At the societal level, food has become more personalized but also more communal in interesting ways. People trust their AI assistants to handle individual health optimization, but they still crave social food experiences. There's a renewed trend of communal dining events, perhaps as a counterbalance to high-tech living. As one futurologist envisioned, even if AI can make all our food, we still highly value human-made meals and gathering [22]. In cities, neighbors come together for periodic "analog dinners" where everyone cooks something from scratch (sometimes deliberately ignoring AI suggestions to experiment). Food culture adapts: cooking by hand might become a hobbyist or artisan activity, akin to how some people today knit or bake bread for pleasure rather than necessity. The best restaurants still have human chefs who use AI as a creative partner but put their personal artistic stamp on dishes. There is even a greater appreciation for the craft of cooking precisely because it's no longer mandatory; it's done for love, not labor.

In this LIFE 4.0 world, sustainability is largely built into the food system. AI coordinates much of the supply chain to minimize waste: unsold produce from grocery AI hubs is automatically rerouted to food banks or converted into longer-lasting forms (dried, pickled) based on demand prediction models. Consumers are far more aware of their environmental "foodprint" because every meal can display an eco-score computed by AI (though by now, many habits have shifted to plant-rich diets by default). Possibly, we have "carbon labels" on menus and packages as normal as nutrition labels, and AI ensures those calculations are accurate and updated in real time with sourcing changes.

Internationally, AI helps buffer regions from famine. If drought hits a breadbasket country, AI forecasting triggers early ramp-ups of alternative grain production elsewhere, preventing a crisis. Developing countries have widely adopted AI-driven farming as the mobile internet has become nearly universal, and devices have become cheaper. A farmer in Sub-Saharan Africa might wear AR glasses that overlay real-time pest and soil data on her fields, guided by an AI trained on both local knowledge and global agronomy research, yielding crops comparable to those from a fully mechanized farm at a low cost. This hopeful scenario sees global food equity improving, with fewer regions suffering chronic hunger. We might not eliminate all disparities, but food insecurity is tackled with the same seriousness as disease outbreaks, aided by AI's predictive and logistical prowess.

However, not everything is utopian. In this scenario, society had to confront the challenges enumerated earlier: workforce disruption meant major job programs to transition workers into new roles (with success in some countries, unrest in others that were slower to adapt). There was a public debate about the right amount of automation in restaurants. Some chains went fully robotic for efficiency, but many reintroduced human servers after finding that customers missed interaction. Governments implemented regulations: requiring transparency (AI systems must explain nutritional advice), mandating data privacy (strict laws protecting personal food/health data akin to medical records), and setting guidelines for algorithmic fairness (ensuring, for instance, that diet AIs serve diverse populations effectively and don't propagate harmful diet fads). Culturally, people learned to question AI suggestions when appropriate, critically. Essentially, digital literacy expanded to include food literacy, teaching people how to use these tools without becoming overly dependent.

Thus, the LIFE 4.0 culinary future is a blend of high-tech convenience and enduring human values. AI becomes an invisible assistant, making food healthier, more accessible, and more sustainable by default, while humans remain in control of defining what food means to us: pleasure, tradition, creativity, and community. Ideally, it's a future where no one goes hungry because AI has optimized distribution and yield, and where lifestyle diseases wane. After all, diets are better managed, and the planet can heal because our food system no longer entails massive waste and emissions. It is a future scenario, of course, and reaching it will require navigating the challenges we've discussed.

The concluding section will summarize these insights and emphasize the key takeaways from our journey from Culinary 1.0 to 4.0, reinforcing how we might steer the age of AI toward truly delivering a better food future for all.

## 8.6 CONCLUSION

From the first controlled fire that roasted meat in prehistoric times to the intelligent algorithms now crafting personalized meal plans, the journey of food and cooking reflects the broader journey of human civilization and technology. We have traveled through Culinary 1.0, where cooking was an elemental human endeavor rooted in tradition; Culinary 2.0, where industrialization brought both abundance and new dilemmas; Culinary 3.0, where digital networks transformed convenience and connectivity in food; and into Culinary 4.0, where AI is poised to redefine what is possible in our relationship with food. The chapter has illustrated that each phase didn't fully replace the previous but built upon it, and now, in the age of AI, we stand at a crossroads with the potential to solve persistent problems and enrich culinary experiences, provided we address the associated challenges thoughtfully.

The integration of AI into food systems offers tremendous transformational potential. It promises to make agriculture more innovative and more sustainable, reducing environmental strain while feeding more people [2, 12]. It can make our diets healthier by tailoring them to individual needs and by offering knowledge that helps us make better choices [15]. It can even spark creativity, uncovering novel recipes and flavors that broaden our gastronomic horizons [16]. Significantly, AI can enhance resilience in a world facing climate change and resource limits. AI is a tool that can help adapt crops, prevent waste, and ensure food reaches those who need it most [2, 3]. In many ways, Culinary 4.0 could be the key to achieving a food system that is simultaneously efficient, equitable, and environmentally sound, a balance that eluded prior eras.

Yet, as we have emphasized, the benefits of Culinary 4.0 are not automatic. They require human stewardship and ethical guardrails. We must consciously direct AI innovation to serve the public good: algorithms should be transparent, inclusive, and aligned with health and sustainability goals rather than merely optimized for profit or convenience. This might mean new policies, such as standards for AI in nutrition advice (to prevent misinformation or harmful diet fads) and labor policies to support workers transitioning from automated roles. It means cross-disciplinary collaboration: technologists, nutritionists, farmers, ethicists, and policymakers need to be at the same table to design systems that work in the real world.

Crucially, the human element in food must remain central. As AI becomes more advanced, eating will always remain a fundamentally human experience tied to culture, emotion, and sensory pleasure. AI cannot (and should not) replace the warmth of breaking bread with others, the heritage of recipes passed through generations, or the creativity and love a chef pours into a dish. As one scientific perspective noted, today's AI cannot fully grasp the nuanced social, ethical, and sensory dimensions of food that are deeply rooted in human culture [1]. This insight should guide us: AI in food is a powerful tool, but it is we who imbue food with meaning. The best outcomes will likely arise when AI handles the quantitative aspects of science, logistics, and optimization, freeing humans to focus on the qualitative aspects of art, community, and the joy of food.

In summary the Future of Food in the Age of AI is not a story of technology versus tradition, but instead technology empowering us to uphold the core values of nourishment, sustainability, and sharing in new ways. If Culinary 4.0 is guided by wisdom, inclusivity, and ethical considerations, it could herald an era in which hunger is largely eliminated, lifestyle diseases are curtailed, and the food on our plate delights us without costing the earth. Achieving this vision will be one of the defining challenges and opportunities of LIFE 4.0. As we stand at this threshold, it is our collective responsibility as academics, professionals, and citizens to ensure that the digital brains we introduce into our kitchens and fields operate in harmony with the enduring heart and soul of food.

## REFERENCES

1. Kuhl, E., AI for Food: Accelerating and Democratizing Discovery and Innovation. *npj Science of Food*, 2025. 9(1): p. 82.
2. Forum, W.E. How AI Can Help Improve Food Systems and Agricultural Yields, in *World Economic Forum Agenda*, June 2025.
3. AgriFood, T., Feeding the Future: 9 Ways AI Is Transforming the Food System from Farm to Fork, in *THRIVE Blog*, November 12, 2025.
4. Rupp, R., A Brief History of Cooking With Fire, in *National Geographic*, September 2, 2015.
5. Helmer, J., Labor Day: How the Industrial Revolution Changed Agriculture, in *FFA History*, 2020.
6. Mulhern, O., Food Production: Recent History and Where We Are Heading, in *Earth. Org*, February 19, 2021.
7. Morgan, K., T. Marsden, and J. Murdoch, *Worlds of Food: Place, Power, and Provenance in the Food Chain*. London: Oxford University Press, 2006.
8. Hut, P., *Pizza Hut Celebrates 20th Anniversary of World's First Online Purchase with 50 Percent Off Online Deal for Hut Lovers Members*. January 2, 2014.
9. Ahuja, K., et al., Ordering In: The Rapid Evolution of Food Delivery. *McKinsey & Company*, 2021. 22: p. 1–13.
10. Wikipedia, *Online Food Ordering – History. Wikipedia: The First Online Food Order Was a Pizza from Pizza Hut in 1994*, 2023.
11. Nanda, R., How AI and IoT Are Shaping the Next Generation of Smart Kitchens, in *Techstrong.ai*, March 18, 2025.
12. Pizzuto, C., How AI Is Revolutionizing Food Systems for a Sustainable Future, in *AI for Good Stories*, 2024.
13. Kingson, J.A., Restaurant Robots Can Cook, Serve, and Bus Your Meal Now, in *Axios*, June 11, 2024.
14. Kim, M., Robot Chefs Take Over at South Korea's Highway Restaurants, in *Mixed Reviews—Rest of World*, May 14, 2025.
15. Tribe AI, Top AI-Powered Nutrition Apps to Watch in 2025, in *Tribe AI Blog*, April 28, 2025.

16. Marr, B., How AI-Enabled Chef Watson Is Bringing Creativity and Innovation to Cooking, in *Bernard Marr & Co. Blog*, 2018.
17. Kleinberg, S., *Debate Weighs Efficacy of AI for Personalized Nutrition Planning—The Official News Source of the 85th Scientific Sessions*, 2024.
18. Bose, P.P.D., *AI-Driven Personalized Nutrition Shows Promise in Improving Gut Health*, 2025.
19. Marwala, T., *Artificial Intelligence Can Transform Global Food Security and Climate Action*, 2024.
20. Agrawal, K., et al., Artificial Intelligence in Personalized Nutrition and Food Manufacturing: A Comprehensive Review of Methods, Applications, and Future Directions. *Frontiers in Nutrition*, 2025. 12: p. 1636980.
21. ses653, Cybersecurity & Food Safety: A Growing Concern for the Food Industry, in *Food Integrity and Safety*, April 7, 2025.
22. Morley, M., Dinner of the Future – What We'll Be Eating in 2050, in *FoodBev Media*, August 1, 2025.

# 9 Sport 4.0

**Abstract**

In the fourth epoch of sports, the Sports 4.0 era, athletic performance and spectator experiences are being fundamentally transformed by artificial intelligence (AI) and allied technologies. This chapter traces the evolution of sports through four major phases. Sports 1.0 encompasses ancient traditions and local competitions, highlighting how early athletic contests (from the Olympic Games of antiquity to medieval tournaments) were deeply rooted in cultural ritual and communal pride. Sports 2.0 charts the impact of industrialization, urbanization, and mass media in the 19th and early 20th centuries, which professionalized sports and turned games into grand spectacles for national audiences. Sports 3.0 covers the digital revolution of the late 20th century, marked by global broadcasting, real-time data analytics, and the emergence of a worldwide sports economy. Building on these foundations, Sports 4.0 is defined by AI-augmented training, innovative equipment, immersive fan engagement, and intelligent infrastructures. In this era, athletes leverage machine learning for personalized coaching and injury prevention. At the same time, fans enjoy augmented reality (AR), virtual reality (VR), and data-rich interactivity that blur the line between physical and virtual arenas. The chapter explores the opportunities and transformational potential of Sports 4.0, ranging from breaking performance barriers to democratizing access to global sports. It also addresses the challenges and ethical considerations, including data privacy, competitive balance, and the preservation of human-centric values in sport. Looking ahead, we envision a future where AI seamlessly integrates into sports to enhance athlete well-being, spectator inclusion, and global connectivity, a LIFE 4.0 scenario in which athleticism and technological innovation co-evolve to enrich the spectacle of sport for all.

## 9.1 INTRODUCTION

Sports have always been a mirror of societal evolution, from the ancient arenas where athletic feats honored gods and community to today's high-tech stadiums that connect millions of fans in real time [1]. In the 21st century, we stand at the brink of a new paradigm often dubbed "Sports 4.0," echoing the broader Fourth Industrial Revolution driven by digital transformations. Just as Industry 4.0 refers to the fusion of artificial intelligence (AI), robotics, Internet of Things (IoT), and data in manufacturing, Sports 4.0 denotes the integration of these cutting-edge technologies into athletics and sports entertainment. This chapter will contextualize Sports 4.0 by first journeying through earlier epochs of sports development (Sports 1.0, 2.0, and 3.0) to illustrate how we arrived at the current AI-augmented age [2].

Historically, each significant shift in sports has been catalyzed by broader technological and social changes. In Sports 1.0, athletic contests were local and often ritualistic, reflecting the values and mythologies of ancient civilizations. In Sports 2.0,



DOI: 10.1201/9781003455219-9

the forces of industrialization and urban growth professionalized sports and standardized rules. They enabled mass spectatorship from packed city stadiums to the first crackling radio broadcasts that carried live commentary across regions. Sports 3.0 unfolded with the rise of global media and the digital revolution: television and satellite broadcasts made the Olympics and World Cup truly worldwide events [3]. Computer analytics and the internet ushered in new ways to evaluate performance and engage fans.

Against this backdrop, Sports 4.0 emerges as a transformative convergence of AI, data analytics, intelligent devices, and immersive media in the sports domain [4]. The hallmark of this epoch is the augmentation of both athleticism and spectacle through technology. Athletes and coaches now have unprecedented tools from machine learning algorithms that optimize training regimens to wearables that provide real-time biometric feedback. Fans, however, experience the spectacle of sports through personalized content, interactive platforms, and augmented (ARs) or virtual realities (VRs) that were inconceivable to earlier generations. The very definition of what constitutes a sports "arena" or "competition" is expanding, as virtual competitions (such as eSports and AR-enhanced games) take their place alongside traditional physical sports.

Crucially, Sports 4.0 is not happening in isolation in one corner of the globe; it is a global phenomenon unfolding in diverse ways across different countries and sports. Elite sports programs in countries like the US, Germany, and China are investing heavily in AI-driven performance analytics, while emerging sports markets are leveraging technology to leapfrog infrastructure gaps and reach audiences in new ways. The chapter will incorporate these global perspectives, recognizing that the AI epoch in sports has varied manifestations and impacts worldwide.

In the sections that follow, we delineate the characteristics of Sports 1.0, 2.0, and 3.0, which serve as the historical foundation for understanding Sports 4.0. We then delve deeply into the components of Sports 4.0, from AI-augmented training and innovative equipment to immersive spectating, illustrating how each is reshaping the athletic experience. Finally, we discuss the significant opportunities this transformation offers, the challenges and ethical dilemmas it raises, and a forward-looking LIFE 4.0 scenario that imagines the future of sports if current trends mature to their full potential (Table 9.1).

**TABLE 9.1**
**Comparative Summary**

| Sports Era | Core Mode | Key Driver | Focus |
|---|---|---|---|
| 1.0 | Community sport | Tradition | Participation |
| 2.0 | Professional sport | Media and industry | Performance |
| 3.0 | Digital sport | Data and platforms | Engagement |
| 4.0 | Intelligent sport | AI and XR | Optimization and immersion |

## 9.2 SPORTS 1.0–4.0

### 9.2.1 Sports 1.0: Ancient Traditions and Local Competitions

Sports in their earliest recorded forms were deeply intertwined with ritual, religion, and community. In ancient civilizations, athletic contests were not viewed as standalone entertainment or professions as they often are today, but rather as ritualized expressions of cultural values and local identity. Participation was typically local or regional, rules were derived from tradition, and spectators were members of nearby communities or worshippers at festivals.

One of the most iconic embodiments of Sports 1.0 is the ancient Olympic Games of Greece. Starting in 776 BCE, the Olympics were held every four years at Olympia to honor Zeus [5]. These early Games featured events such as footraces, wrestling, boxing, and chariot races, which also had religious significance. Athletes competed as representatives of their city-states, and victors like Koroibos of Elis (the first recorded Olympic champion) were immortalized in lore. Over time, additional Panhellenic games (Pythian, Nemean, Isthmian) emerged, attracting athletes from across the Greek world and fostering a sense of shared Hellenic culture. During these festivals, an Olympic truce was declared so that athletes and spectators could travel safely, an ancient recognition of sport's power to bring temporary peace. Prizes in these games were wreaths and glory rather than money, and champion athletes like Milo of Croton became celebrated heroes, sometimes mythologized as demigods in lineage or inspiration. The ethos of ancient competition emphasized honor, excellence (arete), and often a spiritual connection. For instance, Greek athletes would dedicate their victories to the gods, and some events (such as the footrace in armor or the brutal pankration) had origins in military training or mythic exploits.

Other ancient societies had their own sporting traditions. In Ancient Egypt, tomb paintings depict Pharaohs and commoners engaging in activities like wrestling, swimming, archery, and an early form of handball, often tied to military training or ritual demonstrations of strength. In Mesopotamia, artifacts from Sumer (c. 3rd millennium BCE) show wrestlers in combat, and the epic Gilgamesh mentions belt-wrestling bouts as a measure of valor. The people of Mesoamerica developed the famous ballgame (Ōllamaliztli), a ceremonial sport played since at least 1400 BCE, where a rubber ball was passed through stone hoops. This game had such ritual importance that it was associated with cosmology and, at times, human sacrifice. In ancient China, competitive pastimes such as Cuju (an early form of soccer) and archery contests were present by the Zhou dynasty, often linked to military readiness and court pageantry. Across these diverse cultures, sports served as proving grounds for physical excellence and courage, but equally as social and religious gatherings that reinforced community bonds and shared myths [2].

Localism defined sports in the 1.0 era. Events were organized by and for the local populace, be it a city-state, a religious festival, or a feudal lord's court. This meant rules and styles of play could vary significantly from one region to another. For example, the exact format of a wrestling match or footrace could differ between Olympia and Delphi, just as medieval folk games varied from one village to the next. There were no international standards or federations: a "sport" was a communal agreement of what contest was being played and how to score it. In many cases, athletic contests were integrated into cultural festivals (like the Greek Olympics or the Scottish Highland Games later in the medieval period) rather than being standalone events.

It's important to note that while these early sports did not have technology in the modern sense, they were still influenced by the tools and knowledge of their time. The design of ancient javelins, discuses, or bows, and the architecture of arenas (such as the Roman Colosseum or the circuses for chariot racing) reflected the engineering capabilities of those eras. Training methods, albeit primitive, existed among ancient Greek athletes; for instance, they followed special diets and used trainers (paidotribes) to prepare for competition (metmuseum.org) [5]. However, outcomes relied purely on human and animal physical abilities, and the concept of augmenting performance with external devices was limited to basic gear (e.g., padded gloves for boxers, or performance footwear in some cultures).

Another hallmark of Sports 1.0 was the spectator experience: spectators were physically present, often as co-participants in the ritual. There was no remote viewing, of course, the "broadcast" of a great victory was through traveling bards or written records, not instantaneous media. The roars in the stadium (or simple clearing) were the only feedback loop of fan engagement. Nevertheless, ancient sports could draw huge crowds in their context. The Roman gladiatorial games and chariot races, for example, filled massive arenas; the Circus Maximus in Rome could host tens of thousands of spectators, illustrating that the appetite for dramatic sports spectacle is by no means a modern invention.

In summary, Sports 1.0 represents the age when sports were localized competitions steeped in tradition, with minimal technological influence beyond the basic tools of play. Athletes were often amateurs in the true sense, farmers, soldiers, or aristocrats competing for honor, and the notion of a professional sports industry was non-existent. Yet even in this era, we see the fundamental human themes of sports persist: the celebration of physical excellence, the forging of social unity, and the use of sports as a stage for values (be it religious piety, civic pride, or personal honor). These ancient and local roots set the stage for the seismic changes that would come with industrialization and the rise of global media in later centuries.

### 9.2.2 Sports 2.0: Industrialization, Mass Spectatorship, and Professionalization

The dawn of the Industrial Revolution in the 18th and 19th centuries brought dramatic changes to nearly every aspect of human life, and the realm of sports was no exception. Sports 2.0 refers to the period roughly spanning the 19th and early 20th centuries when sports transformed from localized folk activities into organized, codified, and professional institutions. This era was characterized by the formalization of rules, the rise of professional athletes and leagues, and the advent of mass spectatorship, aided by new communication technologies and urban infrastructure.

### 9.2.3 Several Key Developments Define Sports 2.0

#### 9.2.3.1 Codification of Rules and Organized Competition

Many modern sports took their current form during this period. For example, the rules of football (soccer) were standardized in England in 1863 with the formation of The Football Association, rugby split off with its own code in 1871, basketball was invented with a fixed rule set in 1891, volleyball in 1895, and so on. This was a shift from the variable local rules of earlier times to a common rulebook, enabling fair competition on a larger scale. National and international sport governing bodies emerged (e.g., FIFA for soccer in 1904, the International Olympic Committee in 1894), reflecting a new institutionalization of sports.

#### 9.2.3.2 Professionalization

While in Sports 1.0, athletes were primarily amateurs (gentleman competitors or participants in their leisure time), Sports 2.0 saw the birth of the professional athlete and coach. The late 19th century witnessed the first professional sports clubs and leagues. In baseball, for instance, the Cincinnati Red Stockings became the first openly professional team in 1869, paying players a salary. By the turn of the 20th century, professional leagues in baseball (MLB) and football (various regional leagues in soccer, including the English Football League, founded in 1888) were drawing paying crowds. Industrialization had created an urban working class with some disposable income and leisure on weekends, fueling demand for spectator entertainment. Talented players could now make a living from sports, elevating the quality of play and the stakes of competition. This era also saw the rise of sports heroes like cricket's W.G. Grace, boxing's John L. Sullivan, or baseball's Ty Cobb, whose fame spread beyond their local community, hinting at the early seeds of celebrity culture in sports.

#### 9.2.3.3 Infrastructure: Stadiums and Transport

Industrial-era advancements made it feasible to gather large crowds in one place. Railways and later trams allowed teams and fans to travel for regional matches, expanding the geographic scope of competitions. Iconic stadiums were built in this era: e.g., Lord's Cricket Ground (England) was established earlier but expanded in the 19th century, the first modern Olympic stadium in Athens (1896) revived the ancient tradition in a new internationalist form, and large baseball parks and football grounds sprang up in major cities. The concept of a dedicated sports venue with ticketed entry and spectator amenities became normalized. Sports shifted from open fields or informal settings to enclosed arenas where admission could be charged, signifying the commercialization of the spectator experience.

#### 9.2.3.4 Media and Mass Spectatorship

While the 19th century predates radio or television broadcasts, the mass media of the time (newspapers, magazines, and telegraphs) played a crucial role in expanding sports fandom. Newspapers started dedicating sections to sports, and sports journalism was born, allowing fans who couldn't attend events to follow their teams. The first telegraph reports of sports events in the 1890s enabled rapid news dissemination [3]. For example, in 1896, a telegraph line relayed updates of a Stanley Cup hockey series from Montreal to Winnipeg. This was an early form of "live" remote sports experience, albeit via text updates rather than audio/visual. By the early 20th century, radio broadcasting was on the horizon (with the first sports radio broadcast occurring in April 1921, a boxing match broadcast by KDKA in Pittsburgh [6], and the first baseball game broadcast later that year, but even before that, sports had achieved a national following through print media and the telegraph. The Olympic Games, revived in 1896 in Athens, were a hallmark of this era's globalizing impulse, an international multi-sport event inspired by industrial-age ideals of progress and unity, even if early editions were only followed through print reports.

#### 9.2.3.5 Social Change and Democratization of Sports

Industrialization also brought ideas of organized physical education and the use of sports for social development. Schools and universities began formal sports programs (e.g., the birth of collegiate sports in the US and Europe in the late 1800s). Sports clubs were formed not just by elites but also by workers (many European football clubs began as factory teams or workers' clubs). This era saw sports used to bridge social divides (though often they also reflected them; for instance, certain sports were seen as "gentlemanly" and others as working-class). The inclusion of women in sports also progressed slowly during this period, with women's events in tennis, golf, and the Olympics (first women's Olympic events in 1900) making tentative appearances despite prevailing social norms restricting female athletic participation.

In sum, Sports 2.0 marks the transition of sports into a structured societal institution. What had been informal or festival activities became regular-season schedules, championship tournaments, and international contests. The concept of the spectator evolved from being no longer just a neighbor watching a local contest. Still, a paying customer in a crowd of thousands, or a reader eagerly following distant events via the press. Sports became a form of mass entertainment and even soft power; for example, countries began to measure prestige in the Olympic medal counts or the successes of their national teams. The seeds of the sports industry were planted as revenues from tickets, betting, and sponsorship (early forms, such as companies sponsoring trophies or providing kit) entered the picture.

Technologically, the Sports 2.0 era set the stage for broadcast and data revolutions. By the end of this period (early 20th century), the invention of radio would soon allow live sports commentary to be heard in homes, and innovations like electronic scoreboards and lighting (the first night games under electric lights began in the 1930s) hinted at how tech could further enhance sports viewing. Yet, even at this stage, technology's role was mostly limited to infrastructure and communication. The core athletic contest remained a human-versus-human (or team-versus-team) affair decided on the field with modest equipment. That paradigm would begin to shift in the mid-20th century, as we enter the digital age of Sports 3.0.

### 9.2.4 Sports 3.0: Digital Broadcasting, Globalization, and Data Analytics

By the mid-20th century, sports entered a new phase fueled by electronic media and computer technology. Sports 3.0 can be seen as spanning roughly the post-World War

II era through the late 20th and early 21st centuries, characterized by the globalization of sports, the rise of real-time digital broadcasting, and the incorporation of data analytics into competitive strategy and training. This era transformed sports into a truly worldwide cultural phenomenon and a massive commercial industry.

One of the most revolutionary changes was the advent of broadcast technology, first with radio, then with television, and eventually with the internet, which turned sports into a real-time shared experience for global audiences. Radio broadcasts of sports gained popularity in the 1920s–1930s (e.g., millions tuned in to boxing title fights or baseball's World Series), but it was television that had the most profound impact. The first experimental TV sports broadcast occurred in 1939 (a baseball game between Columbia and Princeton universities televised in New York), and by the 1950s, televisions were entering homes en masse. Live TV coverage of events like the 1948 London Olympics and the 1954 FIFA World Cup began to knit together a global community of fans who could watch simultaneously from afar. In 1964, the Tokyo Olympics became the first to have a live satellite broadcast, sending television signals across continents [3]. By the late 20th century, major sports events consistently ranked among the most-watched broadcasts worldwide. For instance, the FIFA World Cup final and the Olympic opening ceremonies drew billions of viewers in aggregate across countries.

Digital broadcasting not only expanded audiences but also enriched the spectator experience with on-screen information and analytics. In the 1990s and 2000s, broadcasts began incorporating digital overlays: the now-ubiquitous score bugs and timers, slow-motion replays, and graphics analyzing plays. A seminal example is the introduction of the virtual "yellow line" for first downs in American football telecasts or the on-screen strike zone in baseball, which uses computer tracking to show pitch locations for viewers. These innovations, powered by technologies such as the Sport Vision system and pitch-tracking software, enhanced viewers' understanding and enjoyment. As noted by sports technology experts, graphics boxes and tracking tools "take the guesswork out" of officiating calls and add drama for fans at home. This trend has accelerated in the 21st century: broadcast enhancements now include real-time statistics, player-tracking data (e.g., showing how far and how fast a soccer player has run), and multi-angle interactive replays, all made possible by digital data capture and processing [7].

Another pillar of Sports 3.0 is globalization. While Sports 2.0 planted the seeds with events like the Olympics and the World Cup, in Sports 3.0, the international exchange of sports culture exploded. The Olympics grew into a mega-event attracting over 200 nations. The FIFA World Cup became the planet's most popular sporting event by viewership. Sports leagues expanded beyond their home countries through both media and physical expansion (e.g., the National Basketball Association and the English Premier League gained huge followings in Asia and Africa through broadcasts and began staging occasional games abroad). International tournaments and circuits proliferated in tennis, golf, motorsports, and other sports, creating a year-round calendar of global sports entertainment. The result was a massive sports economy: by the 2020s, the international sports industry was valued in the hundreds of billions of dollars annually, encompassing media rights, sponsorships, merchandising, and tourism. Sports became a key sector not just culturally but economically; stadiums and events were engines of jobs and urban development, and athletes became global celebrities and brand ambassadors.

Underpinning this global sports boom was the rise of data analytics and sports science as integral parts of competition. Starting in the latter half of the 20th century, teams and athletes began to harness computers and systematic data collection to gain a competitive edge. A famous early milestone was the sabermetrics revolution in baseball, which gained attention in the 1980s and 1990s and was popularized by the Oakland A's "Moneyball" approach in the early 2000s. Instead of relying purely on traditional scouting, teams mined statistical data to evaluate player performance and make strategic decisions, often revealing undervalued players or more effective tactics. This data-driven philosophy soon spread to other sports: basketball teams embraced analytics for shot selection and lineup efficiency, soccer clubs hired data analysts to study player tracking and optimize formations, and so on. By the 2010s, virtually all elite sports organizations had analytics departments.

Concurrently, advancements in sports science and technology provided teams with new kinds of data. Wearable sensors, such as heart-rate monitors, began being used in training in the 1990s, allowing coaches to quantify exertion and recovery. Motion-capture and video analysis tools were developed so that coaches could break down film not just qualitatively but quantitatively (measuring angles, speeds, etc.). The development of GPS trackers and accelerometers small enough for athletes to wear (often worn in a vest between the shoulder blades) has led to detailed in-game data on distance covered, sprint speeds, work rate, and even impact forces in contact sports. Injury prevention and training optimization became more data-driven. For example, by monitoring an athlete's workload metrics and biomechanics, sports scientists could detect fatigue or technique issues and proactively adjust training to prevent injuries [8]. A review of machine learning in sports notes that AI models have been built to predict injury risk by analyzing training load and biomechanical data [9], a concept that began to take hold in the late Sports 3.0 era.

Sports officiating also saw increased tech integration in this period. We see the introduction of technologies such as the instant replay for officiating reviews (the NFL implemented replay review in 1986, albeit in a limited form at first; other sports followed suit in various ways), the Hawk-Eye ball-tracking system in tennis and cricket (2000s) to adjudicate close line calls with high precision, and goal-line technology in soccer (2010s) to confirm if the ball crossed the line. These systems, while not AI in the sophisticated sense, rely on advanced cameras, sensors, and algorithms to assist human referees and add a layer of fairness and accuracy that was not possible before the digital era.

The fan experience in Sports 3.0 also evolved with the internet and mobile revolution. By the late 1990s and early 2000s, fans could follow games via web updates and, soon, streaming audio and video online. Sports leagues launched official websites and then social media accounts, enabling global fans to get news and highlights instantly. Fantasy sports has become a massive subculture (and industry) in which fans use real player statistics (updated in real time) to compete in virtual manager games, an engagement made possible by the widespread availability of data. Social media in the 2010s further changed fan engagement: athletes began communicating directly to fans on Twitter/Instagram, and fan communities became more interconnected globally, sharing memes, analysis, and reactions 24/7.

By the end of the Sports 3.0 era, roughly the 2010s, the convergence of technology and sports was well underway: data analytics had impacted coaching and management, digital media had turned sports into a continuous global spectacle, and fans expected high-tech enhancements as part of their sports consumption. Yet, it is essential to note that the full potential of AI and innovative technology has only begun to be realized at this stage. The use of analytics, while significant, was often still descriptive or diagnostic, telling us what happened and sometimes why, but less usually making autonomous predictions or decisions. The stage was now set for the next leap: applying AI and interconnected technology to augment further both the play and consumption of sports, launching us into the era of Sports 4.0.

### 9.2.5 Sports 4.0: AI-Augmented Training, Smart Equipment, and Immersive Spectatorship

The current era, Sports 4.0, is defined by the infusion of AI, machine learning, and innovative technologies throughout the sports ecosystem. Building on the digital foundations of Sports 3.0, this epoch introduces intelligent automation and enhanced interconnectivity into training, equipment, and fan experiences. In Sports 4.0, the traditional boundaries of sports are extended by data and algorithms, enabling athletes to train smarter, equipment to "think" and adapt, and spectators to engage with events in immersive, personalized ways.

#### 9.2.5.1 AI-Augmented Training and Smart Equipment

At the heart of Sports 4.0 is the concept of AI-augmented training using advanced algorithms to analyze performance data and provide insights or guidance that go far beyond what human observation alone could achieve. Modern athletes and coaches are increasingly supported by intelligent systems that process vast amounts of data from practices, games, and even biometric monitoring. These systems can identify subtle patterns and make tailored recommendations, essentially acting as an ever-watchful data-driven assistant coach.

One application is in performance analysis and skill acquisition. Using computer vision AI, video from training sessions can be analyzed frame-by-frame to assess an athlete's technique. For example, in soccer, deep learning models can track a player's body mechanics when shooting or cutting, detecting inefficiencies or risk factors in their movement. The Chinese Olympic diving team recently demonstrated a striking example of this: they employed a Baidu-developed AI training system that records divers in mid-air and provides instant feedback on their body positions and entry angles [10], comparing each attempt to an ideal model. Coaches and athletes reported that this AI, with "sharp eyes and a strong brain," captures intricate details in the 1.8-second dive that traditional video review might miss, allowing divers like Quan Hongchan and Chen Yuxi to refine their techniques and achieve near-perfect, splash-free dives. Such AI systems exemplify how machine learning can break down elite techniques into quantifiable feedback, accelerating the learning curve.

In team sports, predictive analytics guide strategy and talent development. Major League Baseball's Statcast system, mentioned earlier, is an AI-powered platform that tracks every movement on the field, pitch velocities, swing angles, and fielder positioning. This yields data that AI models can use to predict outcomes or optimal decisions. Coaches now routinely consult data-driven models: in basketball, AI might suggest optimal lineups against a certain opponent by simulating matchups; in American football, machine learning models assess tendencies, so that on 4th down, coaches have a statistical recommendation (sometimes leading to unorthodox, aggressive plays). Similarly, AI scouting tools evaluate player performance over seasons to project future potential, helping identify rising stars or undervalued players while mitigating human bias or oversight.

Central to AI-augmented training is the explosion of wearable and innovative equipment that supplies raw data. Athletes today often train with wearable devices that monitor heart rate, GPS position, acceleration, muscle activation (EMG sensors), and more. These devices form part of the IoT in sports, a network of sensors feeding continuous data. AI algorithms digest this data to manage training loads and prevent injuries. For instance, an AI system might learn an individual athlete's patterns and flag when metrics deviate toward a risk zone (e.g., a decrease in stride length combined with an elevated heart rate could indicate fatigue that presages injury). Indeed, studies show AI models can integrate training load, biomechanical, and even psychological data to alert coaches when an injury risk threshold is reached [9]. Professional teams across sports have adopted such injury-prediction AI. There are reports of European soccer clubs using machine learning to reduce soft-tissue injuries by analyzing GPS and physiological data from practices, and of NBA teams using AI to optimize rest and recovery schedules for players during the long season.

Innovative equipment goes beyond wearables on the athlete's body. We now have "smart balls" and "smart rackets" with embedded sensors. A smart basketball, for example, can record the force and spin of each shot, feeding an app that uses AI to tell a player how to adjust their shooting form. Smart tennis rackets can log the point of contact and swing speed for every stroke, helping players (and their AI assistants) analyze patterns of play. Equipment manufacturers are incorporating AI into design as well for instance, training machines that use AI to adapt to the athlete: a pitching machine that analyzes a batter's weaknesses and adjusts pitch types and locations to challenge them, or a soccer ball launcher that changes trajectory unpredictably to train goalkeepers, informed by an AI learning what reactions are more complex for the keeper to save.

Perhaps one of the most visually striking uses of AI in training is AR and VR for skill practice. VR training systems place athletes in a virtual game situation to practice decision-making at game speed without physical wear and tear. Quarterbacks in American football have used VR to practice reading defensive schemes; baseball hitters use VR to time pitches from virtual replicas of real pitchers. These systems increasingly incorporate AI to generate realistic scenarios – for instance, an AI can simulate an opponent's tendencies (maybe based on real data) in the virtual environment, giving athletes a form of "scrimmage" tailored to upcoming opponents. In motorsports, drivers train on AI-enhanced simulators that not only mimic the physics of their vehicle but also simulate weather, track conditions, and even AI "ghost" racers to compete against. In one example of global outreach, a program in Africa provided VR headsets to young athletes in remote areas to experience high-quality coaching scenarios; paired with AI assessment tools, the system can evaluate their skills and offer personalized training tips. This shows how AI and immersive tech together can democratize access to advanced training, a theme we will revisit in the opportunities section.

Importantly, integrating AI into training doesn't eliminate the human coach; rather, it augments coaching with deeper insights. The best Sports 4.0 practices leverage synergy between human intuition/experience and AI-driven data analysis. A coach might use AI recommendations to adjust a game plan, but apply a human understanding of team psychology to decide how to implement it. The athlete gets immediate, objective feedback from AI, but still relies on their own creativity and the coach's mentorship to internalize improvements. In essence, AI acts like a hyper-intelligent assistant coach or sports scientist, tirelessly crunching numbers and watching film, freeing human coaches to focus on strategy and motivation.

#### 9.2.5.2 Immersive Spectatorship and Smart Venues

While athletes train and compete with AI support, spectators in the Sports 4.0 era are enjoying an increasingly immersive and personalized viewing experience, often blurring the lines between live and virtual, and between observer and participant. Advances in AR, VR, and innovative stadium technology are revolutionizing how fans consume sports, whether they are in the stadium or halfway across the world [11].

One significant development is the rise of immersive media for remote viewing. Fans who cannot attend a game in person can wear a VR headset to feel virtually present at the venue. Modern sports broadcasts and tech companies have experimented with 360-degree VR streams, where multiple 360° cameras in a stadium allow a fan at home to "choose their seat." For example, VR applications for basketball let viewers position themselves courtside or even virtually on the backboard to experience a dunk in first-person. These VR experiences are often coupled with real-time data overlays. Imagine turning your head in the virtual stadium and seeing a player's live stats floating above them, or instantly replaying a pivotal moment from a different angle. Fans have found such experiences so engaging that "virtual watch parties" have emerged: friends meet in a VR environment, each from their own home, to watch a game together as if in the same virtual space. This social aspect of immersive tech underscores a key point: Sports 4.0 tech isn't about isolating fans with gadgets, but rather connecting them in novel ways [12].

AR is also enhancing in-person and at-home viewing. AR can overlay digital content onto the real world through a smartphone, tablet, or AR glasses. In stadiums, some venues have experimented with AR apps that, for instance, allow fans to point their phone camera at players on the field and immediately see contextual stats or player profiles. Others have dedicated AR zones where fans can take pictures with virtual representations of their favorite players or virtually "try on" championship rings and trophies for fun. At home, AR might let a fan project a virtual scoreboard or player holograms onto their coffee table. The convergence of AI and AR enables creative features. One idea is an AR "x-ray" view that could, say, show an ice hockey fan the paths of the puck and players traced on the rink in real time (especially when the puck is hard to see), generated by AI tracking.

Smartphones have become ubiquitous second screens for fans, and AI plays a role in personalizing content delivered to these devices. Sports team and league apps leverage AI to recommend highlights and news based on a fan's preferences (e.g., the app learns which team or player you follow closely and curates content accordingly). Push notifications driven by AI might alert a fan when a game is getting exciting (a close score in the final minutes) or when a record is about to be broken, luring them to tune in. As noted in a GMU sports management insight, franchises analyze fan data, such as favorite players and merchandise purchases, to customize content and even the ticket offers fans receive [13]. The result is a more engaging and loyalty-building experience; a fan feels the team "knows" them. For instance, an app might send a discount on jerseys of the fan's most-watched player, or automatically create a highlight reel of all the goals scored by the fan's supported team that week.

For those attending games in person, the smart stadium incorporates many Sports 4.0 innovations to enhance comfort and engagement. A smart stadium is outfitted with IoT sensors, high-speed connectivity (5G Wi-Fi), and automation systems to create a seamless experience. Fans can use digital tickets for entry, sometimes via facial recognition or biometric scans, to speed up gate security. Inside, smartphone apps connect to the stadium: one can find the shortest line for concessions, order food delivered to their seat, or even adjust the climate around their seat (some venues have climate-controlled sections). Many of these features are coordinated by AI in the background; for example, AI might manage crowd flow by directing fans to less crowded entry points via app notifications, or adjust air conditioning and lighting dynamically based on crowd density and weather. The Mercedes-Benz Stadium in Atlanta, as an example of a state-of-the-art facility, is equipped to handle over 75,000 simultaneous online connections and features a 360-degree "Halo Board" [14] screen that gives every fan an immersive visual of the action. Smart stadiums often incorporate AR on big screens as well; during breaks, fans might play interactive AR games on the jumbotron using their phones as controllers [15].

Smart venues are not just about fan luxury; they also dramatically improve operational efficiency and safety. AI-driven security systems analyze CCTV feeds for unusual behavior or overcrowding, enabling security staff to take pre-emptive action. IoT sensors can detect if a section of seating is empty and redirect resources accordingly, or identify a spill/cleanup need. Energy management is also crucial. AI systems regulate lighting and temperature to save energy when parts of a stadium are not in use, and some stadiums (such as Levi's Stadium in California or M. Chinn Swamy Stadium in India) integrate renewable energy and IoT to significantly reduce their carbon footprints. In Levi's Stadium's case, reports indicate that the use of IoT and data analytics helped reduce energy consumption by 40 percent while enhancing the fan experience, demonstrating the dual benefits of intelligent design.

Another facet of immersive spectatorship in Sports 4.0 is social interactivity and gamification. Fans are not passive; technology allows them to be part of the spectacle. Stadium apps and broadcasts now sometimes feature real-time polls ("Vote for the player of the match" or "predict the next play outcome"), with results shown live, making fans feel their input is part of the event. Some teams have experimented with letting fans use an app to change stadium lighting colors or graphics when a goal is scored, effectively crowdsourcing some of the celebration ambiance. On social media, AI curates global fan content. For instance, AI can compile the best fan reactions from Twitter or TikTok after a big play and present them during the broadcast, highlighting the worldwide chorus of fandom. This phenomenon was notably evident at events like the FIFA World Cup, where broadcasts would show a mosaic of fan selfies celebrating, drawn from an app that uses AI to select enthusiastic submissions from millions.

Underpinning all these spectator-side advances is a heavy dose of AI and data analytics working behind the scenes. Personalization algorithms ensure each fan gets relevant content. Computer vision and sensor analytics keep the venue safe and efficient. Natural language processing may power chatbots that answer fan questions ("Where's the nearest restroom?" asked via an app chat, answered instantly by an AI agent). Even entertainment is getting AI-infused: some arenas have experimented with AI-composed music, robots for halftime shows, and drone swarms for light displays, making the overall spectacle high-tech.

It is worth noting that Sports 4.0's immersive and innovative features are currently more accessible in wealthier markets and at top-tier events, but they are spreading. Simpler versions, like mobile ticketing and basic data overlays on TV, are nearly universal in pro sports. As technology costs drop, we see even smaller events adopting elements of smart operations (for instance, a local marathon using AI for race tracking and an app for spectators to follow runners).

In conclusion, Sports 4.0 transforms the sporting experience on and off the field. Athletes train and compete with unprecedented insights and precision thanks to AI and smart devices, pushing the boundaries of human performance. Fans engage through multiple senses and platforms, from the roar of a high-tech stadium that knows their preferences to the solitude of a VR headset that lets them feel on the pitch alongside their heroes, thereby deepening their connection to the sport. Sports 4.0 is making good on the promise of the AI epoch: augmenting human ability and enjoyment rather than replacing it. But with this great power comes new challenges and responsibilities, which we explore next.

## 9.3 OPPORTUNITIES AND TRANSFORMATIONAL POTENTIAL

The marriage of sports and AI-driven technology in the Sports 4.0 era offers a multitude of opportunities and transformational potential. These range from enhancing athletic performance and extending careers to playing sports more inclusively and accessibly, to fostering entirely new economic and participatory models in the sports domain. Here, we highlight some of the most significant positive impacts that Sports 4.0 can deliver globally.

### 9.3.1 Enhanced Athlete Performance and Health

Perhaps the most immediate opportunity of Sports 4.0 is the ability to push the limits of human performance in a safe and scientifically grounded way. With AI-tailored training regimens, athletes can achieve new personal bests and even world records that might have seemed out of reach. By analyzing biomechanics and suggesting adjustments, AI helps athletes optimize their technique for efficiency and power. For instance, sprinters might refine their gait and start using insights from AI that identify minute asymmetries; weightlifters might use smart wearables to ensure optimal form and avoid injury as they increase loads. The precision training enabled by Sports 4.0 could usher in an era of shattered records across disciplines, as training becomes less about brute force repetition and more about targeted improvements.

More importantly, these performance gains need not come at the cost of athlete health. AI's ability to monitor fatigue and predict injury risk means athletes and coaches can act proactively, adjusting training loads, improving recovery protocols (e.g., sleep, nutrition, and stress management) to prevent injuries and burnout. This can extend athletic careers, allowing top players to maintain high performance longer and retire on their own terms rather than due to chronic injuries. For example, professional sports teams have reported reductions in soft-tissue injuries after implementing AI-based monitoring, keeping star players on the field longer in a season and over their careers. In youth sports, which in some countries face issues of overtraining and injury, AI guidance can ensure training is age-appropriate and safe, thus nurturing talent more sustainably.

### 9.3.2 Data-Driven Decision Making and Strategy

Sports 4.0 provides an unprecedented informational advantage. Teams that fully embrace analytics and AI can make far more informed decisions in recruiting, game strategy, and player development. The democratization of insight is a key aspect: even teams with smaller budgets can tap into open-source analytics tools or affordable sensor tech to gain insights that previously required vast scouting networks or expensive systems. In that sense, Sports 4.0 can level the playing field to some degree; it's not solely about money, but about who uses the information better. We've already seen smaller-market teams in various sports successfully compete with giants by being early adopters of analytics (the "Moneyball" effect). As AI lowers the barrier to advanced analysis, more teams, including even semi-professional or amateur clubs, can use these tools to improve. This could raise the overall standard of play globally.

For fans, data-driven decision-making has a transformational effect, too: it deepens engagement and understanding. Fans today enjoy delving into stats and strategy as part of their fandom (the rise of sports analytics blogs and discussions testifies to this appetite). Sports 4.0 provides rich data and even AI-driven visualizations that help fans appreciate the game's nuances, effectively educating the audience and potentially attracting a new segment of analytically minded fans. It also allows broadcasters to tell more compelling stories (e.g., "Player X has a 43% probability of making this shot from this position, up from 30% last season after his training adjustments."). These narratives, backed by data, add layers to the spectacle.

### 9.3.3 Global Access and Inclusion

One of the most profound opportunities of Sports 4.0 is using technology to make sports more inclusive and globally accessible [1]. AI and digital platforms can overcome geographic, economic, and physical barriers that previously limited participation. For instance, remote coaching platforms powered by AI can bring high-level training to athletes in developing regions or rural areas that lack access to in-person elite coaching. We saw an example with VR training in Africa for youth athletes.

Such initiatives can be scaled to many sports, meaning a talented kid in a remote village might still get noticed and developed through virtual combines or AI talent identification programs that scout via smartphone videos. In fact, some organizations are exploring AI-based talent scouting, in which young players upload footage that AI algorithms analyze to flag prospects for further trials, thus democratizing talent discovery beyond traditional scouting networks.

Inclusivity also extends to people with disabilities or those previously sidelined in sports. AI-driven prosthetics and assistive devices are advancing (e.g., running blades, exoskeletons, and enhanced wheelchairs) and can be tuned to individuals using AI to optimize performance. In the spectator realm, AI provides tools such as real-time captioning and sign-language avatars for deaf fans, as well as descriptive audio commentary for blind fans, making live events and broadcasts more accessible. The concept of "differently-abled athletes" competing in augmented categories (such as the Cybathlon, an Olympic-style competition for bionic athletes) showcases technology's empowering side. Sports 4.0 can validate and amplify these efforts, possibly integrating some events or achievements across Paralympic and Olympic lines as technology enables more equal competition. As noted by sports leaders, there's a push toward integration rather than separation, when possible, to ensure true inclusion.

### 9.3.4 Fan Engagement and Community Building

The immersive and interactive technologies of Sports 4.0 provide opportunities to deepen fan engagement and create new communities. Fans worldwide can come together in virtual spaces, a phenomenon already evident in eSports and now extending to traditional sports, where virtual fan zones can host thousands of fans in an interactive environment. These technologies can particularly engage younger audiences raised on digital interactivity, helping future-proof sports' fanbases. For example, gamified experiences like prediction games, fantasy integrations, or even AR scavenger hunts in stadiums make being a fan more active and participatory. When fans are more engaged, their loyalty and the time they spend following the sport increase, which, in turn, attracts sponsorship and investment, fueling a virtuous cycle for sports leagues and teams.

Furthermore, Sports 4.0 can help sports reach new markets and demographics. A person in a country without a tradition in a particular sport might become a passionate fan through digital exposure and online communities. Social media and AI-curated content mean that someone in a remote part of India can follow the NBA with the same depth of information and community as a person in New York. This truly globalizes fan communities (we see examples such as huge NFL fan clubs in Europe and millions of English Premier League fans in Asia engaging via online forums and AI-translated content). Broadly, this can promote cultural exchange and understanding, as sports have often done, but now amplified by direct fan-to-fan connections worldwide.

### 9.3.5 Economic Growth and New Career Pathways

The intersection of sports and technology is spawning entirely new industries and job roles from data scientists and performance analysts on team payrolls, to entrepreneurs developing sports tech products (apps, wearables, VR experiences). The Sports 4.0 boom contributes to the economy by creating startups, investment opportunities, and educational programs (sports analytics degrees, etc.). Cities that embrace smart sports infrastructure might become new hubs for sports tech innovation. Also, the enhanced reach of sports content means more monetization opportunities: personalized subscriptions, interactive advertising in AR/VR, and NFT collectibles of sports moments (some teams are already selling highlight clips as blockchain-based digital collectibles, creating new revenue streams). All these can inject financial health into sports organizations and even grassroots sports if implemented wisely (for instance, tokenization models where fans invest directly in athletes or clubs could provide funding for under-resourced talent, as is being experimented with via blockchain).

In developing countries, mobile technology has enabled the creation of sports content and even new sports entertainment models (such as mobile-based eSports or fantasy sports leagues), contributing to local economies. One interesting trend is the rise of mobile eSports and sports gaming, which in some places is an entry point to sports fandom and has lower barriers to entry than fielding physical teams. This was noted in an economic context: for example, mobile gaming and eSports are generating revenue and jobs in regions where traditional sports infrastructure is lacking.

### 9.3.6 Sustainability and Optimization

Sports 4.0 also offers the potential to play sports more sustainably and efficiently. Smart stadiums, as discussed, can drastically reduce waste and energy usage. AI can optimize scheduling and logistics for significant events to minimize the carbon footprint (e.g., by optimizing travel routes and loads). Digital ticketing and streaming reduce the need for physical materials. There's even the concept of using sports popularity to promote sustainability messages, such as apps that encourage fans to take public transit to games (with incentives) or stadiums that showcase renewable energy in action (e.g., Solar panels at venues as educational pieces). If the sporting world embraces sustainability tech as part of Sports 4.0, it can influence millions of fans to consider greener behaviors, leveraging sports' inspirational power [16].

Finally, an overarching transformational aspect is the blending of physical and digital to create new forms of sports. We may see entirely new "tech sports" emerge, for example, drone racing (already a professional sport) or AR-enhanced physical competitions (think of virtual obstacles or targets in a real-life race that only exist via AR). There's also interest in mixed-reality sports, where perhaps human athletes compete alongside robotic assistants, or where team strategy includes AI players (this sounds far-fetched, but even now there are competitions like robot soccer world cups, which, while separate, push the envelope of what competition means). Such innovations could draw new audiences and participants who are more inclined toward tech.

In sum, Sports 4.0 holds promise not just for improving existing sports but for transforming the role of sports in society, making it more inclusive, more engaging, and more deeply integrated with global cultural and economic currents. It can amplify the positive values of sports (excellence, teamwork, fair play) by reaching more people in more personalized ways. To fully realize these opportunities, however, stakeholders must navigate the associated challenges and ensure the technology is applied ethically and equitably.

## 9.4 CHALLENGES, RISKS, AND ETHICAL CONSIDERATIONS

The leap into the Sports 4.0 epoch, while exciting and full of promise, is not without significant challenges and ethical dilemmas. As AI and pervasive technology become embedded in sports, stakeholders must grapple with issues ranging from privacy and data security to fairness and the very essence of sport's human spirit. Below, we outline some of the most pressing challenges and risks that accompany the transformation of sports in the AI age.

### 9.4.1 Data Privacy and Security

Sports 4.0 relies heavily on data, athlete performance metrics, biometric data, fan personal information, etc. This raises serious privacy concerns. Athletes are now among the most quantified humans on the planet: teams collect data on their heart rates, sleep patterns, diet, location during training, and even psychological indicators. Who owns this data, and how can we ensure it's not misused? If an athlete's health data (say, a predisposition to injury or an illness) were leaked, it could affect their contract negotiations or public image. Recently, a group of over 850 professional footballers in the UK launched the "Project Red Card" legal action against analytics companies, claiming that their personal performance and tracking data were being collected and sold without proper consent [17]. This case highlights that athletes want control and possibly compensation for the data generated by Sports 4.0 tools from them, framing it as an asset akin to image rights. Ensuring informed consent and fair use of player data is paramount. Leagues may need to establish clear regulations on what data can be collected and how it's stored and shared. There is also the question of cybersecurity; a team's proprietary data (playbooks, strategies drawn from analytics) could be targeted by hackers or rival teams if not adequately protected, analogous to corporate espionage.

For fans, the personalized experiences of Sports 4.0 also mean apps and platforms gather data about their behaviors and preferences. Sports organizations must thus protect user data and comply with privacy laws (such as GDPR in Europe). A data breach in a team's fan app could expose thousands of users' personal information, eroding trust. There's a delicate balance between leveraging fan data to drive engagement and avoiding crossing the line into invasive tracking or unwanted surveillance.

### 9.4.2 Algorithmic Bias and Fairness

AI systems are only as good as the data and assumptions that build them. In sports, if not carefully managed, AI could inadvertently introduce or reinforce biases. For example, an AI scouting program trained mostly on data from male athletes might undervalue female athletes due to subtle differences in how performance metrics translate, or it might have cultural biases (e.g., favoring athletes from regions with more abundant data). If AI is used in referee decisions (say, automated officiating), any bias in those algorithms can directly affect game outcomes. Ensuring transparency in AI decision-making is crucial. Yet AI, like deep learning, often functions as a "black box," making it hard to explain decisions such as why a particular player is injury-prone, according to the model. This opaqueness can lead to mistrust among athletes and coaches if a computer says, "bench this player today." Is the coach comfortable accepting that, and can they explain it to the player in human terms? [18]

In terms of competitive fairness, consider that wealthier teams have more resources to invest in advanced Sports 4.0 tech. This technology gap could widen the performance gap. Rich clubs might afford the latest AI performance optimization and injury-prevention tools, while poorer clubs lag, creating a new form of inequality in competition. Similar concerns have been raised historically about access to training facilities or medical staff; now it extends to algorithms and data scientists on the payroll. Leagues might need to consider guidelines or sharing mechanisms (for instance, making specific baseline data analysis tools available to all teams) to avoid a tech arms race that skews competitive balance.

### 9.4.3 Diminished Human Element and Ethical Boundaries

A more philosophical but prevalent concern is that an over-reliance on data and AI might diminish the human drama and unpredictability that make sports captivating. If coaches make decisions based on algorithms and players follow AI-derived instructions to the letter, do we risk turning sports into a sterile optimization exercise? Many argue that the beauty of sports lies in its spontaneity and the emotional, irrational factors, the hot hand, the underdog impulse, and the inspired risk that defies percentages. If everything becomes "by the book" because the book is written by AI analyzing thousands of scenarios, some fear sports could lose some of their charm. This is somewhat speculative, but one already hears criticism that some sports (like baseball) have become too dominated by analytics, leading to strategies that are practical but possibly less entertaining (e.g., defensive shifts, or, in basketball, the heavy emphasis on three-point shots due to efficiency metrics).

There's also the ethical question of technological doping or augmentation. Sports have long banned performance-enhancing drugs to keep competition fair and health-conscious. Now, where is the line with technology? If one athlete uses advanced neurofeedback AI and another doesn't have access to it, is that an unfair advantage? What if brain-computer interfaces eventually allow athletes to enhance reaction times or focus? Will that be allowed? The governing bodies will have to update regulations to define what constitutes acceptable aid continuously. Already, we saw controversy with the introduction of ultra-performance running shoes (with carbon fiber plates and special foam) that were seen as "tech doping" as numerous records fell; track authorities ultimately regulated shoe designs. We might face similar debates with AI-driven training regimens or equipment. Does an AI-designed training plan constitute an "unfair aid" or just savvy preparation? Like the latter, but the optics matter if some feel only those who can afford the AI get the benefits.

### 9.4.4 Officiating and Integrity of Competition

As we introduce more technology into officiating (like VAR in soccer, which uses video and sometimes semi-automated offside detection), we solve some problems but create others. VAR has clearly corrected many wrong calls, but it's also sparked new kinds of debates among fans and pundits, for example, complaints about the flow of the game being interrupted, or that offside decisions by an automated line are too forensic ("toenail offside") and maybe against the spirit of the rule. In the future, if AI handles more officiating (as the Australian Open tested AI for line calls in tennis, or FIFA explores AI for offsides), the sports community must be comfortable with machine judgment. A significant challenge is maintaining trust in these systems. Any high-profile mistake by an AI referee (and no system is infallible) could cause backlash. And if an AI system decided that players strongly feel it is wrong, yet is unoverridable, could that undermine respect for officiating? Right now, human referees, for all their faults, are an accepted part of the game's fabric, and their authority is usually respected (even when we yell at the TV). Transitioning that authority to machines is non-trivial and must be handled gradually and transparently.

Moreover, as AI helps detect cheating (e.g., pattern analysis for match-fixing or doping signals), cheaters may also use AI to avoid detection, leading to an arms race. Sports organizations will have to invest in AI ethics and governance, for instance, ensuring AI used in anti-doping or gambling fraud detection is accurate and doesn't falsely accuse innocent parties (which could ruin careers unjustly).

### 9.4.5 Digital Divide and Global Inequality

As mentioned earlier, not all countries or sports organizations can equally benefit from advances in Sports 4.0. There's a risk of a digital divide in which wealthy countries' sports leap ahead while poorer regions fall further behind in competitiveness. This also applies to fan access. Rural or poorer fans might not afford VR headsets or 5G connections for fancy experiences, potentially creating a two-tier fan experience. Sports have a unifying power, partly because of their universality; if the best experience of sport (say, fully interactive viewing) is accessible only to some, that challenges inclusivity. Bridging this gap is an ethical imperative: international bodies such as the IOC and global federations may need to facilitate knowledge transfer and subsidize technology for developing sporting nations to ensure the gap doesn't widen. Programs could be introduced to donate equipment or provide open-source AI tools for training that anyone can use at low cost.

### 9.4.6 Mental and Emotional Impact on Athletes

Constant monitoring and data-driven evaluation can be a double-edged sword for athletes' mental health. Knowing that every movement is measured and an algorithm might dissect every mistake could induce stress or rob athletes of the sense of personal agency. Some athletes have voiced that they feel like "lab rats" under exhaustive performance analytics regimes. There is an ethical line in how far to intrude into athletes' personal lives with monitoring. For example, if an AI suggests that an athlete is underperforming due to a lack of sleep (detected via a wearable) and that information is shared with coaches, does it violate personal boundaries? Similarly, as mental health becomes a focal point, we should consider how an always-on data environment affects athletes' mental health. Teams will need to implement these tools with care, ensuring athletes are educated on their benefits and given off-the-grid time if required.

### 9.4.7 Integrity of Fan Experience

On the fan side, a challenge is maintaining authenticity in an era of AR/VR. If a significant portion of fans experience a game in VR with overlays, are we all still watching the "same game"? Also, deepfakes and synthetic media could emerge as an issue, imagine fake clips generated by AI of controversial moments that never happened, spreading on social media to confuse. Ensuring the truthfulness of sports content may become a task (for instance, broadcasters may need to certify certain footage as authentic to combat AI-generated misinformation in betting or propaganda contexts).

### 9.4.8 Environmental and Sustainability Concerns

While technology can improve sustainability, it also carries environmental footprint concerns (e.g., energy use for data centers powering AI and electronic waste from devices). Sports organizations must ensure that the tech they adopt aligns with green practices and that they mitigate any adverse environmental impacts. With fans and society more conscious of climate issues, a league that heavily promotes tech that is seen as environmentally harmful could face public relations issues.

In conclusion, the challenges of Sports 4.0 revolve around maintaining the humanity, fairness, and integrity of sports amidst rapid technological change. Sports have always been governed not just by rules, but by unwritten codes of ethics and sportsmanship. Those need reinterpretation in the new context – from how we consider augmented performance to how we treat data rights as part of athlete rights. Proactive governance is key: leagues and federations should establish committees focused on technology ethics that involve players, coaches, tech experts, and ethicists to update guidelines continuously. The sports tech revolution should be guided by a principle of augmenting the human experience, not undermining it. If done thoughtfully, many of the challenges can be mitigated, but ignoring them could lead to pitfalls where technology erodes trust or fairness in sports.

## 9.5 THE FUTURE OUTLOOK (LIFE 4.0 SCENARIO)

Looking ahead, we consider a LIFE 4.0 scenario a vision of sports in a mature AI-integrated society, where technology is ubiquitous and deeply woven into daily life. In this future, the distinctions between physical and digital, human and machine, are even more blurred, and sports continue to be a vibrant domain reflecting those shifts. While speculative, this outlook is informed by current trends extrapolated to their potential outcomes in, say, a couple of decades.

### 9.5.1 Hyper-Personalized Athlete Support

In LIFE 4.0, every athlete might have an AI-driven personal performance companion from early development through their professional career. This AI, fed by lifelong data on the athlete's body and performance, would act almost like a second self, constantly learning and refining training strategy to suit the athlete's evolving condition and style. Young athletes worldwide could access virtual coaching assistants that are as effective as having a personal elite coach, thus flattening the talent development landscape. With advancements in biotech and AI, training could become hyper-targeted to the genetic and molecular levels, for instance, with real-time blood and genetic analysis informing nutrition and recovery protocols with unprecedented precision. Ethical considerations will swirl around gene editing and augmentation. By LIFE 4.0, society may have drawn red lines (e.g., no gene doping in sports) but also possibly embraced specific safe enhancements (perhaps allowing genetic therapies for injury recovery). The definition of "natural" talent may evolve as athletes use brain-computer interfaces to train mental aspects such as focus and visualization. We might see early forms of neuro-athletes who practice not just physically but also in simulated environments via direct neural stimulation, dramatically reducing injury risk in practice and accelerating learning.

### 9.5.2 Cyborg and Robotic Sports

A striking element of future sports could be the normalization of competitions involving augmented humans or robots. There could be a Paralympic/Olympic convergence in which athletes with bionic limbs compete directly with able-bodied athletes if technology makes their capabilities comparable, raising philosophical questions about what constitutes an advantage. Alternatively, separate categories might flourish: for example, an X-League where enhancement is allowed, and athletes push beyond current human limits, akin to Formula 1 but with the human body. On the robotic side, spectator-friendly robot sports leagues may become mainstream (imagine robot soccer or drone combat leagues on primetime TV, with AI players whose "brains" are products of the best AI engineering teams, much like Formula 1 is a competition between car manufacturers). Human-AI collaboration in sports could emerge as well, perhaps with mixed teams of humans and AI-controlled robots competing to emphasize synergy. While this sounds far-out, it could showcase the positive narrative of humans and AI working in tandem, each leveraging their strengths [19, 20].

### 9.5.3 Immersive Spectatorship Reimagined

For fans, the sports of LIFE 4.0 might be experienced through immersive media indistinguishable from reality. Holo-stadiums could project full 3D holographic games into local arenas, so fans in multiple cities watch the players "live" in front of them via holograms, feeling almost like the real thing. Those at home might use advanced VR with haptic feedback suits so a fan feels the rumble of the crowd, the breeze of a fast break, maybe even the impact of a tackle (to a safe extent!). AI concierge systems will curate each fan's experience: for instance, an AI could dynamically choose the best camera angle for you at each moment based on your past preferences, or even render the game with a preferred commentary style (using trained voice models of famous commentators, or perhaps your own avatar providing analysis). Entertainment might be deeply interactive, so fans could virtually step into a replay and walk around the field to inspect what happened from any angle, essentially becoming their own broadcasters.

Moreover, fandom might become even more participatory. We see early signs of fan tokens granting voting rights on minor team decisions; in the future, decentralized autonomous clubs could emerge, where large fan collectives literally run a team via blockchain-based voting on decisions (within legal frameworks). Fantasy sports might blur reality. For example, fans could influence real games through crowd-sourced strategy suggestions fed to coaches via AI analysis of the crowd's wisdom (though purists might resist), and experimental exhibition games could toy with this concept. Virtual sports – entirely AI-generated competitions might captivate fans too, especially for matchups that can't happen (think simulations of historical dream teams playing each other, presented with lifelike graphics and AI that emulate the players' styles). Already, games and simulations are approaching realism; by LIFE 4.0, a simulated game could be so convincing that it itself becomes a form of sport consumption, raising questions of whether the simulation is its own sport.

### 9.5.4 Global Unity and Cultural Exchange

Optimistically, LIFE 4.0 sports could be a force for global unity. With translation AI, a Brazilian soccer fan and a Chinese basketball fan can communicate seamlessly in real time, breaking language barriers in fan communities. Global cross-sport events might arise, such as an "AI World Cup" in which each nation's AI competes in various tasks, bridging sports and AI as a showcase of national pride. Traditional sports might also see a revival and preservation aided by tech, e.g., indigenous games and ancient sports could be supported by global platforms, helping keep cultural heritage alive and even blending old and new (imagine an ancient game like traditional archery being augmented with AR for broader appeal).

### 9.5.5 Ethical and Regulatory Evolution

In a LIFE 4.0 scenario, sports governance will likely have evolved elaborate regulations around tech. Perhaps a "Technology Use Charter" akin to anti-doping codes becomes standard, specifying which enhancements and AI uses are allowed. We might see "integrity AI" overseeing competitions, detecting anomalies in play that suggest cheating (almost like a referee AI in the sky, assuring fairness in real time). Athlete welfare might be improved. Maybe an AI mediator monitors workload and can even override coaching decisions if data indicate risk (would we allow an AI to say "this pitcher must be pulled now to prevent injury" regardless of the coach's will?). Such measures could better protect athletes, albeit amid debate over autonomy.

### 9.5.6 New Sports and Hybrid Experiences

LIFE 4.0 could yield entirely new sports that are hybrids of physical exertion and e-sports. For example, AR could turn any environment into a sports field with dynamic obstacles (think AR parkour where the cityscape digitally changes). Or exoskeleton races where competitors in robotic suits run at 60 km/h, blurring vehicle racing with human racing. These innovations will challenge our notion of what a sport is, but they may gain followings, especially among youth who seek novel experiences beyond classic sports.

It is also plausible that climate and societal changes will affect sports. If extreme heat becomes common, indoor or night events might dominate, with technology enabling more indoor arenas that simulate outdoor atmospheres (maybe even projecting a realistic sky). If travel is restricted for sustainability reasons, perhaps holographic team presence could reduce the need for physical travel. Teams could play "away" games from their home stadium via hologram projection to the other site, an idea that sounds crazy now but could technically preserve home-field dynamics without incurring carbon costs.

In sum, the LIFE 4.0 sports scenario is one of deep integration of AI and humans co-creating sport experiences that are richer, more inclusive, but also more complex to manage ethically. Sports would likely maintain their core appeal: competition, stories of human (and maybe transhuman) effort, and community celebration. But the medium and methods of those stories will expand dramatically. Crucially, the spirit of play and fair contest must be carefully safeguarded. Even as we imagine AI coaches or bionic players, we hold onto the essence of sports as a test of will, skill, and heart, qualities that, even when enhanced by AI, remain fundamentally human (or conscious, in the case of advanced AI entities that become part of the games).

The challenge for the sports world will be embracing innovation without losing the values that make sports meaningful. If done right, the future can be an era in which sports are more accessible, exciting, and entwined with our lives than ever, truly exemplifying "LIFE 4.0," where living, working, and playing are all enhanced by intelligent technology.

## 9.6 CONCLUSION

Sports have continually evolved alongside humanity, reflecting our social structures, technological advances, and collective aspirations. From the sacred games of antiquity (Sports 1.0) to the industrial-age rise of professional leagues and mass audiences (Sports 2.0) to the globally televised and data-informed competitions of the late 20th century (Sports 3.0), each era of sport has built upon the last. Today, as we navigate the Sports 4.0 epoch, we are witnessing perhaps the most rapid and profound transformation yet: the infusion of AI and innovative technology into nearly every facet of athletic performance and spectatorship.

The chapter has examined how AI-augmented training and innovative equipment are enabling athletes to reach new heights while striving to stay healthier and prolong careers. It has explored how immersive media and smart venues are revolutionizing the fan experience, making it more interactive, personalized, and global than ever before. The opportunities presented by these changes are immense, from more inclusive participation and democratized talent development to richer fan engagement and new economic models in sports. At the same time, we have delved into the challenges and ethical considerations that accompany Sports 4.0: ensuring data privacy and fair play, avoiding technology-driven disparities or dehumanization, and preserving the fundamental unpredictability and fairness that give sports their drama and integrity.

As we look toward a future LIFE 4.0 scenario, it's clear that the boundaries of sports will continue to expand. The notion of what constitutes an athlete, a competition, or a spectator will be redefined by technologies that allow us to augment human abilities and create hybrid realities. Yet, amidst all the change, certain constants persist. People will always be drawn to the stories and emotions of sport, the underdog victories, the record-breaking feats, the split-second moments of decision that turn games around. AI can provide analysis and even anticipation of these moments, but the emotional impact remains a profoundly human experience. The roar of a crowd (whether physical or virtual), the camaraderie among fans across continents, and the inspiration we draw from athletes pushing their limits are anchored in our human psyche.

The task for stakeholders in the era of Sports 4.0 and beyond is to harness technology in the service of sport's timeless values. That means crafting regulations and ethical norms that keep competition fair and inclusive, investing in tools that enhance athlete welfare as much as performance, and leveraging new platforms to bring the joy of sports to broader audiences without alienating traditional fans. It requires collaboration between technologists, athletes, coaches, leagues, and fans to shape the evolution of sports in a deliberate, thoughtful way.

In conclusion, "Sports 4.0: Athleticism and Spectacle in the AI Epoch" encapsulates a paradigm shift where the synergy of human athletic spirit and AI creates a spectacle more dynamic and far-reaching than ever. Sports are not being overtaken by technology; they are being transformed by it in ways that can elevate the best of what sports offer to individuals and societies. If we manage the transition well, the result will be a richer sports ecosystem, one that still celebrates the thrill of human competition and achievement, now amplified through the tools of our most advanced technologies. In the AI epoch, life and sport–4.0 can flourish together, continuing to unite and inspire us in new and unexpected ways [15, 19].

## REFERENCES


1. Guarda, D., Key Technologies of 4IR in Sports & Opportunities for Sports. *Sports DNA*, December 19, 2024.
2. Yellow brick, Evolution of Sports History: From Ancient Games to Modern Sporting Events, in *Yellow Brick Blog*, November 2, 2023.
3. Wikipedia, Broadcasting of Sports Events, in *Wikipedia*, last modified August 2024.
4. Guarda, D., The 4IR in Sports: Transforming Performance, Engagement, and Socio-Economic Impact, in *Sports DNA*, December 19, 2024.
5. Belis, A., The Ancient Olympics and Other Athletic Games, in *Metropolitan Museum of Art – Perspectives*, July 23, 2021.
6. History.com Editors, *First Live Sporting Event Broadcast on Radio*, April 11, 1921.
7. Hernandez, J., AI 3D Tracking Is Changing How Coaches Prevent Injuries, in *Wall Street Journal*, April 10, 2023.
8. Reel Mind, *Soccer Player Injury on Field: AI for Sports Safety (Biomechanical Analysis & Fatigue Monitoring)*, in *Reel Mind Blog*, 2023.
9. Claudino, J.G., et al., Artificial Intelligence and Machine Learning Approaches in Sports: Concepts, Applications, Challenges, and Future Perspectives. *Brazilian Journal of Physical Therapy*, 2024. 28: p.45–62.
10. *China Daily*, Chinese AI Making All the Smart Moves at Paris Olympics, August 8, 2024.
11. IBM., IBM and Wimbledon Launch 'Catch Me Up' Feature with AI Highlights, in *IBM Newsroom*, July 14, 2024.

12. Mwareya, P., Marshall Islands Air Historic Match with AI Cameras, in *Reuters*, June 2023.
13. Aylsworth, J., How Is Technology Amplifying the Sports Fan Experience? in *George Mason University – CEHD Features*, October 6, 2025.
14. Aylsworth, J., "Smart" Stadiums, in *George Mason University – CEHD Features*, October 6, 2025.
15. Liang, X., Smart Stadiums and the Future of Sports Entertainment: Leveraging IoT, AI, and Blockchain for Enhanced Fan Engagement and Venue Management. *Journal of Sports Technology*, 2025. 12(1): p. 345–364.
16. Deloitte, Fan Experience Survey: Connecting with Fans through Technology, in *Deloitte Sports Industry Report*, 2020.
17. Card, P.R., Legal Action over Players' Data Rights, in *Sports Law & Taxation Reports*, 2020.
18. Guarda, D., Challenges and Ethical Considerations in the 4IR Era, in *Sports DNA*, December 19, 2024.
19. Mastro Georgiou, Y., et al., Sports Industry 5.0: Reimagining Sport through Technology, Humanity, and Sustainability, in *Frontiers in Sports Technology*, 2025: p. 134–142. 1.
20. Yao, Z., et al., Technology Perception, Convenience, and Smart Venue Experience: A Study on Spectator Behavior. *Journal of Smart Sports Management*, 2025. 3(4): p. 76–92.

# 10 Wellness 4.0


## Abstract

Medicine and wellness have evolved in tandem with human civilization, reflecting broader transformations in knowledge systems, technology, social organization, and ethical priorities. Historical analyses of healthcare reveal that dominant medical paradigms have consistently been shaped by prevailing worldviews, ranging from spiritual and communal interpretations of illness to mechanistic and data-driven models of the human body. This chapter traces the historical trajectory of healthcare and wellness paradigms from Wellness 1.0, characterized by ancient holistic and community-based healing traditions, through Wellness 2.0, marked by the rise of scientific medicine, institutionalized healthcare systems, and biomedical specialization, to Wellness 3.0, defined by digital health technologies, preventive care, and population-level data analytics.

At the center of the chapter is Wellness 4.0, an emerging paradigm in which artificial intelligence (AI), precision medicine, genomics, and interconnected innovative health ecosystems converge to reshape healthcare delivery and wellness management. Wellness 4.0 represents a structural shift from reactive, episodic treatment toward continuous, predictive, and personalized health optimization, enabled by machine learning algorithms, AI-driven diagnostics, digital biomarkers, wearable and implantable sensors, and real-time health monitoring platforms. In this model, individuals transition from passive recipients of care to active participants in adaptive health systems that learn continuously from biological, behavioral, environmental, and social data streams.

Drawing on global perspectives from international health organizations and peer-reviewed research, the chapter synthesizes evidence demonstrating how AI-enabled technologies are already transforming diagnostics, therapeutics, and public health governance. Machine learning models have shown high accuracy in medical imaging interpretation, early disease detection, clinical risk stratification, and drug discovery pipelines, in some cases equaling or surpassing human expert performance in narrowly defined tasks. At the population level, AI-powered surveillance and modeling tools are increasingly used to support epidemic forecasting, healthcare resource optimization, and health system resilience, particularly in complex and resource-constrained settings.

However, the rise of Wellness 4.0 also introduces profound ethical, social, and political challenges. Concerns surrounding data privacy, algorithmic bias, transparency, accountability, and unequal access have emerged as central risks in AI-driven healthcare systems. Empirical studies have shown that poorly governed algorithms can unintentionally reinforce racial, socioeconomic, and gender-based disparities in health outcomes, especially when trained on biased or incomplete datasets. Moreover, the increasing reliance on digital infrastructures raises questions about surveillance, informed consent, data ownership, and the commodification of health information. At the same time, uneven global access to AI technologies risks creating a new "digital divide" in healthcare quality and outcomes.

Looking forward, the chapter outlines plausible future scenarios in which integrated AI health assistants, genomic medicine, and ubiquitous monitoring technologies


DOI: 10.1201/9781003455219-10

support a LIFE 4.0 society oriented toward proactive well-being rather than disease management. In such scenarios, healthcare systems evolve into intelligent, adaptive networks that continuously anticipate risk, personalize interventions, and support lifelong wellness across diverse populations. The chapter concludes by identifying key policy, governance, and research priorities required to ensure that AI's transformative potential is directed toward inclusive, ethical, and human-centered wellness, rather than technological determinism. Ultimately, Wellness 4.0 is framed not merely as a technological upgrade, but as a civilizational transition that will shape how societies define health, care, and human flourishing in the AI epoch.

## 10.1 INTRODUCTION

Human healthcare systems have always reflected the dominant social, technological, and philosophical contexts of their time [1–10]. From antiquity to the present, medicine has evolved not only through scientific discovery but also through changing understandings of the human body, disease causation, responsibility for care, and the relationship between individuals and institutions [1, 11]. In early civilizations, medical practice was inseparable from culture, spirituality, and community life. Ancient healing systems, including Egyptian and Greco-Roman medicine, Ayurveda, Traditional Chinese Medicine (TCM), and indigenous medical traditions, emphasized holistic balance, natural remedies, and the interconnectedness of physical, mental, spiritual, and environmental well-being [12, 13]. These approaches, described here as Wellness 1.0, were community-based, locally adapted, and transmitted through tradition rather than formal scientific institutions.

The Enlightenment and subsequent Industrial Revolutions marked a decisive rupture with earlier paradigms, ushering in Wellness 2.0. Advances in anatomy, microbiology, chemistry, and experimental science led to vaccines, germ theory, anesthesia, antisepsis, and the rise of hospitals as centralized sites of care, transforming medicine into a professionalized and institution-driven enterprise [1, 14]. These developments produced dramatic gains in life expectancy, reductions in infant mortality, and improved control of infectious diseases, particularly in industrialized nations [15]. However, they also generated highly specialized and reactive healthcare systems that prioritized treatment over prevention and often overlooked social and environmental determinants of health. By the late 20th century, rising chronic disease burdens, demographic aging, escalating costs, and persistent inequities exposed the limitations of this model, prompting the emergence of Wellness 3.0, characterized by digital health innovations such as electronic health records (EHRs), telemedicine, mobile health applications, wearables, and large-scale health data analytics [3, 4]. These trends were rapidly accelerated by the COVID-19 pandemic, which forced widespread adoption of telehealth, remote monitoring, and artificial intelligence (AI)-assisted tools, highlighting both the potential and fragility of digitally mediated healthcare systems [16, 17]. Today, healthcare is entering a new epoch: the AI era, or Wellness 4.0, in which AI, machine learning, big data analytics, genomics, and smart connected devices converge to enable predictive, personalized, and continuously adaptive models of care [4, 5]. AI is increasingly used in medical imaging interpretation, clinical decision support, disease risk prediction, drug discovery, and personalized treatment planning, enabling a shift from episodic care toward lifelong health optimization. Wellness 4.0 is not a localized or experimental phenomenon but a global transformation, with adoption advancing most rapidly in countries with strong digital infrastructure while increasingly spreading through international collaboration and open innovation [18]. Global institutions such as the World Health Organization, OECD, and World Economic Forum emphasize that AI-driven health systems must be governed ethically, inclusively, and securely to support Universal Health Coverage and public trust [7, 19–21]. Building on this historical and global context, this chapter examines how Wellness 4.0 is redefining medicine and health in the AI epoch, assessing both its transformative potential and the ethical, social, and governance challenges that will shape health and well-being in a LIFE 4.0 society (Table 10.1).

**TABLE 10.1**
**Comparative Summary**

| Wellness Era | Core Orientation | Primary Driver | Focus |
|---|---|---|---|
| 1.0 | Holistic and communal | Tradition | Balance |
| 2.0 | Biomedical and institutional | Science | Treatment |
| 3.0 | Digital and preventive | Data | Self-management |
| 4.0 | Intelligent and predictive | AI | Integrated well-being |

## 10.2 WELLNESS 1.0–4.0

### 10.2.1 Wellness 1.0: Ancient Medicine and Traditional Healing

Before the rise of modern biomedical science, human societies worldwide developed sophisticated systems of traditional medicine rooted in holistic understandings of health, disease, and the human condition. These early medical traditions viewed the body not as an isolated biological machine, but as an integrated system in which physical, mental, spiritual, social, and environmental dimensions were inseparably linked. Healing was embedded within cultural norms, cosmologies, ethical frameworks, and communal practices, rather than centralized institutions or standardized scientific protocols. Illness was commonly interpreted as a disruption of balance within the body, between the individual and nature, or between humans and the spiritual realm rather than the result of a single pathogenic cause [1, 11]. This foundational worldview, referred to here as Wellness 1.0, represents humanity's earliest organized approach to health and well-being.

One of the most influential Wellness 1.0 traditions is Ayurveda, which originated in ancient India more than 2,000 years ago and remains one of the world's oldest continuously practiced medical systems. Ayurveda conceptualizes health as a dynamic balance among three fundamental life energies, or doshas – Vata, Pitta, and Kapha – each governing different physiological and psychological functions. Disease arises when these doshas become imbalanced due to factors such as diet, lifestyle, emotional stress, or environmental conditions. Treatment strategies emphasize personalized care, incorporating herbal medicines, dietary regulation, yoga, breathing practices, meditation, and ethical living to restore harmony between body and mind [22]. Similarly, TCM developed over millennia around the principles of qi (vital energy), yin–yang balance, and the Five Elements theory. Health is understood as the smooth flow of qi through meridians, while illness results from stagnation, deficiency, or excess. Therapeutic modalities such as acupuncture, herbal pharmacology, massage (tui na), and qigong aim to rebalance energy flows rather than suppress symptoms, reflecting a preventative and systems-oriented approach to medicine [23].

In the Western tradition, ancient Greek medicine, particularly the work attributed to Hippocrates in the 5th century BCE, laid the ethical and observational foundations of clinical practice. Hippocratic medicine rejected purely supernatural explanations for disease and instead emphasized careful observation of symptoms, environmental factors, diet, and lifestyle. The theory of the four humors (blood, phlegm, yellow bile, and black bile) dominated Western medical thought for centuries, framing health as a balance of bodily fluids analogous to natural elements. While scientifically obsolete, this framework reinforced the principle that disease reflects systemic imbalance rather than isolated malfunction. Equally important was the ethical dimension of Hippocratic medicine, including the emphasis on physician responsibility, patient-centered care, and the moral duty to "do no harm," principles that continue to underpin modern medical ethics [24]. Beyond these literate traditions, indigenous and shamanic healing systems worldwide, from Native American sweat lodge ceremonies to African herbalism and Amazonian plant medicine, integrated spiritual rituals, community participation, and deep ecological knowledge. Healing often involved the collective, reinforcing social cohesion and shared responsibility for well-being [12, 13].

Despite their diversity, Wellness 1.0 systems share several defining characteristics. They rely heavily on natural remedies derived from plants, minerals, and animal products; prioritize prevention and lifestyle regulation over crisis intervention; and emphasize the therapeutic relationship between healer and patient. Health was understood as a state of harmony rather than merely the absence of disease, and healing was a continuous process rather than a discrete medical event. In modern times, many of these practices have been subjected to scientific investigation, leading to renewed interest in integrative medicine, which combines evidence-based elements of traditional healing with contemporary biomedical approaches. Acupuncture, herbal pharmacology, and mind–body therapies are now recognized by institutions such as the World Health Organization and incorporated into healthcare systems worldwide [13]. As such, Wellness 1.0 did not disappear with scientific modernity; instead, it provided the conceptual groundwork for holistic, preventative, and patient-centered models of health that continue to influence medicine in the AI era. By foregrounding balance, lifestyle, and the interconnectedness of human life, Wellness 1.0 established enduring principles that remain essential as healthcare evolves toward increasingly technological futures.

### 10.2.2 Wellness 2.0: Scientific Revolution and Institutional Healthcare

The emergence of Wellness 2.0 represents a decisive shift in the history of medicine, as healthcare moved from tradition-based healing toward a scientific, experimental, and institutional model. This transformation unfolded during the Enlightenment and Industrial eras, when empirical observation, laboratory experimentation, and technological innovation began to dominate medical thought. Disease was increasingly understood as a biological phenomenon governed by natural laws rather than spiritual imbalance or humoral disharmony. This epistemological shift laid the groundwork for evidence-based medicine, redefining health as something that could be systematically studied, measured, and controlled through scientific inquiry [1].

A foundational breakthrough of Wellness 2.0 was the establishment of the germ theory of disease, pioneered by figures such as Louis Pasteur, Robert Koch, and Joseph Lister. Their work demonstrated that specific microorganisms caused specific diseases, enabling medicine to move from symptomatic relief to causal intervention. Lister's introduction of antiseptic surgery in the 1860s drastically reduced postoperative mortality, while Koch's identification of pathogens such as Mycobacterium tuberculosis provided a framework for targeted treatment and prevention [25, 26]. These discoveries also catalyzed the rise of modern public health, informing sanitation systems, clean water initiatives, vaccination campaigns, and hygiene practices that dramatically reduced infectious disease mortality at the population level [27].

Equally transformative was the institutionalization of healthcare and medical education. The late 19th and early 20th centuries saw the rise of university-affiliated hospitals that integrated patient care, laboratory research, and clinical training. Johns Hopkins Hospital, founded in 1889, became a global model by embedding scientific research into bedside medicine and setting rigorous standards for physician training. This model was reinforced by regulatory reforms such as the Flexner Report (1910), which standardized medical curricula, strengthened licensing requirements, and professionalized the medical field across the US and beyond. Medicine thus became firmly embedded within hospitals, universities, and state-regulated systems, gaining institutional authority and public trust [11, 28].

The 20th century marked the consolidation and expansion of Wellness 2.0 through biomedical breakthroughs and the development of large-scale healthcare systems. Antibiotics, vaccines, organ transplantation, medical imaging, and pharmaceuticals for chronic diseases transformed once-fatal conditions into manageable or preventable illnesses. Global public health initiatives, most notably the World Health Organization's smallpox eradication campaign, demonstrated the power of coordinated institutional healthcare on a worldwide scale [29]. These advances significantly increased life expectancy and improved population health, firmly establishing the paradigm that disease could be prevented and treated through science, technology, and institutions. At the same time, the growing specialization and systematization of care revealed limitations, particularly in addressing chronic illness, mental health, and holistic well-being, setting the stage for subsequent wellness paradigms that seek to reintegrate the human experience into scientific medicine.

### 10.2.3 Wellness 3.0: Digital Health, Telemedicine, and Preventive Care

Wellness 3.0 builds on institutional medicine, infusing it with digital connectivity and a preventive focus. Beginning in the late 20th century, breakthroughs in information technology, telecommunications, and biotechnology reshaped healthcare delivery.

EHRs, genome sequencing, smartphones, and wearable sensors transformed health data into a continuous, real-time stream. At the same time, medical devices became increasingly networked through the Internet of Medical Things (IoMT) [4]. Patients gained unprecedented access to online health information and the ability to communicate with healthcare providers remotely, shifting medicine toward more distributed, data-driven care models. A defining feature of Wellness 3.0 is the rise of telemedicine and telehealth. By the 2010s, video consultations, remote triage, and digital monitoring were increasingly common, but adoption accelerated dramatically during the COVID-19 pandemic. A U.S. Centers for Disease Control and Prevention (CDC) analysis found that telemedicine use among office-based physicians rose from 16.0 percent in 2019 to 80.5 percent in 2021, reflecting emergency-driven normalization of virtual care [30]. Similarly, telehealth visits surged from negligible levels before the pandemic to approximately 15 percent of all outpatient claims in 2020, according to national claims data [31]. While these figures are U.S.-centric, parallel trends were observed globally as lockdowns and infection risks necessitated remote care. International organizations, including the World Health Organization (WHO), actively promoted telehealth adoption during and after the pandemic, emphasizing its role in maintaining access to care for rural and underserved populations [32, 33].

Beyond access, Wellness 3.0 emphasizes patient empowerment and preventive care. Wearable technologies such as fitness trackers, smartwatches, and continuous glucose monitors allow individuals to track vital signs, sleep patterns, and physical activity in daily life. Mobile health applications support medication adherence, symptom monitoring, behavioral nudges, and lifestyle modification. This shift aligns with what has been described as a "left shift" in medicine, moving care upstream from late stage treatment toward early detection and prevention. A landmark review characterizes this model as "P4 medicine" predictive, preventive, personalized, and participatory, enabled by large-scale health data and digital analytics [34]. Examples include population-level risk algorithms that identify individuals at high risk for diabetes or cardiovascular disease before symptoms emerge, as well as genetic and biomarker screening programs that support targeted prevention, such as earlier cancer surveillance for individuals with BRCA mutations.

Importantly, Wellness 3.0 is a global phenomenon. Digital connectivity has expanded rapidly in low- and middle-income countries (LMICs), with approximately 65 percent of the worldwide population having internet access by the mid-2020s. In contrast, mobile broadband, 5G, and satellite internet continue to bridge remaining gaps [35]. WHO surveys indicate that most member states now maintain national digital health strategies, reflecting widespread institutional commitment to digital transformation in healthcare [7]. Social media and online platforms further demonstrate global engagement: an analysis of the 2024 Digital Health Public Health Symposium (dHPSP2024) shows that recorded contributions came from participants across 38 countries, highlighting active discourse in Asia, Europe, North America, and beyond [36]. At the population level, digital epidemiology, such as smartphone-based disease surveillance and tools like Google Flu Trends, has enabled near real-time monitoring of outbreaks, enhancing global health preparedness [37].

Nevertheless, Wellness 3.0 also faces significant challenges. Digital literacy gaps, particularly among the elderly, low-income, and marginalized populations, limit equitable access to digital health tools. Fragmented data systems and poor interoperability hinder information flow across providers and platforms. Privacy and cybersecurity concerns intensify as vast repositories of electronic health data become attractive targets for misuse or attack. Without deliberate policy interventions, digital inequities such as a lack of broadband access or affordable devices risk widening existing health disparities. Despite these limitations, Wellness 3.0 plays a foundational role in the evolution of modern healthcare by generating the data infrastructure, connectivity, and preventive orientation upon which AI-driven Wellness 4.0 systems will ultimately depend.

### 10.2.4 Wellness 4.0: AI-Driven Diagnosis, Personalized Medicine, and Smart Health Ecosystems

Wellness 4.0 centers on AI and advanced technology, fundamentally reshaping care. In this phase, machines learn from data to aid diagnosis, genomics guides individualized therapies, and pervasive devices form intelligent health networks [4, 7]. Key components include:

#### 10.2.4.1 AI Diagnostics

Machine learning algorithms can now analyze complex medical data, including images, genetic data, lab test results, and even patient speech, to detect disease. For instance, deep neural networks can interpret X-rays, CT scans, and MRIs with accuracy rivaling expert radiologists [4, 38, 39]. Studies confirm that AI can spot tumors, fractures, or hemorrhages faster and sometimes more accurately than humans. In the adjacent figure, an AI model's confusion matrix shows its high true-positive rates in classifying multiple lung pathologies from chest X-rays. Such systems sift through subtleties (tiny lesions or patterns indiscernible to the eye) and flag abnormalities for clinician review. The algorithms process vast patient data, including imaging, vital signs, genomics, and EHR histories, to provide a comprehensive view. As research notes, "AI algorithms can analyze medical images … and assist healthcare providers in identifying and diagnosing diseases more accurately and quickly" [6]. In practice, AI diagnostic tools are being deployed for diabetic retinopathy screening, skin lesion classification, and pathology slide analysis, among others [6, 40, 41].

#### 10.2.4.2 Personalized (Precision) Medicine

Building on genetic and molecular data, Wellness 4.0 delivers treatments tailored to the individual. Genetic sequencing of tumors, for example, now guides the choice of targeted cancer drugs. AI systems can integrate a patient's genome, microbiome, environment, and lifestyle to predict drug responses and disease risk. Pharmacogenomic algorithms help choose medications with minimal side effects. Over time, this means medicine moves away from a "one-size-fits-all" approach [4, 42, 43]. In this era, treatments are customized: a diabetic patient might receive a bespoke insulin regimen based on genetic markers and real-time glucose readings, while a cardiology patient's therapy is adjusted continuously by AI analyzing wearables and imaging. By correlating vast population genetics data with outcomes, AI also accelerates the discovery of new treatments and repurposes existing drugs for new indications [44].

#### 10.2.4.3 Smart Health Ecosystems

Beyond individual tools, Wellness 4.0 envisions an interconnected healthcare environment. The "Internet of Medical Things" (IoMT) connects devices (smart monitors, ingestible sensors, robotic assistants) so that data flows seamlessly. Hospitals become smart facilities with automated patient flow, real-time inventory tracking, and AI-driven staffing [4, 32]. Even the home is a mini clinic: smart scales, blood pressure cuffs, and sleep trackers automatically send data to doctors. AI coordinates this ecosystem. Keragon's industry analysis describes workflows where IoT devices collect real-time health data, AI analyzes it, and actionable alerts are sent to providers. For example, a wearable may detect an atrial fibrillation episode and immediately notify a cardiologist. These systems can predict health events before they occur: continuous monitoring combined with predictive modeling could warn of a pending heart attack or migraine hours in advance, allowing preventive action. On the public health side, aggregated data feeds powerful surveillance: AI can spot emergent disease patterns (as in pandemic tracking) and optimize resource allocation across regions [45].

The practical effect of Wellness 4.0 is profound. Clinical workflows are evolving: AI serves as an assistant or second opinion, flagging issues and reducing diagnostic errors. Digital twin models of patients (virtual physiological replicas driven by real data) could simulate disease progression and test therapies in silico. Advanced robotic systems, guided by AI, perform minimally invasive surgeries with micro-precision. Overall, healthcare is moving toward a proactive model: continuous data and intelligent algorithms catch and correct disease trajectories before acute crises, shifting care from reaction to prevention.

Examples from around the world: in China, hospitals are implementing AI radiology platforms that screen millions of scans. In India, start-ups use AI to analyze smartphone pictures for skin cancer risk. In the US and Europe, algorithms have been FDA-approved to detect stroke from CT scans in seconds. The UK's NHS has launched AI labs to integrate AI into routine practice. Meanwhile, low-resource settings leapfrog traditional barriers: mobile phone networks and simple AI apps are used in parts of Africa and Asia to screen for eye disease and maternal health risks remotely. Across nations, systems are converging on a model where personalized, AI-augmented care is standard.

## 10.3 OPPORTUNITIES AND TRANSFORMATIONAL POTENTIAL

Wellness 4.0 presents unprecedented opportunities to transform healthcare delivery by improving diagnostic precision, personalizing treatment, expanding access, reducing costs, and accelerating innovation. By leveraging AI, continuous data streams, and digital connectivity, healthcare systems can move from reactive, episodic models toward proactive, predictive, and equitable care. Key opportunities include:

### 10.3.1 Enhanced Diagnostic Accuracy and Speed

AI-driven diagnostic systems can analyze complex, high-dimensional medical data far faster than human clinicians while maintaining or exceeding expert-level accuracy.

Deep learning models now interpret medical imaging (X-rays, CT scans, MRIs), electrocardiograms (ECGs), pathology slides, and multimodal clinical data within milliseconds, significantly reducing time to diagnosis [4, 38]. This speed enables earlier detection of disease, often before symptoms become clinically apparent, which is particularly critical for cancer, cardiovascular disease, stroke, and rare genetic disorders. Earlier diagnosis increases treatment success rates, reduces complications, and lowers long-term healthcare costs. Studies demonstrate that AI-assisted imaging can detect conditions such as intracranial hemorrhage or lung cancer more consistently than human readers alone, serving as a powerful second opinion and reducing diagnostic error rates [6, 46].

### 10.3.2 Personalized Treatment Plans

Wellness 4.0 enables a shift from standardized protocols to individualized, precision-based therapies. By integrating genomic information, lifestyle data, environmental exposures, and continuous physiological monitoring, AI systems can tailor treatment choices and dosages to each patient's unique biological profile [42, 43]. Pharmacogenomic models help clinicians select medications that maximize efficacy while minimizing adverse drug reactions, improving patient safety and satisfaction. For example, oncology treatments are increasingly guided by tumor-specific genetic mutations. At the same time, chronic disease management, such as diabetes or hypertension, can be dynamically adjusted using real-time wearable data and predictive analytics. Over time, these personalized approaches are expected to improve cure rates, reduce trial-and-error prescribing, and enhance long-term patient outcomes [4].

### 10.3.3 Continuous and Preventive Care

A defining opportunity of Wellness 4.0 is the transition from episodic, clinic-based care to continuous, preventive health management. Wearable devices, home monitoring systems, and IoMT platforms generate real-time data on vital signs, activity levels, sleep patterns, and disease biomarkers. When combined with AI analytics, these data streams allow clinicians to detect early warning signals and intervene before conditions escalate into emergencies [47]. Chronic diseases such as diabetes, heart failure, and hypertension can thus be managed proactively, reducing hospital admissions and improving quality of life. This preventive orientation shifts healthcare's focus from crisis response to long-term wellness maintenance, aligning clinical care with public health goals [7].

### 10.3.4 Global Access and Equity

AI and digital health technologies significantly expand access to healthcare for underserved and remote populations. Telemedicine platforms, smartphone-based diagnostics, and lightweight AI models enable rural clinics to transmit data for expert interpretation or perform on-device analysis when specialists are unavailable. According to the WHO Digital Health Strategy, approximately 40 percent of surveyed countries have reached advanced levels of digital health maturity (Phase 3 or 4), indicating widespread adoption of digital tools even in resource-constrained settings [48, 49]. AI-assisted screening tools can empower community health workers to detect malnutrition, infectious diseases, or maternal health risks without on-site physicians, helping address global shortages of medical professionals. By lowering geographic, economic, and infrastructural barriers, Wellness 4.0 offers a pathway toward more equitable global healthcare delivery.

### 10.3.5 Efficiency and Cost Savings

Automation and AI-assisted workflows reduce administrative burden and optimize resource utilization across healthcare systems. Virtual triage, AI-supported clinical documentation, predictive bed management, and automated scheduling free clinicians to focus on patient care rather than paperwork [4]. Predictive analytics help hospitals anticipate demand for intensive care beds, staff, and equipment, improving operational efficiency. Moreover, AI systems enable scalable expertise, allowing a single radiologist or specialist to oversee diagnostics across multiple facilities through AI-assisted review. Over time, these efficiencies can reduce healthcare expenditures while maintaining or improving care quality, contributing to the sustainability of national health systems [50].

### 10.3.6 Accelerated Research and Innovation

The data-rich environment of Wellness 4.0 dramatically accelerates biomedical research and innovation. AI can mine vast patient databases to identify novel disease patterns, therapeutic targets, and biomarkers. At the same time, in silico clinical trials and digital twin models simulate treatment effects before real-world deployment [51]. Drug discovery timelines are shortened as AI models rapidly screen chemical compounds and identify candidates for repurposing, reducing development costs and failure rates [44]. Global collaboration platforms further enable clinicians and researchers to share anonymized data, fostering collective learning and faster translation of discoveries into clinical practice.

In summary, Wellness 4.0 holds transformative potential for healthcare delivery. By harnessing AI to analyze vast medical datasets, healthcare systems can achieve earlier diagnosis, more personalized treatment, broader access, and greater efficiency. As one widely cited assessment notes, AI's ability to process complex medical data at scale offers the opportunity to transform the field of medicine, leading to improved patient outcomes and more efficient and effective healthcare systems [4]. If implemented responsibly and equitably, Wellness 4.0 promises a future in which healthcare is more innovative, more responsive, and more patient-centered than ever before.

## 10.4 CHALLENGES, RISKS, AND ETHICAL CONSIDERATIONS

Despite its transformative promise, Wellness 4.0 introduces substantial challenges, risks, and ethical dilemmas that must be addressed to ensure that AI-driven healthcare advances are safe, equitable, and trustworthy. Many of these concerns echo long-standing issues in public health ethics, fairness, autonomy, privacy, and nonmaleficence, but are amplified by the scale, speed, and opacity of digital systems. Without deliberate governance, inclusive design, and global coordination, AI-enabled health technologies risk reinforcing existing inequities and introducing new forms of harm.

A primary challenge is data bias and algorithmic fairness. AI systems learn from historical data, and when that data underrepresents specific populations, algorithmic outputs may systematically disadvantage them. Studies have shown that some clinical algorithms underestimate illness severity in patients of color due to biased proxy variables, leading to delayed or inadequate care [10]. Similarly, non-AI medical technologies reveal structural bias: pulse oximeters have been shown to overestimate oxygen saturation in darker-skinned patients, a disparity now acknowledged by regulators and medical associations [52]. In AI-driven diagnostics, similar risks arise when training datasets, such as dermatology images or radiological scans, disproportionately represent lighter skin tones or high-income populations. As a result, Wellness 4.0 must contend not only with technical accuracy but with ethical dataset design, continuous bias auditing, and explicit inclusion of diverse demographic groups to prevent the amplification of health inequities [53].

### 10.4.1 Privacy and Trust

Privacy and trust are foundational to any healthcare system, yet they are particularly vulnerable in Wellness 4.0 due to the scale and sensitivity of data involved. AI systems rely on vast datasets, including EHRs, genomic sequences, wearable sensor data, behavioral logs, and even social media signals. The aggregation and reuse of such data raise profound concerns about consent, surveillance, and misuse. Patients may hesitate to share sensitive information such as mental health status or genetic predispositions if they fear discrimination, stigmatization, or unauthorized access [54].

Questions of data ownership and control further complicate data governance. It remains contested whether health data belong primarily to patients, healthcare providers, technology companies, or states. Regulatory frameworks vary widely across regions: the European Union's GDPR emphasizes individual data rights and consent, while other jurisdictions adopt more permissive or fragmented approaches. In response, global organizations have begun articulating ethical principles for AI in health. The World Health Organization's Ethics and Governance of AI for Health (2021) stresses human rights, transparency, accountability, and explainability as essential foundations for trust. Without robust safeguards and transparent governance, privacy breaches or opaque data practices could undermine public confidence and stall the adoption of beneficial AI technologies [7].

### 10.4.2 Digital Divide

The digital divide poses a significant risk to the equitable deployment of Wellness 4.0. While AI-driven healthcare promises global reach, its benefits are unevenly distributed due to disparities in infrastructure, digital literacy, and financial resources. High-income countries and urban centers often adopt advanced technologies rapidly.

In contrast, rural regions and LMICs face barriers such as limited broadband access, unreliable electricity, and shortages of technical expertise. Although global AI adoption reached approximately 35 percent of organizations by 2022, this growth has been concentrated in wealthier regions, potentially widening existing health disparities [8]. Without targeted interventions, Wellness 4.0 could inadvertently reinforce a two-tier health system, one characterized by AI-enhanced precision care, and another reliant on under-resourced traditional services. To counter this risk, experts advocate an "AI catch-up" strategy in LMICs that involves international partnerships, technology transfer, open-source tools, and local capacity-building initiatives [55]. When designed inclusively, AI can also enable leapfrogging: mobile-based diagnostics and lightweight algorithms can bring specialist-level insights to underserved areas. However, achieving this outcome requires intentional policy alignment and global investment, rather than relying solely on market forces.

### 10.4.3 Clinical Safety and Validation

Clinical safety and validation remain central concerns in AI-enabled healthcare. AI diagnostic tools that are insufficiently tested, poorly validated, or deployed outside their intended contexts can mislead clinicians and harm patients. Unlike traditional medical devices, AI systems may evolve as they encounter new data, complicating regulatory oversight. Regulatory bodies such as the U.S. Food and Drug Administration (FDA) and the European Medicines Agency (EMA) are actively developing frameworks for AI-based medical devices, yet global standards remain fragmented and evolving [46].

Liability presents another unresolved issue: when an AI system contributes to a clinical error, responsibility may be unclear. Does it lie with the developer, the clinician, or the healthcare institution? Trust in AI will depend on transparency, explainability, and rigorous clinical trials demonstrating safety and efficacy across diverse populations. Additionally, integrating AI into clinical workflows demands substantial investment in workforce training. Healthcare professionals must acquire new digital competencies, and facilities must be designed to support human–AI collaboration. Without adequate training and institutional readiness, AI tools may be resisted, misused, or over-relied upon, undermining their potential benefits [56].

### 10.4.4 Ethical Use

Ethical use of AI in medicine extends beyond technical performance to encompass autonomy, informed consent, and moral decision-making. Patients may question whether to accept AI-generated recommendations, particularly when algorithms function as "black boxes" whose reasoning is difficult to explain. This raises philosophical and practical concerns about accountability and shared decision-making in clinical care [57]. Moreover, ethical dilemmas emerge regarding how outcomes are prioritized: should AI systems optimize for longevity, quality of life, cost-effectiveness, or societal benefit?

At the societal level, large-scale health data analytics may be perceived as intrusive if used for purposes beyond care delivery. For instance, insurers or employers could seek to use AI-generated risk scores to adjust premiums or employment decisions, raising concerns about discrimination and coercion. Psychological effects also warrant attention: constant health monitoring may induce anxiety or erode the human dimensions of care. The use of AI in mental health, such as chatbots or emotion-recognition systems, introduces further ethical questions about empathy, confidentiality, and the limits of automation in deeply human domains [58].

### 10.4.5 Security and Robustness

Security and robustness are critical in a healthcare ecosystem increasingly dependent on connected devices and algorithms. The proliferation of networked medical devices, from insulin pumps to pacemakers, expands the attack surface for cyber threats and can have potentially life-threatening consequences. Documented vulnerabilities in medical devices highlight the risks of hacking, data manipulation, or system disruption [59]. AI systems themselves are also susceptible to adversarial attacks, in which subtly altered inputs can deceive diagnostic models and produce incorrect outputs.

Ensuring cybersecurity in hospitals and patient-facing technologies requires continuous investment, rigorous testing, and coordinated policy responses. As healthcare becomes more digital, cybersecurity must be treated not as an auxiliary IT concern but as a core patient safety issue. Policymakers, regulators, and technology developers must collaborate to establish resilient health data infrastructures capable of withstanding evolving threats [60].

In summary, Wellness 4.0 introduces risks related to bias, inequity, privacy, safety, ethical governance, and security, mirroring long-standing public health ethics challenges but intensified by digital complexity and scale. Addressing these challenges requires globally coordinated standards, such as the WHO's ethical guidelines for AI in health, alongside inclusive data practices, robust validation frameworks, and meaningful stakeholder engagement. Communities must be involved in the design and deployment of AI systems that affect them. If guided responsibly, many of these risks can be mitigated; if ignored, AI-driven health innovations risk undermining public trust and exacerbating the very disparities they promise to resolve.

## 10.5 THE FUTURE OUTLOOK (LIFE 4.0 SCENARIO)

Looking ahead, LIFE 4.0 envisions a world where Wellness 4.0 technologies mature and integrate into everyday life. In this scenario, most people have a personal AI health assistant accessible via a phone or other ambient device that continually analyzes health data and provides preventive advice. For example, a person's digital health record, including their genome, microbiome profile, continuous glucose monitor data, and daily activity, feeds into an AI system that alerts them to looming health issues days or months in advance. Imagine an AI-driven health hub at home: a bright mirror or toilet that monitors vital signs, a virtual coach in your earpiece nudging you to hydrate or do breathing exercises, and AR glasses guiding a rehab exercise set prescribed by your physiotherapist (all personalized via AI). Such technologies are already in prototype form and could be commonplace by mid-century.

Medical care becomes more distributed and continuous. Routine check-ups might be AI-led, with triage handled by algorithms and tele-consults for most cases. In hospitals, robots could perform routine surgeries or assist in elderly care, with human caregivers focusing on empathic tasks. Precision therapies (gene editing via CRISPR, immunotherapies) become widespread and affordable, effectively curing many once-intractable diseases. Continuous monitoring with nanotech sensors inside the body could detect diseases at the molecular level. AI models predict pandemics before they start by scanning global travel, climate, and animal disease data, allowing pre-emptive measures.

Importantly, preventive wellness is deeply embedded in education and culture. Schools teach children about personalized nutrition and stress management based on their biology. Cities are designed around health: pollution control, green spaces, and IoT networks that remind citizens to exercise. Healthcare extends outside hospitals: local community health workers use AI tools to serve villages and slums, dramatically raising care standards in low-resource settings. Smartphones or even drones might deliver AI diagnostics to remote areas. If current efforts succeed, the digital divide lessens, so LIFE 4.0 does not become a tale of "haves" versus "have-nots."

By mid-century, many expect improvements in longevity and quality of life. Life expectancy could be extended by 10–20 years if age-related diseases (Alzheimer's disease, heart disease) can be delayed or cured through AI-accelerated research. Some even imagine "digital immortality" experiments where minds are partially uploaded or simulated (though this is speculative). Mental health benefits are significant too: AI-trained therapy bots could provide 24/7 psychological support, alleviating global depression and anxiety burdens.

However, this future also demands balance. Society will have had to navigate trust in AI and establish that human values guide health decisions. We may see new regulations ensuring AI therapies are equitable. Privacy norms will evolve, possibly giving individuals greater ownership of their health data, as exemplified by emerging proposals such as personal data ownership frameworks. Insurance and ethical boards will have settled many early dilemmas (for example, how to weigh life extension technologies fairly).

One concrete future framework might be precision population health: global health organizations could use AI to allocate resources where they are needed most, predicting outbreaks or health crises. Perhaps a World Health AI Center is established to analyze aggregated data from around the world to advise on policy, detect emerging diseases, and coordinate research. This cooperative model would reflect LIFE 4.0's spirit – networked intelligence working for human health everywhere.

Ultimately, the LIFE 4.0 scenario imagines a symbiosis of humans and AI in health: not AI replacing doctors, but augmenting society's ability to understand and manage our biology at a level never possible before. The emphasis shifts from reacting to illness toward proactive well-being. As Bajwa et al. conclude, AI could make "personalized, precise, predictive, and portable" healthcare a reality, delivering high-quality care to anyone, anywhere [61]. If we address the ethical and equity challenges along the way, LIFE 4.0 could herald unprecedented improvements in global wellness.

## 10.6 CONCLUSION

The chapter has examined the evolution of medicine and wellness as a mirror of humanity's changing relationship with knowledge, technology, and the human body. Across history, healthcare systems have never been neutral or purely technical enterprises; they have always embodied deeper assumptions about causality, responsibility, authority, and what it means to live well. From the holistic, spiritually grounded, and community-oriented traditions of Wellness 1.0, through the scientific rationalism and institutional dominance of Wellness 2.0, to the digitally mediated, preventive focus of Wellness 3.0, each paradigm has expanded human capability while simultaneously introducing new limitations and tensions. Wellness 4.0 emerges from this lineage not as a rupture from the past, but as a convergence of long-standing aspirations: to understand health more precisely, to intervene earlier, and to support human flourishing more comprehensively across the lifespan.

What distinguishes Wellness 4.0 from earlier paradigms is the scale, speed, and adaptiveness of intelligence embedded within healthcare systems. AI transforms medicine from a predominantly reactive enterprise into a continuously learning ecosystem capable of anticipating risk, personalizing interventions, and coordinating care across settings and populations. AI-driven diagnostics augment human perception, revealing patterns invisible to clinicians alone; precision medicine replaces population averages with individualized biological insight; and innovative health ecosystems dissolve the boundaries between clinic, home, and community. In this emerging model, health is no longer managed primarily through episodic encounters with institutions, but through ongoing, data-informed relationships between individuals and adaptive systems that evolve alongside them. This represents a profound redefinition of care, shifting medicine's center of gravity from disease treatment to proactive wellness stewardship.

Yet the chapter has also underscored that technological capability alone does not guarantee social benefit. The same systems that promise unprecedented precision and efficiency also introduce new vulnerabilities and moral dilemmas. Algorithmic bias, rooted in historical inequities and incomplete data, risks reproducing or amplifying disparities in diagnosis and treatment. Expanding data collection raises urgent questions about privacy, consent, ownership, and surveillance, particularly as health data increasingly intersect with commercial, governmental, and insurance interests. Cybersecurity threats transform digital infrastructure into a matter of patient safety, while opaque algorithms challenge traditional notions of accountability and professional responsibility. These challenges reveal that Wellness 4.0 is not only a technical transition, but a governance and ethical transition that tests the capacity of societies to align innovation with justice, transparency, and public trust.

The global dimension of Wellness 4.0 further complicates this transition. While AI-enabled healthcare offers powerful tools to expand access in underserved and remote settings, unequal infrastructure, digital literacy gaps, and resource constraints risk creating a bifurcated health future. Without intentional policy design and international cooperation, advanced AI-driven care could remain concentrated in high-income regions, deepening existing global health inequities. Conversely, when paired with inclusive strategies such as open-source technologies, capacity building, and community-centered deployment, AI can enable leapfrogging, allowing LMICs to bypass legacy limitations and deliver higher standards of care. The direction Wellness 4.0 ultimately takes will depend less on technological inevitability than on collective political, institutional, and ethical choices.

Viewed within the broader LIFE 4.0 framework, Wellness 4.0 exemplifies how AI reshapes fundamental human systems rather than isolated sectors. Health, like work, relationships, governance, and everyday life, becomes increasingly mediated by intelligent technologies that blur boundaries between human agency and machine support. This raises both existential and practical questions: How much decision-making should be delegated to algorithms? How can human judgment, empathy, and moral reasoning remain central in AI-augmented care? What forms of literacy and institutional adaptation are required for individuals and societies to stay active shapers, rather than passive subjects, of intelligent health systems? Addressing these questions is essential if Wellness 4.0 is to enhance, rather than erode, human dignity and autonomy in the AI epoch.

Ultimately, Wellness 4.0 should be understood as a civilizational opportunity rather than a predetermined outcome. It offers the possibility of extending both lifespan and health span, reducing preventable suffering, and aligning healthcare systems more closely with the lived realities of individuals and communities. At the same time, it exposes the limits of purely technological solutions to deeply human challenges. The future of medicine in the AI era will be shaped not only by algorithms and data, but by values, governance structures, and social commitments that determine how intelligence is designed, deployed, and constrained. If guided by ethical foresight, inclusivity, and human-centered principles, Wellness 4.0 can serve as a cornerstone of LIFE 4.0 – supporting a future in which technology amplifies humanity's capacity to care for itself, rather than redefining health solely through the logic of machines.

## REFERENCES


1. Porter, R., *The Greatest Benefit to Mankind: A Medical History of Humanity*. London: HarperCollins, 1997.
2. WHO, *WHO Global Strategy on People-Centred and Integrated Health Services*. Geneva: WHO, 2015.
3. Raghupathi, W., and V. Raghupathi, Big Data Analytics in Healthcare: Promise and Potential. *Health Information Science and Systems*, 2014. 4(2): p. 98–123.
4. Topol, E., *Deep Medicine: How Artificial Intelligence Can Make Healthcare Human Again*. New York: Basic Books, 2019.
5. National Academies of Medicine, *Artificial Intelligence in Health Care: The Hope, the Hype, the Promise, the Peril*. Washington, DC: National Academies Press, 2019.
6. Esteva, A., B. Kuprel, R.A. Novoa, et al., Dermatologist-Level Classification of Skin Cancer with Deep Neural Networks. *Nature*, 2017. 542(7639): p. 115–118.
7. WHO, *Ethics and Governance of Artificial Intelligence for Health*. Geneva: WHO, 2021.
8. McKinsey Global Institute, *The AI-Powered Organization: Unlocking the Potential of AI at Scale*, 2021.
9. European Commission, *Ethics Guidelines for Trustworthy AI*. Brussels: European Commission, 2019.

10. Obermeyer, Z., B. Powers, C. Vogeli, and S. Mullainathan, Dissecting Racial Bias in an Algorithm Used to Manage the Health of Populations. *Science*, 2019. 366(6464): p. 447–453.
11. Rosenberg, C.E., *The Care of Strangers: The Rise of America's Hospital System. 2002*. Baltimore, MD: Johns Hopkins University Press, 2013.
12. Leslie, C., *Asian Medical Systems: A Comparative Study. 1976*. Berkeley: University of California Press, 2015.
13. WHO, *WHO Traditional Medicine Strategy 2014–2023*. Geneva: WHO, 2013.
14. Mokyr, J., *The Enlightened Economy: An Economic History of Britain 1700–1850*. New Haven, CT: Yale University Press, 2009.
15. Riley, J.C., *Poverty and Life Expectancy*. Cambridge: Cambridge University Press, 2005.
16. Keesara, S., A. Jonas, and K. Schulman, Covid-19 and Health Care's Digital Revolution. *New England Journal of Medicine*, 2020. 382: p. e82.
17. Whitelaw, S., M.A. Mamas, E. Topol, and H.G.C. Van Spall, Applications of Digital Technology in COVID-19 Pandemic Planning and Response. *The Lancet Digital Health*, 2020. 2(8): p. e435–e440.
18. OECD, *Health in the 21st Century: Putting Data to Work for Stronger Health Systems*. Paris: OECD Publishing, 2023.
19. WHO, *Global Strategy on Digital Health 2020–2025*. Geneva: WHO, 2020.
20. OECD, *Artificial Intelligence in Society*. Paris: OECD Publishing, 2019.
21. WEF, *Global Governance of Artificial Intelligence for Health*. Geneva: WEF, 2020.
22. Sharma, R.K., and B. Dash, *Charaka Samhita: Text with English Translation*. Varanasi: Chow Khamba Sanskrit Series Office, 2013.
23. Kaptchuk, T.J., *The Web That Has No Weaver: Understanding Chinese Medicine*. New York: McGraw-Hill, 2000.
24. Nutton, V., *Ancient Medicine*, Vol. 2nd ed. London: Routledge, 2013.
25. Lister, J., On the Antiseptic Principle in the Practice of Surgery. *The Lancet*, 1867. 90(2299): p. 353–356.
26. Brock, T.D., *Robert Koch: A Life in Medicine and Bacteriology*. Madison: University of Wisconsin Press, 1999.
27. Szreter, S., The Importance of Social Intervention in Britain's Mortality Decline. *Social History of Medicine*, 2002. 1: p. 1–38.
28. Flexner, A., *Medical Education in the United States and Canada*. New York: Carnegie Foundation, 1910.
29. WHO, *The Global Eradication of Smallpox*. 1980, Geneva: WHO Press, 2103.
30. Centers for Disease Control and Prevention, *Trends in the Use of Telehealth Among Health Care Providers*, 2022.
31. Koonin, L.M., et al., Trends in the Use of Telehealth During the Emergence of the COVID-19 Pandemic. *MMWR*, 2020. 69(43): p. 1595–1599.
32. WHO, *Telemedicine: Opportunities and Developments in Member States*. Geneva: WHO, 2022.
33. Boston Consulting Group, *Telehealth: A Quarter-Trillion-Dollar Post-COVID-19 Reality?* McKinsey & Company Original Publication, Washington, DC, 2020.
34. Hood, L., and M. Friend, Predictive, Personalized, Preventive, Participatory (P4) Cancer Medicine. *Nature Reviews Clinical Oncology*, 2011. 8: p. 184–187.
35. World Bank, *Individuals Using the Internet (% of Population)*, 2023.
36. Exploration Pub, Digital Health Public Health Symposium 2024 (dHPSP2024), in *Social Media Analysis*, 2024.
37. Salathé, M., et al., Digital Epidemiology. *PLoS Computational Biology*, 2012. 8(7): p. e1002616.
38. Rajpurkar, P., et al., Deep Learning for Chest Radiograph Diagnosis. *PLoS Medicine*, 2018. 15(11): p. e1002686.

39. McKinney, S.M., et al., International Evaluation of an AI System for Breast Cancer Screening. *Nature*, 2020. 577: p. 89–94.
40. Gulshan, V., et al., Development and Validation of a Deep Learning Algorithm for Detection of Diabetic Retinopathy. *JAMA*, 2016. 316(22): p. 2402–2410.
41. Campanella, G., et al., Clinical-Grade Computational Pathology Using Weakly Supervised Deep Learning. *Nature Medicine*, 2019. 25: p. 1301–1309.
42. Collins, F.S., and Harold Varmus, A New Initiative on Precision Medicine. *New England Journal of Medicine*, 2015. 372(9): p. 793–795.
43. Ashley, E.A., Towards Precision Medicine. *Nature Reviews Genetics*, 2016. 17: p. 507–522.
44. Zhavoronkov, A., et al., Artificial Intelligence for Drug Discovery, Biomarker Development, and Generation of Novel Chemistry. *Nature Reviews Drug Discovery*, 2019. 18: p. 463–477.
45. Keragon, Healthcare IoT Integration and Automation Workflows, 2023.
46. FDA, *Artificial Intelligence and Machine Learning in Software as a Medical Device*, 2022.
47. Perez, M.V., et al., Large-Scale Assessment of a Smartwatch to Identify Atrial Fibrillation. *New England Journal of Medicine*, 2019. 381(220): p. 1909–1917.
48. WHO, *Global Digital Health Monitoring Report*. Geneva: WHO, 2023.
49. Statista. Digital Health Adoption Worldwide, 2024.
50. Institute of Medicine., *Crossing the Quality Chasm: A New Health System for the 21st Century*. Washington, DC: National Academies Press, 2001.
51. Viceconti, M., et al., In Silico Trials Using Digital Twins. *Nature Reviews Materials*, 2021. 6: p. 112–126.
52. Sjoding, M.W., et al., Racial Bias in Pulse Oximetry. *New England Journal of Medicine*, 2020. 383(25): p. 2477–2478.
53. Benjamin, R., *Race after Technology*. Cambridge: Polity Press, 2019.
54. Floridi, L., et al., AI4People—An Ethical Framework for a Good AI Society. *Minds and Machines*, 2018. 28: p. 689–707.
55. UNESCO, *Recommendation on the Ethics of Artificial Intelligence*. Paris: UNESCO, 2021.
56. Jiang, F., et al., Artificial Intelligence in Healthcare. *Stroke and Vascular Neurology*, 2017. 2: p. 230–243.
57. Mittelstadt, B.D., et al., The Ethics of Algorithms. *Big Data & Society*, 2016. 3(2): p. 345–356.
58. Luxton, D.D., *Artificial Intelligence in Behavioral and Mental Health Care*. San Diego: Academic Press, 2016.
59. Ransford, B., et al., Security and Privacy in Implantable Medical Devices. *IEEE Security & Privacy*, 2014. 12(1): p. 30–39.
60. Coventry, L., and Dawn Branley, Cybersecurity in Healthcare. *Future Healthcare Journal*, 2018. 5(3): p. 184–189.
61. Bajwa, J., U. Munir, A. Nori, and B. Williams, Artificial Intelligence in Healthcare: Transforming the Practice of Medicine. *NPJ Digital Medicine*, 2021. 4: p. 100.

# 11 Relationships 4.0

**Abstract**

This chapter traces the evolution of human relationships from their ancestral roots through the Industrial Age and the digital era, culminating in a Relationships 4.0 paradigm shaped by artificial intelligence (AI) and immersive technologies. We survey how kinship and tribal bonds (Relationships 1.0) provided the original framework for social cohesion, and how social norms, institutions, and economic changes of the Industrial era (2.0) transformed family and community ties. We then examine the Digital Age (3.0), where mobile connectivity, social media, and globalization created new forms of community and communication. In Relationships 4.0, AI-driven companionship, virtual reality (VR) relationships, and hybrid human-machine social systems introduce unprecedented modes of connection. We discuss empirical findings and case studies, ranging from elderly care robots to VR courtship, that illustrate both the opportunities (e.g., reduced loneliness and global empathy-building) and the challenges (e.g., privacy, authenticity, and mental health) of this transformation. Cultural and policy perspectives are woven throughout to highlight how different societies adapt. Finally, we sketch an outlook (LIFE 4.0) scenario where human AI partnerships are commonplace, underscoring the need for ethical frameworks and proactive stewardship.

## 11.1 INTRODUCTION

The story of human connection is one of continual change. From the genealogical ties of ancient tribes to the digital networks of the 21st century, each technological and social revolution has reframed how we bond. This chapter explores the next phase of the AI Epoch and considers "Relationships 4.0," in which artificial intelligence (AI) and virtual environments become integral to human intimacy and community. We trace this evolution in four stages:

- **Relationships 1.0 (Kinship and Tribal Bonds)**: In early human societies, extended family and clan relations were the basis of trust and cooperation. Everyone's social identity and obligations were defined by birth and bloodlines.
- **Relationships 2.0 (Social Institutions, Industrial Age)**: Industrialization and urbanization detached many people from extended kin networks, replacing them with formal institutions (schools, corporations, nation-states) and new social norms. Family structures shifted toward nuclear units, and concepts such as romantic love and individual rights gained prominence.
- **Relationships 3.0 (Digital Networks and Globalization)**: The internet and mobile technology connected billions worldwide. Social media, instant messaging, and global travel enabled friendships that span continents, diluting the importance of proximity and traditional community. Local customs

DOI: 10.1201/9781003455219-11

mix with international trends, and people can belong to multiple overlapping communities (e.g., online gaming clans and diaspora associations).

- **Relationships 4.0 (AI and Virtual Companionship)**: Today, conversational AI, social robots, and VR/AR are creating hybrid relationships that blend the virtual and the real. People form emotional attachments to chatbots and digital avatars, while robotics promises to become a caretaker or a partner. This AI-driven era raises fundamental questions: How do we define friendship and love when one party is an algorithm? What new forms of community will emerge?

In what follows, we elaborate on each stage with empirical studies, global examples, and interdisciplinary insight. We examine both optimistic possibilities (e.g., AI as a tool for empathy and inclusion) and cautionary issues (e.g., privacy and emotional dependence). Throughout, we consider cultural variations, for instance, how Japanese society's embrace of companion robots contrasts with Western skepticism and policy implications, such as the need for AI ethics guidelines. By the end, we look ahead to a LIFE 4.0 scenario: a future world where AI is fully integrated into the social fabric (Table 11.1).

**TABLE 11.1**
**Comparative Summary**

| Relationships Era | Core Mode | Scale | Key Driver |
|---|---|---|---|
| 1.0 | Communal | Local | Tradition |
| 2.0 | Mediated | Regional/national | Institutions |
| 3.0 | Networked | Global | Digital platforms |
| 4.0 | Hybrid and AI-mediated | Adaptive | Artificial intelligence |

## 11.2 RELATIONSHIPS 1.0–4.0

### 11.2.1 Relationships 1.0: Kinship and Tribal Bonds

Human beings evolved as intensely social primates, and for most of human history, kinship constituted the primary organizing principle of social life. Long before formal states, markets, or legal institutions emerged, social order was structured through networks of biological descent, marriage, and ritualized affiliation. Anthropologists have long emphasized that "the single most important aspect of people's in-group is their kin relations," because kinship defined identity, obligation, authority, and belonging in early human societies [1, 2]. In this sense, Relationships 1.0 refers to a social world in which trust, cooperation, and moral responsibility were overwhelmingly grounded in family and clan ties.

In small-scale societies such as hunter-gatherer bands, pastoral groups, and early agrarian villages, nearly everyone was connected through blood or marriage. These kinship ties prescribed rights and duties: who shared food, who cared for children and elders, who married whom, and how conflicts were resolved. Marriage rules, including exogamy (marrying outside one's clan), were essential for forging alliances between groups and reducing intergroup conflict. At the same time, incest taboos helped formalize social boundaries and stabilize group identity [2, 3]. Kinship thus functioned as a social operating system, coordinating cooperation without written laws or centralized authority.

Although kinship systems varied widely across cultures, patrilineal versus matrilineal descent, patrilocal versus matrilocal residence, and nuclear versus extended households, their functional role was remarkably consistent. Kinship networks provided economic security, reproductive continuity, and social welfare in environments where survival depended on collective effort. In tightly knit kinship regimes, cooperation was primarily restricted to in-group members, and moral obligations were strongest toward relatives. Comparative studies of pre-industrial societies show that cultures with vigorous kinship intensity tend to emphasize communal norms, collective punishment, and family-based enforcement of behavior. In contrast, societies with weaker kinship structures rely more on universal rules, formal institutions, and impersonal trust mechanisms [4].

#### 11.2.1.1 Real-World Anthropological Examples

Ethnographic evidence illustrates how Relationships 1.0 operated across diverse ecological and cultural contexts:

- The Kung San of the Kalahari and the Hadza of Tanzania, both hunter-gatherer societies, organize life around small, mobile bands where food sharing, childcare, and labor are communally distributed. In these societies, nearly all adults are related by blood or marriage, and extensive sharing norms function as a survival strategy in unpredictable environments [5, 6]. Cooperation is enforced not through formal authority but through kinship obligations, reputation, and social sanction.
- Among the Māori of New Zealand and the Navajo (Diné) of North America, extended family structures such as whānau, hapū, or clans remain central to social organization, even in modern contexts. Genealogical descent determines land rights, leadership roles, inheritance, and ceremonial responsibilities, demonstrating the enduring influence of kinship beyond pre-industrial settings [7, 8].
- Many tribal societies practiced fictive kinship, in which outsiders were ritually adopted into the group as "brothers," "sons," or "clan members." Apprentices, war captives, or slaves could be symbolically incorporated into kinship networks, extending trust and obligation beyond biological ties. Anthropologists note that fictive kinship served as a powerful mechanism for social integration, alliance-building, and labor organization in societies lacking bureaucratic institutions [9].

#### 11.2.1.2 Strengths and Limitations of Kin-Based Social Systems

Kinship-based social organization generated extraordinary cohesion. Trust flowed naturally along family lines, reducing the need for surveillance, contracts, or enforcement mechanisms. Emotional bonds reinforced cooperation, while shared ancestry created strong collective identities. These systems enabled humans to survive harsh environments, raise offspring cooperatively, and transmit cultural knowledge across generations.

However, the same features that made Relationships 1.0 effective at small scales also imposed constraints. Strong clan loyalties often fostered out-group suspicion, inter-tribal conflict, and cycles of revenge. Chieftaincies and early states frequently experienced warfare driven by kin-based alliances and vendettas. Anthropological analyses suggest that very tight kinship structures can hinder large-scale cooperation, innovation, and institutional development, because trust rarely extends beyond the clan [4, 10]. As populations grew and interactions increasingly involved non-kin, societies were forced to develop new mechanisms, such as customary law, religious universalism, markets, and eventually formal states, to regulate cooperation among strangers. In this broader historical arc, Relationships 1.0 laid the evolutionary foundations of human sociality. It provided the earliest templates for trust, obligation, and moral order within small groups, while simultaneously revealing the limits of kinship as societies scaled. These limitations set the stage for later relationship paradigms in which bonds would increasingly be mediated by institutions, technologies, and abstract norms rather than by blood alone. Thus, Relationships 1.0 represents both the origin of human cooperation and the threshold beyond which humanity would gradually expand social trust beyond kin.

### 11.2.2 Relationships 2.0: Social Norms, Institutions, and Industrial Age Interactions

The transition from subsistence agriculture to the Industrial Age (roughly the late 18th to early 20th centuries) marked one of the most profound transformations in human relationships. While Relationships 1.0 were grounded primarily in kinship and clan-based cooperation, Relationships 2.0 emerged as social life became increasingly mediated by institutions, formal norms, and impersonal systems. The most fundamental shift occurred as economic production shifted from the household to factories, offices, and markets, reshaping family structure, social roles, and the basis of interpersonal trust. In pre-industrial societies, the household functioned as a unit of both production and reproduction. Families worked collectively on farms, in workshops, or in small-scale trade, pooling labor, land, tools, and income. According to Encyclopedia Britannica, a typical 18th-century rural household operated as a "domestic enterprise," in which parents, children, and often extended relatives contributed economically to shared survival (Britannica, "Family"). Skills and occupations were transmitted across generations, and multiple generations frequently lived under one roof, reinforcing dense interdependence and authority structures within the family.

#### 11.2.2.1 Industrialization and the Reconfiguration of Family Life

Industrialization disrupted this model by centralizing production in factories and urban centers, drawing individuals, particularly men, and later women, out of the household and into wage labor. As production relocated to factories, the family gradually ceased to be a primary economic unit and instead became a unit of consumption and socialization. Income increasingly belonged to individuals rather than to the household, weakening the financial basis of extended family living arrangements [11, 12].

Governments and emerging nation-states progressively assumed roles once fulfilled by kin networks. Public education systems replaced parental apprenticeship, welfare programs and pensions reduced dependence on extended family support, and hospitals and professional medicine displaced household-based care. By the late 19th century, particularly in Western Europe and North America, the nuclear family consisting of two parents and their children had become the dominant family form, supported by state institutions and legal frameworks [13]. Scholars describe this transformation as a shift from "the family as a mini-society" to "the family as a private emotional refuge," with primary responsibility for child-rearing rather than economic production [14].

#### 11.2.2.2 Institutions and Ideologies Reshaping Social Bonds

Simultaneously, new institutions and ideologies emerged that reorganized social relationships beyond kinship:

- Religion and caste/class systems played a central role in structuring marriage, friendship, and social networks. In Europe, the Catholic Church's prohibition of close-kin marriage weakened extended clan structures and encouraged broader social alliances, contributing to the gradual decline of kin-based solidarity [15]. In South Asia, caste endogamy strictly regulated marriage and social interaction, embedding relationships within hierarchical institutional frameworks that extended far beyond the family.
- Nation-states and nationalism reoriented loyalty from clan or tribe toward imagined political communities. Citizenship, military service, taxation, and legal equality fostered a sense of shared identity among strangers. Benedict Anderson famously described nations as "imagined communities," held together not by kinship but by shared symbols, institutions, and narratives [16]. This marked a decisive expansion of social trust beyond blood ties.
- Education systems, workplaces, and public institutions created new arenas for relationship formation. Schools grouped children by age rather than kinship, while factories and bureaucracies organized adults by occupation and rank. Friendships, professional networks, and collective identities are increasingly formed around shared education, ideology, or employment rather than family lineage. A factory worker, civil servant, or soldier developed bonds based on role-based cooperation, discipline, and shared experience, reflecting a new relational logic grounded in institutions.

#### 11.2.2.3 Individualization, Gender, and Intimacy

Relationships 2.0 also coincided with profound changes in gender roles and intimate life. Industrial-era reforms gradually expanded women's rights, including access to education, property ownership, and paid employment. Although progress was uneven and contested, these changes altered power dynamics within families and expanded individual autonomy [17].

Marriage itself was transformed. While pre-industrial marriages were often arranged for economic or kinship advantage, industrial societies increasingly promoted romantic love and emotional fulfillment as legitimate foundations for marriage. Sociologist Anthony Giddens argues that the rise of the "companionate marriage" reflected broader processes of individualization, in which people gained greater freedom to choose partners, careers, and life trajectories [14]. Emotional intimacy became central to personal identity, while relationships were increasingly evaluated in terms of personal satisfaction rather than collective obligation.

#### 11.2.2.4 Cross-Cultural Variation and Ongoing Tensions

Notably, the transition to Relationships 2.0 was neither uniform nor universal. In East Asian, Middle Eastern, and African societies, extended family structures and filial obligations remained influential well into the 20th century. Confucian norms in China and Korea, for example, continued to emphasize ancestor worship, filial piety, and intergenerational co-residence, even as industrialization progressed [18]. Similarly, in many African contexts, kinship networks retained economic and social significance alongside wage labor and state institutions.

Nevertheless, globalization, urbanization, and economic development have progressively drawn most societies into the Relationships 2.0 paradigm. The resulting tension between traditional norms (such as extended family obligations or arranged marriage) and modern ideals (such as nuclear families, individual choice, and gender equality) remains a defining social theme in countries ranging from India to Brazil. These tensions highlight that Relationships 2.0 is not merely a historical phase but an ongoing negotiation between institutional modernity and inherited social norms.

In summary, Relationships 2.0 represents the expansion of human social bonds beyond kinship into institutional, legal, and ideological frameworks. Trust increasingly flowed through laws, contracts, education systems, and shared national identities rather than blood ties alone. While this transformation enabled large-scale cooperation, mobility, and economic growth, it also introduced new forms of alienation, individualism, and social fragmentation. These dynamics set the stage for subsequent relationship paradigms, in which digital technologies would further reshape how humans form, maintain, and experience social connections in the modern world.

### 11.2.3 Relationships 3.0: Digital Social Networks, Mobile Communication, and Globalization

By the early 21st century, human relationships entered a new phase marked by pervasive digital connectivity, mobile communication, and unprecedented global reach. By early 2025, global digital access had reached historic levels: approximately 5.78 billion people (around 70.5 percent of the world's population) owned mobile phones, 5.56 billion (67.9 percent) were connected to the internet, and 5.24 billion (63.9 percent) actively used social media platforms [19]. These figures underscore a structural shift in how relationships are formed, maintained, and experienced. While Relationships 2.0 relied on physical institutions such as workplaces, schools, and nation-states, Relationships 3.0 are increasingly mediated through digital platforms that operate continuously, globally, and at scale.

#### 11.2.3.1 Digital Platforms as Social Infrastructure

Social media platforms, messaging apps, and mobile networks have become the core social infrastructure of contemporary life. Platforms such as Facebook, WhatsApp,

WeChat, Instagram, Telegram, and TikTok function as digital "town squares," enabling communication across vast geographic and cultural distances. These platforms facilitate not only personal interaction but also news sharing, emotional support, political mobilization, and cultural production. For diaspora communities in particular, digital communication has fundamentally reshaped relational continuity. Migrant families can now sustain daily contact through text, voice, and video calls, reducing the emotional costs of migration and sustaining transnational family identities [20].

Unlike earlier communication technologies, mobile internet access is persistent and portable, embedding social interaction into everyday routines. Relationships are no longer confined to specific places or times; instead, they unfold asynchronously across time zones and contexts. A single individual may simultaneously maintain familial ties in one country, professional collaborations in another, and friendship networks distributed across multiple continents. This constant connectivity marks a decisive break from earlier relationship paradigms rooted in physical co-presence.

#### 11.2.3.2 Expansion of Relational Scope

The rise of digital networks has expanded the scope of relationships in several important ways:

- **Geographic dispersion**: Maintaining long-distance friendships, romantic relationships, and professional ties has become routine. During the COVID-19 pandemic, video calls and messaging platforms became essential lifelines for separated families and social groups, normalizing virtual intimacy and remote sociality [21].
- **New communities and identity groups**: Online environments enable the formation of "families of interest" based on shared hobbies, identities, or causes rather than geography or kinship. Gaming communities, fandoms, online support groups, and activist networks often foster deep emotional bonds that can rival offline relationships in intensity and longevity. Scholars note that for many young people, online communities function as primary sites of belonging and identity formation [22].
- **Information networks and collective action**: Digital platforms dramatically accelerate the spread of information and ideas. Social movements such as #MeToo and #BlackLivesMatter illustrate how localized experiences of injustice can rapidly scale into global solidarity networks through social media amplification [23]. At the same time, algorithm-driven content curation can produce echo chambers, reinforcing ideological polarization and fragmenting shared public discourse [24].
- **Cultural exchange and virtual cosmopolitanism**: Technologies such as live streaming, automatic translation, and global content platforms expose users to diverse cultural practices in real time. A student in Europe can attend a live-streamed music event in Brazil or collaborate with peers in Asia, fostering what scholars describe as "networked cosmopolitanism," a form of global cultural awareness mediated through screens [25].

#### 11.2.3.3 Empirical Indicators of Digital Social Saturation

Empirical data underline the ubiquity of Relationships 3.0. Over two-thirds of the global population now uses social media every month, and the average user maintains accounts on nearly seven platforms, spending approximately 18.6 hours per week engaging in online social activity [19]. In practical terms, this means that many individuals now spend the equivalent of a full waking day each week socializing, consuming media, or participating in online communities.

Mobile technology has been pivotal in this transformation. Smartphones account for approximately 87 percent of mobile handsets worldwide, making internet access continuous rather than episodic and often replacing desktop computing altogether [26]. The rapid diffusion of affordable smartphones and data plans has accelerated digital inclusion across the Global South. In sub-Saharan Africa, smartphone penetration has more than doubled over the past decade, while in India, internet access surpassed 50 percent of the population by the mid-2020s [27]. These trends demonstrate that Relationships 3.0 is not limited to high-income societies but represents a genuinely global transformation.

#### 11.2.3.4 Real-World Examples of Digitally Mediated Relationships

The social effects of digital connectivity are evident in countries with large migrant or diaspora populations. In the Philippines, overseas workers routinely use video calls and messaging apps such as Viber and WhatsApp to sustain family life across continents. Grandparents participate in daily childcare rituals virtually, while family decisions are negotiated across borders in real time. Similar patterns are observed among migrant laborers in the Gulf States, who maintain close cultural and emotional ties with home communities through language-specific social media groups and messaging platforms [28].

At a global scale, shared digital experiences now generate ephemeral but meaningful social bonds. Major sporting events, political elections, and even television series finales attract massive "co-viewing" audiences online, where viewers collectively comment, meme, and react in real time. These moments illustrate how digital platforms enable synchronized emotional experiences among strangers, forming temporary but powerful relational communities.

#### 11.2.3.5 Transformation of Community and Identity

While digitalization has amplified connection, it has also redefined the nature of community. The concept of social space has expanded to include virtual "third spaces," neither home nor workplace, where screens, profiles, and avatars mediate interaction. Individuals may accumulate hundreds or thousands of online "friends" or followers without ever meeting them offline. For younger generations in particular, online identity can carry social weight comparable to, or even exceeding, offline presence [29].

This transformation introduces new tensions. Online interactions often privilege visibility, performance, and metrics such as likes, shares, and reshares, reshaping how social value and recognition are perceived. Relationships may feel simultaneously expansive and fragile: easily initiated yet easily abandoned. As scholars note, digital sociality produces a paradox of hyperconnection and loneliness, in which individuals are constantly networked but not always deeply bonded [30].

In summary, Relationships 3.0 are characterized by digitally mediated, mobile, and global networks of interaction. Trust and belonging increasingly flow through platforms, algorithms, and shared digital experiences rather than through institutions or physical proximity alone. This paradigm dramatically expands the scale and diversity of human connection, enabling new forms of community, identity, and collective action. At the same time, it introduces challenges related to misinformation, polarization, identity fragmentation, and the commodification of social life.

Relationships 3.0 thus represents a transitional phase: a world in which human connection is no longer bounded by place but not yet fully integrated with intelligent systems. These dynamics set the stage for Relationships 4.0, in which AI, immersive environments, and human–machine interaction further reshape how relationships are formed, sustained, and valued in the AI epoch.

### 11.2.4 Relationships 4.0: AI-Driven Companionship, Virtual Relationships, and Hybrid Social Systems

In Relationships 4.0, the boundary between human and machine in social life begins to blur. AI, embodied robotics, and immersive virtual environments introduce new forms of companionship and interaction that were once confined to science fiction. Unlike Relationships 3.0, where digital platforms mediated human-human interaction, Relationships 4.0 are defined by the presence of artificial agents as active social participants. These systems do not merely transmit communication; they simulate conversation, respond emotionally, and adapt to individual users over time. Three interrelated trends characterize this emerging relational paradigm.

#### 11.2.4.1 AI Chatbots and Virtual Assistants as Social Companions

Natural language AI agents such as chatbots, conversational assistants, and virtual therapists are increasingly designed not only to provide information, but also to engage users socially and emotionally. Early conversational programs, such as ELIZA in the 1960s, demonstrated that even simple rule-based systems could evoke emotional responses in users [31]. Today's large language model (LLM) based companions, such as ChatGPT, Replika, XiaoIce, and therapeutic chatbots, can sustain nuanced, context-aware conversations that approximate empathy, humor, and social reciprocity.

Empirical studies suggest that these systems can reduce perceived loneliness, particularly among individuals with limited social contact. A randomized controlled trial reported that an AI conversational agent produced a statistically significant short-term reduction in loneliness, comparable to the effect of brief human social interaction [32, 33]. Users frequently report feeling "heard" or emotionally supported, even when they are aware that the companion is artificial. This highlights a central feature of Relationships 4.0: social presence is increasingly defined by responsiveness rather than biological origin.

However, researchers also emphasize essential limitations. Intensive reliance on AI companions has been associated with increased emotional dependence and reduced motivation for offline social engagement in some studies [29, 34]. Public health experts caution that while AI chatbots may serve as valuable supplements, particularly in mental health triage or elder support, they are not substitutes for deep human belonging, which involves mutual vulnerability, shared embodiment, and long-term reciprocity [35]. Relationships 4.0 thus introduces a fundamental tension between accessibility and authenticity in social connections.

#### 11.2.4.2 Social and Companion Robots

Beyond software, physically embodied social robots are entering domestic, clinical, and institutional environments. Companion robots such as ElliQ, designed to support older adults living alone, and PARO, a robotic seal used in dementia care, illustrate how physical presence enhances emotional engagement. Studies in eldercare settings show that interaction with PARO can significantly reduce symptoms of depression, anxiety, and loneliness among patients with dementia, while increasing social interaction and positive effects [36, 37].

Similarly, evaluations of ElliQ deployments in the US indicate that older adults find the robot highly engaging and report improvements in perceived quality of life and adherence to daily routines [38]. Pet-like robots and robotic dogs or cats have also been shown to stimulate social behavior and mobility in nursing home residents, especially where human caregiving resources are limited.

These systems illustrate how AI and robotics extend the reach of caregiving and companionship, particularly in aging societies facing shortages of healthcare workers. Countries such as Japan and South Korea, which face acute demographic aging, have actively integrated social robots into eldercare policy frameworks, often through initiatives such as "Society 5.0," which emphasize human-centered technology integration [39]. While critics warn against "outsourcing care" to machines, proponents argue that such robots can enhance dignity, autonomy, and emotional well-being when used alongside human caregivers rather than as replacements.

#### 11.2.4.3 Hybrid Social Systems and Human–AI Co-Participation

Taken together, AI chatbots, companion robots, and immersive digital environments give rise to hybrid social systems, in which humans and artificial agents coexist as participants in shared social spaces. In online forums, multiplayer games, and virtual worlds, AI agents may act as moderators, collaborators, or even perceived peers. In workplaces, "digital colleagues" powered by AI assist in summarizing meetings, translating conversations, or managing workflows. In homes, smart speakers such as Alexa or Google Assistant engage in casual conversation, respond to emotional cues, and become integrated into daily routines.

In these systems, the social categories of "friend," "assistant," or "community member" expand to include entities that lack human consciousness but exhibit socially meaningful behavior. Scholars describe this as a shift from exclusively human social networks to socio-technical assemblages, in which agency is distributed across people, algorithms, and devices [40]. Trust and familiarity may develop through repeated interaction, personalization, and emotional responsiveness rather than shared human experience.

##### *11.2.4.3.1 Illustrative Scenario*

Consider an elderly individual living alone. She interacts daily with a robotic companion dog for comfort and routine, while a smart home assistant greets her each morning and manages reminders. She video-calls her children abroad and regularly visits a virtual environment where an AI-generated grandchild avatar interacts playfully with her. When she feels lonely, she converses with a culturally attuned chatbot in her native language. She also belongs to an online knitting circle with members across continents. This blended ecology of relationships, human, artificial, and virtual, keeps her socially engaged across time zones and contexts.

Such scenarios are no longer speculative; they are emerging realities in pilot programs across Asia, Europe, and North America. In Europe, trials of VR-based therapy and AI-assisted care are expanding, while in many developing regions, basic AI chatbots and voice assistants in local languages are growing rapidly due to their low cost and scalability [41].

#### *11.2.4.3.2 Global Variation, Ethics, and Social Implications*

Adoption of Relationships 4.0 is uneven worldwide. Japan and South Korea have embraced social robots more readily due to cultural attitudes toward technology and urgent demographic pressures. In contrast, Western societies often express greater ethical concern about emotional attachment to machines. Meanwhile, in low- and middle-income countries, lightweight AI companions delivered via smartphones are more common than advanced robotics, reflecting infrastructure constraints.

Ethically, Relationships 4.0 raises critical questions:

- Should machines be designed to simulate emotional attachment?
- Who is responsible if users develop dependency or distress?
- How do we preserve human dignity, consent, and autonomy in AI-mediated intimacy?

Global organizations such as the World Health Organization and UNESCO emphasize that AI systems in social and caregiving roles must remain human-centered, transparent, and accountable, with clear boundaries between support and substitution [35, 41].

In summary, Relationships 4.0 reflect a pluralistic and hybrid social order, where companionship, care, and interaction extend beyond exclusively human relationships to include AI-driven agents and virtual entities. Face-to-face human bonds remain central for most people, but a growing diversity of social ties now includes artificial and virtual co-inhabitants of everyday life. Relationships are increasingly defined by continuity, responsiveness, and emotional attunement rather than shared biology or physical presence.

This paradigm marks a profound civilizational shift: from societies organized around kinship (1.0), institutions (2.0), and digital networks (3.0), toward ecosystems of human–AI coexistence. How societies govern, integrate, and ethically constrain these emerging relationships will shape not only social life but also fundamental understandings of connection, care, and what it means to belong in the AI epoch.

## 11.3 OPPORTUNITIES AND TRANSFORMATIONAL POTENTIAL

The convergence of AI, immersive virtual technologies, and digital platforms introduces substantial opportunities to augment, diversify, and democratize human connection. While earlier relationship paradigms were constrained by geography, institutions, or physical presence, Relationships 4.0 enables new modalities of companionship, communication, and collective participation. When designed and governed responsibly, these technologies have the potential not only to mitigate long-standing social challenges such as loneliness, exclusion, and access barriers but also to reimagine how individuals and communities connect across distance, difference, and ability.

### 11.3.1 Alleviating Loneliness and Social Isolation

One of the most widely documented benefits of AI-driven companionship is its capacity to reduce loneliness and social isolation. Conversational agents and companion robots can provide consistent social interaction for individuals who live alone, are geographically isolated, or experience reduced mobility. Experimental studies show that interacting with AI companions can produce short-term reductions in perceived loneliness comparable to brief human interactions [33]. In eldercare settings, social robots such as PARO and ElliQ have been associated with improved mood, increased engagement, and enhanced social interaction among older adults, particularly those with dementia [36–38].

For individuals with disabilities or chronic illnesses, AI assistants and virtual reality (VR) experiences can bring social encounters into the home, mitigating the isolating effects of physical constraints. These tools do not replace human relationships, but they can serve as bridging mechanisms, sustaining emotional stimulation and social rhythm where human contact is limited. Public health scholars increasingly recognize such technologies as potential complements to traditional social care in aging societies [35].

### 11.3.2 Augmenting Long-Distance Relationships

AI, VR, and augmented reality (AR) significantly enhance the quality of long-distance relationships by enabling shared experiences rather than mere communication. Couples, families, and friends separated by geography can engage in virtual travel, collaborative games, or immersive environments that simulate co-presence. Educational and professional contexts also benefit: international families can attend weddings or graduations via VR. At the same time, distributed teams collaborate in virtual meeting spaces that restore non-verbal cues and spatial awareness absent from traditional video calls.

Empirical research suggests that shared virtual experiences increase relationship satisfaction and perceived intimacy compared to text- or video-only communication [42]. As immersive technologies mature, they may partially offset the relational costs of globalization, migration, and remote work by restoring a sense of "being together" across distance.

### 11.3.3 Cross-Cultural Communication and Understanding

AI-powered translation and natural language processing tools are transforming cross-cultural communication. Real-time translation enables individuals to converse in their native languages while maintaining mutual comprehension, significantly lowering linguistic barriers to friendship, collaboration, and community participation. These tools facilitate global online communities in which members from diverse cultural backgrounds can interact more fluidly, fostering cross-cultural empathy and cosmopolitan awareness [25].

Exposure to diverse AI-mediated personalities, global content, and intercultural dialogue can also broaden perspectives and reduce prejudice. Research on mediated contact suggests that sustained interaction across group boundaries, even when digitally mediated, can promote mutual understanding and reduce intergroup bias under appropriate conditions [43, 44]. In this way, Relationships 4.0 holds promise for cultivating empathy on a global scale.

### 11.3.4 Expanded Social Platforms and Intelligent Matching

AI-driven algorithms extend social platforms beyond traditional romantic matching toward friendship formation, mentorship, and peer support. Intelligent systems can identify shared interests, complementary skills, or emotional needs, recommending connections that might not emerge organically in offline contexts. For example, AI could connect newly relocated individuals with local support networks, match students with mentors, or suggest peer groups for mental health or caregiving support.

In community settings, AI moderation and content curation tools can improve group cohesion by reducing harassment, filtering harmful content, and encouraging constructive participation. Studies show that effective moderation increases user retention and perceived safety in online communities, enabling more inclusive digital public spaces [45]. When carefully designed, such tools can strengthen collective norms rather than undermine them.

### 11.3.5 Accessibility and Social Inclusion

Relationships 4.0 technologies offer powerful mechanisms for social inclusion, particularly for individuals who have historically been marginalized due to communication or sensory barriers. AI-driven speech recognition, text-to-speech, and alternative communication interfaces enable participation for people with speech or hearing impairments. Virtual avatars can provide social presence for individuals with social anxiety, allowing gradual engagement in group interactions at a self-determined pace.

Educational applications of VR and adaptive AI have shown promise in supporting children with autism by creating structured, interactive social learning environments tailored to individual needs [46]. More broadly, these technologies challenge traditional assumptions about what forms of interaction "count" as social participation, expanding the boundaries of inclusion.

### 11.3.6 Enhanced Empathy Training and Perspective-Taking

Immersive technologies also offer novel tools for empathy training and moral imagination. VR experiences have been used to place participants in simulated perspectives, for example, experiencing homelessness, visual impairment, or racial discrimination, producing measurable increases in empathic concern and prosocial attitudes [47]. As VR content becomes more sophisticated, such experiences could be integrated into education, diversity training, and cross-cultural exchange programs.

At scale, these applications suggest the possibility of experiential empathy, where understanding is cultivated not only through information but through simulated lived experience. While ethical safeguards are necessary to avoid manipulation or emotional harm, such tools may deepen social understanding in ways traditional media cannot.

### 11.3.7 Mental Health Support and Emotional Resilience

AI chatbots and digital mental health applications provide low-cost, on-demand emotional support, particularly in regions with limited access to trained therapists. Clinical trials indicate that well-designed conversational agents delivering cognitive behavioral therapy techniques can produce small but significant improvements in symptoms of depression and anxiety [32]. While these tools are not replacements for professional care, they can function as first-line support, triage mechanisms, or supplements to traditional therapy.

For populations facing stigma around mental health or logistical barriers to care, AI counselors may lower thresholds for seeking help. Global health organizations increasingly recognize digital mental health tools as potential components of scalable mental health strategies, provided they meet standards of safety, transparency, and accountability [35].

### 11.3.8 Institutional and Societal Transformation

Beyond individual relationships, Relationships 4.0 technologies have implications for institutions. Educational systems may incorporate shared VR experiences to foster collaborative learning across borders. Workplaces increasingly rely on hybrid human–AI teams, reshaping professional relationships and organizational culture. Governance may benefit from AI-moderated community forums that facilitate large-scale civic participation while maintaining deliberative quality.

In this context, an emerging "dual partnership" model becomes apparent: humans maintain rich, irreplaceable human relationships while also engaging artificial companions and systems for specific emotional, cognitive, or practical functions. For example, individuals may confide deeply in trusted friends while using AI companions to rehearse difficult conversations, manage stress, or combat boredom. Rather than displacing human bonds, AI augments the flexibility and reach of social life.

In summary, Relationships 4.0 holds transformative potential to expand the scale, accessibility, and diversity of human connections. AI and virtual technologies can alleviate loneliness, strengthen long-distance bonds, bridge cultural divides, enhance inclusion, and support mental health. When integrated thoughtfully, they enable a more pluralistic social ecosystem in which human and artificial relationships coexist, each serving distinct but complementary roles.

The realization of this potential, however, depends on ethical design, inclusive governance, and cultural sensitivity. If guided responsibly, Relationships 4.0 may not diminish human connection, but rather extend its reach, making social life more resilient, inclusive, and adaptable in a LIFE 4.0 society.

## 11.4 CHALLENGES, RISKS, AND ETHICAL CONSIDERATIONS

Despite the significant opportunities offered by Relationships 4.0, the integration of AI and virtual technologies into intimate social life raises profound ethical, psychological, cultural, and governance challenges. Many of these risks are not entirely new but are intensified by the scale, intimacy, and opacity of AI-mediated relationships. As artificial agents increasingly enter domains traditionally reserved for human connection, friendship, care, intimacy, and emotional support, societies must confront difficult questions about autonomy, authenticity, trust, and power.

### 11.4.1 Privacy and Surveillance

Increased digital intimacy inevitably produces vast quantities of highly sensitive data. AI companions, voice assistants, and immersive VR platforms can capture not only conversational content but also emotional signals, biometric data, movement patterns, and psychological cues. Unlike traditional social media data, this information often reflects users' inner emotional states, vulnerabilities, and private reflections. Scholars warn that such "intimate data exhaust" is among the most sensitive categories of personal information ever collected at scale [48].

Data ownership and consent remain deeply contested. While regulations such as the European Union's General Data Protection Regulation (GDPR) establish principles of data minimization, purpose limitation, and informed consent, enforcement is uneven, and cross-border data flows complicate accountability [49]. An EU citizen interacting with an AI companion hosted on servers in another jurisdiction may have limited legal recourse if data is misused. Unchecked, companies could exploit relational data for targeted advertising, emotional manipulation, or political influence, extending practices already observed in social media ecosystems [45, 48].

### 11.4.2 Erosion of Human Social Skills

A central concern is that over-reliance on AI companions may weaken core human social competencies. Human relationships are inherently complex, involving misunderstanding, negotiation, conflict, and reconciliation, which some sociologists describe as "necessary friction" for emotional development. In contrast, most AI companions are designed to be agreeable, non-confrontational, and responsive to user preferences. A widely cited Time magazine essay argues that technologies optimized to please users may deprive them of opportunities to practice patience, empathy, and conflict resolution [29].

Developmental psychologists and educators express particular concern about children and adolescents. Prolonged screen-mediated interaction has been associated with reduced face-to-face communication skills, diminished attention spans, and difficulties interpreting social cues, though causality remains debated [50]. If young people increasingly default to AI for social rehearsal or emotional regulation, there is a risk that messy but essential human interactions are displaced rather than supplemented.

### 11.4.3 Mental Health Risks and Emotional Dependency

While AI companions can provide short-term relief from loneliness, evidence suggests a more complex long-term picture. Some studies find that heavy emotional reliance on chatbots correlates with increased loneliness, anxiety, and depressive symptoms over time, particularly among vulnerable users [34]. A public health analysis from George Mason University warns that individuals who substitute AI companionship for human relationships may experience a reinforcing cycle of isolation, in which reliance on AI reduces motivation to seek human contact.

More extreme cases have been reported in clinical literature, including users developing delusional or quasi-delusional attachments to AI systems, sometimes described using the psychiatric metaphor folie à deux, a shared emotional fixation in which the AI reinforces distorted beliefs [29, 51]. While such cases are rare, they highlight the need for safeguards, particularly for users with mental health vulnerabilities.

### 11.4.4 Authenticity, Trust, and Parasocial Relationships

AI companions raise fundamental questions about authenticity and trust. Human–AI relationships often resemble parasocial relationships, one-sided emotional attachments traditionally associated with media figures, where users project emotions onto entities incapable of reciprocal understanding [52]. Although users may intellectually recognize that AI lacks consciousness, emotional engagement can blur this distinction, leading to a perception that the AI "cares" or "understands."

This raises ethical concerns about emotional deception, even if unintended. Is it morally acceptable for a system to simulate empathy without the capacity for genuine concern? In eldercare and dementia settings, critics worry that vulnerable individuals may be misled into believing robots such as PARO are alive, potentially undermining informed consent and dignity [53]. Trust, once eroded by perceived deception, may extend beyond AI to broader technological systems.

### 11.4.5 Cultural and Societal Impacts

Cultural norms significantly shape how Relationships 4.0 are perceived and adopted. In Japan, for example, social robots and virtual partners are often viewed as pragmatic responses to demographic aging and social isolation. In contrast, many Western societies regard intimate human–AI relationships as unsettling or socially deviant [54]. Religious traditions may further complicate acceptance, particularly where doctrines emphasize human uniqueness or prohibit simulated intimacy.

Unequal access to AI companionship technologies also risks exacerbating social inequalities. Wealthier families may rely on AI tutors, nannies, or social enrichment tools, while poorer communities lack access to equivalent support. This "relational digital divide" could compound existing educational and social disparities, reinforcing stratification rather than alleviating it [41].

### 11.4.6 Power Dynamics, Manipulation, and Bias

At the core of many ethical concerns lies the question of power: who designs, controls, and benefits from AI companions? Poorly governed systems could exploit users' emotional vulnerabilities for profit, persuasion, or surveillance. An AI companion designed to encourage consumption, political ideology, or behavioral compliance would represent a profound ethical breach, particularly for lonely or marginalized users.

Bias also poses risks. AI systems trained on culturally narrow datasets may implicitly privilege certain norms, personalities, or values, marginalizing others. If companion AIs reinforce dominant cultural scripts, they may subtly shape users' identities and preferences in non-transparent ways. Ethical AI frameworks emphasize the need for fairness, explainability, and accountability in systems that interact socially with humans [55].

### 11.4.7 Regulatory and Legal Uncertainty

Legal frameworks have yet to adapt to Relationships 4.0 fully. Questions surrounding AI companionship challenge existing categories of law:

- If someone claims to "marry" an AI, does this have any legal standing?
- If a social robot is damaged or causes harm, who is liable: the user, manufacturer, or software provider?
- Should advanced social robots possess any form of legal status?

Governments are beginning to respond. The EU AI Act proposes classifying specific AI systems, particularly those that interact intimately with humans, as high-risk and requiring transparency, human oversight, and strict safeguards. However, global consensus remains elusive, and regulatory approaches vary widely across jurisdictions [56].

In sum, while Relationships 4.0 opens remarkable new horizons for connection, it also repackages age-old ethical questions in unprecedented technological forms. What does it mean to be connected? How do we balance convenience with authenticity, support with dependency, and innovation with human dignity? The current evidence suggests caution: the integration of AI into intimate social spheres has outpaced long-term research, ethical reflection, and regulation.

Psychologists, ethicists, and policymakers broadly agree that AI companions should be designed as augmentative tools rather than emotional replacements. As one survey indicates, a significant minority of young adults already believe AI could replace human romance, underscoring the urgency of public education, ethical standards, and robust governance [57].

The future of Relationships 4.0 will depend not only on technological capability but on collective decisions about values, limits, and responsibility. Without thoughtful oversight, AI-mediated intimacy risks deepening isolation, inequality, and manipulation. With careful design and regulation, however, it may yet enrich human social life without eroding its core meaning.

## 11.5 FUTURE OUTLOOK (LIFE 4.0 SCENARIO)

Looking ahead to LIFE 4.0, the maturation of AI, immersive technologies, and hybrid social systems suggests that human relationships will not be replaced, but reconfigured. A widely discussed and credible future model is the "dual partnership" paradigm, in which individuals maintain both human and AI-mediated relationships, each fulfilling distinct emotional, cognitive, and practical roles. Rather than competing, these forms of connection coexist within layered social ecosystems shaped by culture, policy, and individual choice.

In this scenario, people may continue to reserve deep vulnerability, moral accountability, and long-term commitment for trusted human relationships, while engaging AI companions for rehearsal, guidance, emotional regulation, or routine social interaction. A person might confide existential concerns to a close friend, while using an AI coach for daily emotional "check-ins," communication practice, or stress management. As futurists argue, this reflects a broader pattern in technological history: tools that initially appear disruptive often become embedded as background support within human life rather than its core replacements [58, 59].

### 11.5.1 Domestic, Educational, and Workplace Futures

In domestic life, AI partners may become normalized household members by the 2030s and 2040s, much as televisions, smartphones, and voice assistants were rapidly domesticated in earlier decades. Families may "adopt" social robots or AI companions to support eldercare, childcare routines, or emotional well-being, potentially gaining formal recognition as assistive devices under healthcare or family law. In some jurisdictions, such systems may even be acknowledged as contributing to emotional welfare, though not as legal persons [56].

In education, children of the mid-21st century may grow up socializing alongside AI classmates, tutors, and robotic playmates. AI agents could scaffold collaborative learning, model prosocial behavior, or facilitate inclusion for students with special needs. Immersive metaverse environments may blur boundaries between physical classrooms and virtual communities, enabling students to shift seamlessly between in-person and digital collaboration. Early research suggests that when designed appropriately, AI tutors can enhance learning outcomes without undermining peer interaction, provided they are explicitly framed as supports rather than substitutes [60].

Workplaces are likely to evolve into hybrid human–AI social systems, where AI colleagues moderate discussions, summarize meetings, resolve language barriers, and facilitate cross-cultural teamwork. Global corporations may rely on AI moderators in online forums to sustain respectful, productive dialogue across borders, time zones, and cultures. In such environments, relational competence will increasingly include the ability to collaborate effectively not only with humans but with intelligent systems.

### 11.5.2 Language, Globalization, and Social Scale

One of the most transformative implications of LIFE 4.0 is the near-elimination of language barriers. Advances in real-time translation and natural language understanding suggest that conversational fluency across languages may become routine. This could enable friendships, professional collaborations, and communities that are truly global in scale, reshaping how identity, solidarity, and cultural exchange are experienced [25].

At the same time, global social integration will likely coexist with local differentiation. While some individuals embrace fluid, transnational digital identities, others may respond by reasserting local traditions, face-to-face rituals, and place-based belonging. LIFE 4.0 is therefore unlikely to produce a single relational future; instead, it will generate plural and sometimes competing social models.

### 11.5.3 Possible Trajectories and Governance Pathways

The shape of Relationships 4.0 in a LIFE 4.0 society will depend heavily on policy choices, cultural values, and ethical governance. Several plausible trajectories stand out.

#### 11.5.3.1 Regulated Integration

In an optimistic but realistic scenario, governments establish clear ethical standards and regulatory frameworks for AI companionship. These may include mandatory disclosure that an AI is not human, strong user data rights, limits on emotional manipulation, and safeguards for vulnerable populations such as children and the elderly. Multidisciplinary oversight bodies drawing on psychology, sociology, medicine, and law could evaluate "AI well-being impact" much as drugs or medical devices are assessed today [35, 41].

Countries may introduce age-appropriate usage guidelines for relationship-oriented AI, like existing regulations for television, gaming, or social media. Under this model, AI companionship is normalized but bounded, with explicit recognition of its supportive rather than substitutive role.

#### 11.5.3.2 Persistent Human Primacy and Counter-Movements

Even as technology advances, many societies are likely to reaffirm the primacy of in-person human connection. Counter-movements emphasizing "analog life," digital minimalism, and intentional disconnection are already visible. Practices such as digital sabbaths, "no-phone" dinners, offline schools, or tech-restricted communities may expand as cultural correctives to pervasive connectivity [61].

Public health experts increasingly emphasize that regular face-to-face interaction is essential for psychological well-being, regardless of technological sophistication [62]. Policymakers and educators may therefore promote balance rather than immersion, framing AI as a supplement to, not a replacement for, embodied social life.

#### 11.5.3.3 Cultural Hybridization

LIFE 4.0 will also be marked by cultural hybridization. Some societies may accept mixed human–AI social units, incorporating robots or AI agents into rituals, caregiving practices, or online celebrations. Others may reject such practices as violating moral or religious norms. Rituals, holidays, and family structures may adapt unevenly, producing a patchwork of acceptance, resistance, and reinterpretation.

This diversity will likely intensify global debates over personhood, rights, and moral status. Importantly, these debates will not be settled solely by technology, but by collective cultural negotiation.

### 11.5.4 Human-Centered AI and Relational Design

A key determinant of positive outcomes will be design philosophy. One promising trajectory emphasizes AI systems explicitly designed to strengthen human–human relationships rather than replace them. Examples include AI tools that encourage users to reconnect offline, practice difficult conversations before having them in person, or identify moments when human support is preferable to artificial interaction.

Emerging research suggests that AI companions attuned to healthy emotional boundaries can enhance resilience rather than undermine it, by making users feel supported without fostering dependency [32, 33]. Such systems prioritize empowerment, reflection, and connection rather than emotional capture, aligning with a broader vision of human-centered AI.

### 11.5.5 Global Cooperation and Collective Flourishing

In a fully realized LIFE 4.0 scenario, the governance of human–AI relationships becomes a global concern. Cross-border challenges from eldercare robots in aging societies to digital wellness platforms used by millions require international coordination. Organizations such as the United Nations, World Health Organization, and UNESCO are likely to play central roles in establishing norms, sharing research, and supporting capacity-building across regions [35, 41]. Ultimately, the LIFE 4.0 future of relationships is not technologically predetermined. Choices about values, limits, and responsibility will shape it. If societies anticipate risks, invest in research, and guide innovation with ethical clarity, AI can remain a tool for collective flourishing rather than relational erosion.

In this vision, Relationships 4.0 does not signal the end of human connection, but its expansion into a richer, more flexible social ecology, one in which technology supports belonging, care, and understanding without displacing the uniquely human capacities that give relationships their meaning.

## 11.6 CONCLUSION

The evolution of human relationships from kinship-based bonds to AI-mediated connections reflects a broader pattern in human history: social life continually adapts to the dominant technologies, institutions, and values of each era. As the chapter has shown, Relationships 1.0 grounded trust and belonging in family and clan; Relationships 2.0 expanded social bonds through institutions, norms, and nation-states; and Relationships 3.0 globalized connection through digital networks and mobile communication. Relationships 4.0 represents the next inflection point, where AI, social robots, and immersive virtual environments become active participants in the social world rather than mere conduits for human interaction.

Crucially, the rise of AI-driven companionship does not signal the disappearance of human relationships, but their diversification. AI companions, chatbots, and virtual spaces can alleviate loneliness, strengthen long-distance ties, enhance accessibility, and support mental health, particularly in contexts of aging populations, migration, and social fragmentation. At the same time, these technologies introduce serious challenges related to privacy, emotional dependency, authenticity, inequality, and governance. The evidence reviewed in the chapter underscores a central tension of the AI epoch: systems designed to simulate empathy and responsiveness can offer meaningful support, yet they lack the moral agency, mutual vulnerability, and accountability that define deep human relationships. Navigating this tension will require deliberate choices about design, regulation, and cultural norms.

Looking toward a LIFE 4.0 future, the most plausible and ethically defensible path is neither technological rejection nor uncritical adoption, but balanced integration. In a dual-partnership model, human relationships remain the core of moral responsibility and identity, while AI systems function as augmentative tools that extend care, connection, and capability. Whether Relationships 4.0 ultimately enrich or erode social life will depend less on technical sophistication than on collective stewardship. By prioritizing human dignity, transparency, inclusion, and relational well-being, societies can ensure that digital heartbeats amplify, rather than replace, the fundamentally human need to belong, care, and be understood.

## REFERENCES


1. Fortes, M., *Kinship and the Social Order.* Chicago, IL: Aldine, 1969.
2. Lévi-Strauss, C., *The Elementary Structures of Kinship.* Boston, MA: Beacon Press, 1969.
3. Fox, R., *Kinship and Marriage: An Anthropological Perspective.* Cambridge: Cambridge University Press, 1967.
4. Henrich, J., *The WEIRDest People in the World.* New York: Farrar, Straus and Giroux, 2020.
5. Lee, R.B., *The Kung San: Men, Women, and Work in a Foraging Society.* Cambridge: Cambridge University Press, 1979.
6. Woodburn, J., Egalitarian Societies. *Man*, 1982. 17(3): p. 431–451.
7. Metge, J., *New Growth from Old: The Whānau in the Modern World.* Wellington: Victoria University Press, 1995.
8. Witherspoon, G., *Language and Art in the Navajo Universe.* Ann Arbor: University of Michigan Press, 1977.
9. Sahlins, M., *What Kinship Is—and Is Not.* Chicago, IL: University of Chicago Press, 2011.
10. Diamond, J., *Guns, Germs, and Steel: The Fates of Human Societies.* New York: W. W. Norton, 1997.
11. Polanyi, K., *The Great Transformation.* New York: Farrar & Rinehart, 1944.
12. Laslett, P., *Household and Family in Past Time.* Cambridge: Cambridge University Press, 1972.
13. Shorter, E., *The Making of the Modern Family.* New York: Basic Books, 1975.
14. Giddens, A., *The Transformation of Intimacy.* Stanford, CA: Stanford University Press, 1992.

15. Goody, J., *The Development of the Family and Marriage in Europe*. Cambridge: Cambridge University Press, 1983.
16. Anderson, B., *Imagined Communities: Reflections on the Origin and Spread of Nationalism*. London: Verso, 1983.
17. Davidoff, L., and C. Hall, *Family Fortunes: Men and Women of the English Middle Class 1780–1850*. Stanford, CA: Stanford University Press, 1987.
18. Yan, Y., *Private Life under Socialism: Love, Intimacy, and Family Change in a Chinese Village*. Stanford, CA: Stanford University Press, 2003.
19. Datar portal, Digital 2025 Global Overview Report, 2025.
20. Madianou, M., and D. Miller, *Migration and New Media*. London: Routledge, 2012.
21. Keesara, S., A. Jonas, and K. Schulman, Covid-19 and Health Care's Digital Revolution. *New England Journal of Medicine*, 2020. 382: p. e82.
22. Boyd, d., *It's Complicated: The Social Lives of Networked Teens*. New Haven, CT: Yale University Press, 2014.
23. Tufekci, Z., *Twitter and Tear Gas*. New Haven, CT: Yale University Press, 2017.
24. Pariser, E., *The Filter Bubble*. New York: Penguin, 2011.
25. Appadurai, A., *Modernity at Large: Cultural Dimensions of Globalization*. Minneapolis: University of Minnesota Press, 1996.
26. GSMA, *The Mobile Economy*, 2024.
27. Bank. W., *World Development Indicators: Digital Connectivity*, 2023.
28. Madianou, M., Ambient Co-Presence. *Global Networks*, 2016. 16(2): p. 183–201.
29. Turkle, S., *Alone Together*. New York: Basic Books, 2011.
30. Putnam, R.D., *The Upswing*. New York: Simon & Schuster, 2020.
31. Weizenbaum, J., ELIZA—A Computer Program for the Study of Natural Language Communication. *Communications of the ACM*, 1966. 9(1): p. 36–45.
32. Fitzpatrick, K.K., A. Darcy, and M. Vierhile., Delivering Cognitive Behavior Therapy to Young Adults with Symptoms of Depression and Anxiety Using a Fully Automated Conversational Agent. *JMIR Mental Health*, 2017. 4(2): p. 246–254.
33. Ta, V., et al., User Experiences of Social Support From Companion Chatbots. *Journal of Medical Internet Research*, 2020. 22.
34. Skjuve, M., et al., My Chatbot Companion: A Study of Human–Chatbot Relationships. *International Journal of Human–Computer Studies*, 2021. 5(1): p. 149–158.
35. WHO, *Ethics and Governance of Artificial Intelligence for Health*. Geneva: Pinguan, 2021.
36. Wada, K., and T. Shibata, Living With Seal Robots—Its Sociopsychological and Do Physiological Influences on the Elderly. *IEEE Transactions on Robotics*, 2007. 23(5): p. 972–980.
37. Moyle, W., et al., Effect of a Robotic Seal on Engagement and Quality of Life in People With Dementia. *Journal of the American Medical Directors Association*, 2013. 14(9): p. 661–667.
38. Broadbent, E., et al., Robots in Healthcare: A Review. *Journal of Healthcare Engineering*, 2018.
39. Japan's Cabinet Office (C.O.o.), *Society 5.0*. Tokyo: Himalia Press, 2019.
40. Latour, B., *Reassembling the Social*. Oxford: Oxford University Press, 2005.
41. UNESCO, *Recommendation on the Ethics of Artificial Intelligence*. Paris: Xlibiris, 2021.
42. Nowak, K.L., and F. Biocca, The Effect of the Agency and Anthropomorphism on Users' Sense of Telepresence. *Presence*, 2003. 12(5): p. 481–494.
43. Allport, G.W., *The Nature of Prejudice*. Reading, MA: Addison-Wesley, 1954.
44. Pettigrew, T.F., and L.R. Tropp, A Meta-Analytic Test of Intergroup Contact Theory. *Journal of Personality and Social Psychology*, 2006. 90(5): p. 751–783.
45. Gillespie, T., *Custodians of the Internet*. New Haven, CT: Yale University Press, 2018.

46. Parsons, S., and S. Cobb, State-of-the-Art of Virtual Reality Technologies for Children on the Autism Spectrum. *European Journal of Special Needs Education*, 2011. 26(3): p. 355–366.
47. Ahn, S.J., Jeremy Bailenson, and Dooyeon Park, Short- and Long-Term Effects of Embodied Experiences in Immersive Virtual Environments on Environmental Locus of Control and Behavior. *Computers in Human Behavior*, 2014. 39: p. 235–245.
48. Zuboff, S., *The Age of Surveillance Capitalism*. New York: PublicAffairs, 2019.
49. European Commission, *General Data Protection Regulation (GDPR)*. Brussels, 2018.
50. Twenge, J.M., *iGen*. New York: Atria Books, 2017.
51. Danaher, J., Robot Betrayal: A Philosophical Analysis. *Ethics and Information Technology*, 2020. 22: p. 117–130.
52. Horton, D., and R.R.Wohl, Mass Communication and Para-Social Interaction. *Psychiatry*, 1956. 19(3): p. 215–229.
53. Sharkey, A., and N. Sharkey, Granny and the Robots. *Ethics and Information Technology*, 2012. 14: p. 27–40.
54. Robertson, J., *Robo Sapiens Japanicus*. Berkeley: University of California Press, 2018.
55. Floridi, L., et al., AI4People—An Ethical Framework for a Good AI Society. *Minds and Machines*, 2018. 28: p. 689–707.
56. European Commission, *Proposal for a Regulation Laying Down Harmonised Rules on Artificial Intelligence (AI Act)*. Brussels: European Commission, 2021.
57. YouGov, *Public Attitudes Toward AI and Relationships*, 2023.
58. Floridi, L., *The Fourth Revolution*. Oxford: Oxford University Press, 2014.
59. Brynjolfsson, E., and A. McAfee, *Machine, Platform, Crowd*. New York: W. W. Norton, 2017.
60. Holmes, W., M. Bialik, and C. Fadel. *Artificial Intelligence in Education*. Boston: Center for Curriculum Redesign, 2019.
61. Newport, C., *Digital Minimalism*. New York: Portfolio, 2019.
62. Holt-Lunstad, J., Why Social Relationships Are Important for Physical Health. *American Psychologist*, 2018. 73(2): p. 187–197.

# 12 Assistants 4.0

**Abstract**

Everyday LIFE 4.0 marks a profound transformation of routine tasks through artificial intelligence (AI), robotics, and intelligent systems. This chapter traces the evolution from manual pre-industrial living to an era of intelligent assistance. It examines how smart appliances, autonomous mobility, digital assistants, AI-driven commerce, and personalized health and finance services reshape daily life globally. Key trends include the rapid adoption of voice assistants (billions of devices in use) and AI recommendations (with over 70 percent of consumers expecting personalization). We analyze the opportunities for productivity, safety, and well-being, as well as challenges such as privacy, inequality, and ethical design. By integrating a global perspective, from Singapore's ~60 percent AI adoption to low-income regions with <10 percent AI adoption, the chapter outlines a LIFE 4.0 scenario in which predictive, context-aware systems become seamless parts of family routines and work. Throughout, empirical evidence is cited to ground assertions, and figure suggestions (e.g., smart home ecosystem diagrams and charts on autonomous mobility adoption) are provided to illustrate concepts.

## 12.1 INTRODUCTION

The current era of LIFE 4.0, defined by the widespread diffusion of artificial intelligence (AI), robotics, and intelligent automation, marks a decisive shift in how everyday activities are performed, coordinated, and experienced. Unlike earlier historical stages characterized by manual labor (LIFE 1.0), mechanization (LIFE 2.0), or digitization (LIFE 3.0), LIFE 4.0 is distinguished by intelligent assistance embedded directly into daily routines. AI systems now mediate tasks ranging from household management and transportation to healthcare, commerce, and personal productivity, often operating continuously in the background of daily life.

Global adoption of AI-enabled tools has accelerated rapidly. Recent estimates suggest that 1.7–1.8 billion people worldwide will interact with AI-powered systems by 2025, with between 500 and 600 million daily active users engaging with virtual assistants, recommendation systems, or automation services on a routine basis [1, 2]. In the US alone, approximately 153.5 million people, around 46 percent of the population, use voice assistants, representing an 8 percent increase since 2022. These figures underscore not only the scale of adoption but also the normalization of AI as a practical utility rather than a novelty.

Crucially, however, adoption of LIFE 4.0 is uneven. Advanced economies and urban centers tend to outpace rural and low-income regions, reflecting disparities in infrastructure, digital literacy, and institutional capacity. In the Global North, AI is increasingly integrated into smart homes, autonomous mobility systems, and digital health platforms. By contrast, in many parts of the Global South, AI adoption remains concentrated in mobile-based services such as chatbots, mobile payments,

 DOI: 10.1201/9781003455219-12

and language translation tools. This uneven diffusion highlights that LIFE 4.0 is not a single universal condition, but a patchwork of lived realities shaped by geography, wealth, and policy [3].

Understanding this transformation, therefore, requires both historical context and a broad analytical lens. Technological systems have always shaped everyday life, but LIFE 4.0 differs in that AI systems do not merely extend human effort; they observe, predict, and increasingly decide. Tasks such as home care, commuting, scheduling, shopping, health monitoring, and family coordination are now mediated by algorithms that learn from user behavior and adapt over time. This introduces new efficiencies, but also raises questions about dependency, autonomy, and control that echo yet go beyond earlier technological revolutions.

### 12.1.1 The Importance of a Global and Demographic Perspective

A global perspective is essential for understanding how LIFE 4.0 unfolds in practice. For example, Singapore reports AI tool usage approaching 59 percent, driven by national digital strategies, high connectivity, and public trust in technology. In contrast, large portions of sub-Saharan Africa and rural South Asia remain below 10 percent AI usage, primarily due to connectivity gaps, cost barriers, and limited institutional support. Cultural factors also play a significant role: societies differ in their tolerance for automation, data sharing, and machine decision-making.

Demographic variation further complicates the picture. Surveys indicate that older adults express ambivalence toward AI: while approximately 80 percent of Americans over 60 believe technology makes life more convenient, only around 37 percent expect AI to improve their healthcare meaningfully [4]. Younger generations, by contrast, are more likely to view AI as a natural extension of daily life, particularly in areas such as navigation, communication, entertainment, and learning. These generational differences suggest that LIFE 4.0 will be experienced unevenly across regions, age groups, and life stages.

### 12.1.2 Everyday Life as the Core Arena of LIFE 4.0

This chapter, therefore, focuses on everyday life as the primary arena in which LIFE 4.0 becomes tangible. While policy debates and technical advances often dominate discussions of AI, it is in mundane activities, such as ordering food, commuting to work, caring for family members, managing health, and organizing time, that AI exerts its most immediate and lasting influence. These everyday interactions accumulate, shaping habits, expectations, and social norms in subtle but powerful ways.

Building on the relational analysis developed in Chapter 11, the chapter examines how intelligent assistance reshapes tasks rather than relationships, while still carrying significant social consequences. We trace how each "version" of everyday life, manual, mechanical, digital, and now intelligent, reconfigures human agency, labor, and responsibility. In doing so, we draw on survey data, regional case studies, and industry analyses to map how LIFE 4.0 is emerging differently across households, workplaces, and cultures.

Ultimately, understanding LIFE 4.0 at the level of everyday tasks allows us to move beyond abstract discussions of AI toward a human-centered assessment of its impact. By situating intelligent assistance within daily practices of care, movement, productivity, and family life, the chapter sets the foundation for evaluating not only what AI can do but also how it reshapes lived experience and what forms of guidance, policy, and ethical stewardship are required as intelligent systems become routine companions in ordinary life (Table 12.1).

**TABLE 12.1**
**Comparative Summary**

| Assistant Era | Daily Life Mode | Key Technology | Core Orientation |
|---|---|---|---|
| 1.0 | Manual and communal | Simple tools | Survival |
| 2.0 | Mechanized and schedules | Machines | Efficiency |
| 3.0 | Digital and on-demand | Internet and mobile | Convenience |
| 4.0 | Intelligent and adaptive | AI systems | Optimization |

## 12.2 EVERYDAY LIFE 1.0–4.0

### 12.2.1 Everyday LIFE 1.0: Manual Routines and Human Dependency

Everyday LIFE 1.0 refers to pre-industrial and early industrial societies in which nearly all daily activities depended directly on human or animal muscle power. Before the widespread adoption of electricity, mechanized tools, and modern infrastructure, households were labor-intensive units in which survival and comfort required continuous physical effort. Daily routines were shaped by natural cycles of daylight, seasons, and bodily endurance rather than by machines or automated systems.

Household labor dominated everyday life. Cooking was performed over wood or coal fires, requiring time-consuming fuel collection and constant supervision. Clothes were washed by hand, often in rivers or communal wash areas, and dried outdoors. Lighting relied on candles or oil lamps, limiting productive hours after sunset. Farming and food production were similarly manual: fields were tilled with hand tools or animal-drawn plows, crops harvested by hand, and food preserved through salting, drying, or smoking rather than refrigeration [5, 6].

Without plumbing or electricity, routine domestic tasks consumed most of the day. Drinking water had to be fetched from wells or streams; waste disposal was manual and often unsanitary; and heating required constant tending of fires. Historical time-use studies show that in 19th-century Europe and North America, households spent most daylight hours on basic maintenance activities, cooking, cleaning, hauling water, repairing tools, and tending animals, leaving little discretionary time for leisure or education [7]. Commuting, when it occurred, was slow and physically demanding, typically by foot, horse, or cart, reinforcing highly localized patterns of work and social life.

#### 12.2.1.1 Labor, Gender Roles, and Household Organization

Everyday LIFE 1.0 was characterized by high human dependency and rigid labor division, particularly along gender and age lines. In most agrarian and early industrial societies, women bore primary responsibility for domestic labor, cooking, cleaning, childcare, textile production, and food preservation. At the same time, men were more likely to engage in agricultural labor, craft production, or wage work when available. Economic historians estimate that, as late as 1900, only about 5 percent of married women in industrializing societies worked outside the home, with the remainder of their time spent on unpaid household labor [8].

Children and the elderly were also integral contributors to household productivity. Children assisted with farming, animal care, fuel collection, and domestic chores from an early age, while older family members performed lighter tasks or childcare. The household thus functioned as a multi-generational labor unit, pooling physical effort to meet basic needs. Formal schooling, where it existed, was often secondary to labor demands, particularly in rural areas [9].

#### 12.2.1.2 Constraints on Productivity and Mobility

The productivity of Everyday LIFE 1.0 was fundamentally constrained by human stamina and physical limits. Tasks that now take minutes, such as preparing a meal, cleaning floors, and transporting goods, often require hours of sustained labor. Transportation infrastructure was rudimentary, making travel slow, costly, and exhausting. As a result, most economic and social activity remained highly local, and cities were dense out of necessity rather than choice, minimizing travel distances between home, work, and markets [10].

Education, healthcare, and leisure were similarly limited. Access to formal education was uneven and often restricted to elites. Medical care relied on home remedies or local practitioners, with minor specialization or institutional support. Leisure, when it existed, was typically embedded in communal or religious life rather than individualized recreation. Time scarcity was a defining feature of Everyday LIFE 1.0, shaping values of thrift, endurance, and duty [11].

In summary, human dependency was total in Everyday LIFE 1.0. Survival, productivity, and comfort depended on uninterrupted physical labor and close coordination within households and communities. There was minimal buffering against illness, fatigue, or environmental shocks; a failed harvest, injury, or loss of labor capacity could threaten household survival. Technology existed, but it primarily extended muscle power rather than replacing it.

This mode of everyday life established the baseline from which later transformations emerged. The burdens of manual routines created strong incentives for mechanization, infrastructure development, and labor-saving innovations. As subsequent phases of Everyday Life unfolded through mechanization, electrification, digitization, and ultimately intelligent automation, each step progressively reduced direct human dependency, freeing time, redistributing labor, and reshaping social roles. Understanding Everyday LIFE 1.0 is therefore essential for appreciating the magnitude of change introduced by LIFE 4.0, where intelligent systems increasingly replace not muscle power, but cognitive and organizational effort.

### 12.2.2 Everyday LIFE 2.0: Mechanical Aids and Early Automation

Everyday LIFE 2.0 emerged during the Second Industrial Revolution (roughly the late 19th to early 20th centuries), a period marked by the widespread diffusion of electricity, internal combustion engines, and mass-produced mechanical devices. Unlike Everyday LIFE 1.0, which relied almost entirely on human and animal muscle, LIFE 2.0 introduced mechanical assistance as a routine feature of daily life. These technologies did not eliminate human labor, but they amplified physical effort, standardized tasks, and dramatically reduced time intensity across domestic, industrial, and transportation domains.

At the core of this transition was electrification. The electric motor enabled machines to operate reliably within homes and small businesses, decoupling productive activity from centralized steam power. Indoor plumbing further transformed domestic routines by automating water delivery and waste removal tasks that had previously required hours of manual hauling. Household appliances became emblematic of this era's promise. The electric washing machine, commercialized around 1910, replaced hand-cranked and washboard-based laundering, eliminating heavy lifting and repetitive scrubbing. Contemporary advertisements proclaimed that such devices would "dispense the gloom of washday," framing mechanization as both a physical and emotional liberation (Cowan 1983).

Other appliances followed rapidly. Electric irons, widely adopted by the 1920s, replaced heavy charcoal irons and reduced both preparation time and risk of injury. Vacuum cleaners, introduced in the early 1900s, mechanized floor cleaning and reshaped domestic standards of cleanliness. By the mid-20th century, dishwashers, first marketed in the 1910s and diffusing more widely after World War II, transformed dishwashing from an exhausting manual chore into a largely automated process, often described in popular media as making dishwashing "a pleasant task." Even outdoor and utility labor was transformed as water pumps, chainsaws, and mechanized tools replaced hand-powered alternatives [6].

#### 12.2.2.1 Time Liberation and the Reorganization of Daily Life

The cumulative effect of these innovations was a dramatic restructuring of daily time use. Economic and historical time-use studies show that average weekly housework per household in industrialized countries fell from approximately 58 hours per week in 1900 to around 18 hours by the mid-1970s [12]. This reduction did not occur overnight, nor was it evenly distributed, but it represented one of the most significant reallocations of human time in modern history.

Freed time had profound social consequences. One of the most notable shifts was the increased participation of women in paid labor outside the home. While only about 5 percent of married women worked outside the house in 1900, participation rose steadily throughout the 20th century, reaching over 60 percent by 2000 in many industrialized societies [13]. Mechanical aids thus indirectly reshaped gender roles, family dynamics, and economic structures, even as domestic labor itself remained disproportionately feminized.

Refrigeration provides a clear illustration of how technology reshaped everyday rhythms. The widespread adoption of electric refrigerators by the 1930s allowed households to store food safely for extended periods, reducing the need for daily shopping trips and seasonal dependence. Gas and electric cookstoves replaced wood-fired hearths, eliminating the need for fuel collection and enabling faster, more controllable meal preparation. Sewing machines, introduced earlier but refined during this era, accelerated clothing production and repair, blending household labor with early forms of mechanized productivity [14].

#### 12.2.2.2 Mobility and Early Automation beyond the Home

Everyday LIFE 2.0 was not confined to domestic space. Transportation technologies, particularly the automobile, redefined mobility and spatial organization. The Model T, introduced by Ford in 1908, made car ownership accessible to middle-class households and revolutionized commuting, leisure travel, and supply chains. Although early automobiles required significant manual operation, hand-cranking engines, manual gear shifting, and constant driver attention, they nonetheless represented an important step toward automating movement itself [15]. Similarly, agricultural life underwent partial mechanization. Tractors, mechanical harvesters, and engine-powered pumps spread across farms worldwide, reducing reliance on animal labor and increasing productivity. Although adoption lagged in many rural and developing regions, the global diffusion of mechanical engines marked a decisive shift toward labor-saving technologies as a normative expectation, even outside industrial centers [16].

#### 12.2.2.3 Limits and Inequalities of LIFE 2.0

Despite its transformative effects, Everyday LIFE 2.0 remained mechanical rather than intelligent. Machines required direct human operation, monitoring, and maintenance. Automation was limited to physical processes; cognitive tasks such as planning, scheduling, decision-making, and coordination remained entirely human responsibilities. Moreover, access to these technologies was uneven. By the mid-20th century, industrialized nations rapidly adopted household appliances and mechanized transport, while many rural or developing regions relied on simpler tools for decades longer. Electrification itself progressed unevenly, reinforcing global and regional inequalities [17].

Nonetheless, the conceptual shift introduced by Everyday LIFE 2.0 was profound. Daily life was no longer defined solely by human stamina, but by machine-assisted efficiency. The expectation that technology should reduce drudgery and free time became culturally entrenched, setting the stage for later phases of digitization and intelligent automation.

In summary, Everyday LIFE 2.0 represents the transition from total human dependency to machine-assisted routine life. Mechanical aids and early automation transformed domestic labor, mobility, and productivity, significantly reducing physical burdens and reconfiguring social roles. While these technologies did not think, learn, or adapt, they established the principle that everyday tasks could and should be offloaded to machines.

This principle would later be extended in Everyday LIFE 3.0, where digital technologies automated information processing, and ultimately in LIFE 4.0, where intelligent systems increasingly automate cognitive and organizational labor. Understanding

LIFE 2.0 is therefore essential for appreciating how contemporary expectations of convenience, efficiency, and assistance were historically constructed through mechanical innovation.

### 12.2.3 Everyday LIFE 3.0: Digital Convenience and Internet Integration

Everyday LIFE 3.0 spans the late 20th and early 21st centuries and is defined by the pervasive integration of digital technologies, networked computing, and internet connectivity into daily routines. While Everyday LIFE 2.0 mechanized physical labor, LIFE 3.0 primarily automated information processing, communication, and coordination. The emergence of the personal computer in the 1980s, followed by the public diffusion of the internet in the 1990s, fundamentally altered how individuals accessed information, managed tasks, and interacted with institutions.

Early digital tools transformed office work, education, and finance. Email replaced postal correspondence, word processors displaced typewriters, and spreadsheets automated accounting tasks. Online banking and e-commerce emerged in the late 1990s, allowing consumers to manage finances and purchases remotely. These innovations gradually shifted everyday expectations: information became instant, searchable, and portable, reducing reliance on physical records, in-person transactions, and fixed locations [18].

The mobile phone accelerated this transformation. First introduced widely in the 1990s and later converging into smartphones in the 2000s, mobile devices placed computing power directly in users' pockets. By the 2010s, smartphone penetration in high-income countries approached saturation, and internet access increasingly became mobile-first. This enabled constant connectivity, real-time messaging, GPS navigation, entertainment streaming, and social networking, blurring boundaries between work, leisure, and domestic life [19].

#### 12.2.3.1 Digitization of Domestic Life

Domestic routines were significantly reshaped during Everyday LIFE 3.0 as appliances and household systems became digitally connected. Early "smart home" technologies allowed users to remotely monitor energy consumption, adjust thermostats, or control lighting through mobile applications. Although initial systems were fragmented and often incompatible across brands, they introduced a new expectation: household environments could be monitored and controlled digitally.

A pivotal shift occurred with the introduction of voice-based digital assistants, such as Apple's Siri (2011) and Amazon's Alexa (2014). These systems transformed human–machine interaction from keyboard-based input to natural language commands, making digital assistance accessible to broader populations, including children and older adults. Tasks such as checking the weather, setting timers, playing music, or managing calendars became conversational rather than procedural. By the mid-2020s, an estimated 8.4 billion voice assistants embedded in smartphones, smart speakers, vehicles, and appliances were in use globally, exceeding the world's human population [20].

This shift normalized the idea that digital systems should be responsive, always available, and user-friendly, reinforcing reliance on software-mediated convenience.

Everyday language began to reflect this integration, with phrases like "just ask Alexa" or "Google it" becoming routine problem-solving expressions.

#### 12.2.3.2 Digital Commerce, Services, and Platform Integration

Commerce and service provision were radically transformed during LIFE 3.0 through the rise of platform-based digital ecosystems. Major retailers such as Amazon and Alibaba leveraged data analytics, recommendation algorithms, and global logistics networks to deliver personalized shopping experiences at scale. Studies indicate that by the early 2020s, over 70 percent of consumers expected customized recommendations, a demand increasingly met through algorithmic curation and targeted advertising [21].

Service industries underwent similar transformations. Ride-hailing platforms such as Uber and Didi integrated GPS, mobile payments, and algorithmic matching to automate transportation services. Travel and accommodation platforms like Airbnb digitized booking, pricing, and reputation systems, reshaping tourism and urban housing markets. These platforms reduced transaction costs and increased convenience but also introduced new forms of platform dependency and labor precarity [22].

Social media platforms Facebook, Twitter/X, and Instagram, and video streaming services have become dominant forms of leisure, social interaction, and information consumption. For many individuals, daily routines increasingly included scrolling feeds, consuming algorithmically curated content, and maintaining digital social identities alongside offline ones. This marked a shift toward attention-based economies, in which user engagement itself became a central commodity [23].

#### 12.2.3.3 Challenges, Inequalities, and Systemic Limits

Despite its conveniences, Everyday LIFE 3.0 introduced significant challenges. The expansion of digital services relied on extensive data collection, raising concerns about privacy, surveillance, and data ownership. Early smart home systems were criticized for security vulnerabilities, interoperability issues, and opaque data practices. Users often lacked a clear understanding of how personal data was stored, shared, or monetized [24]. Global adoption of LIFE 3.0 technologies was also highly uneven. While high-income and urban regions achieved near-universal internet and smartphone access, many rural and low-income areas, particularly in parts of sub-Saharan Africa and South Asia, remained constrained by limited connectivity, affordability barriers, and infrastructure deficits. This digital divide shaped who could benefit from online services, remote work, digital education, and e-commerce, reinforcing existing inequalities [3].

Technologically, intelligence in LIFE 3.0 systems remained relatively narrow. Most automation relied on predefined scripts, rule-based software, and early forms of machine learning. Digital tools executed tasks efficiently but lacked contextual understanding, emotional sensitivity, or adaptive reasoning. Users still bore responsibility for planning, decision-making, and coordination; digital systems primarily accelerated execution rather than cognition.

In summary, Everyday LIFE 3.0 brought the internet into the fabric of daily routines, integrating digital connectivity across home, work, commerce, and leisure.

Information access became instantaneous, services became platform-mediated, and convenience became a dominant expectation. However, while digital tools reduced friction and expanded reach, they did not fundamentally replace human judgment or relational labor.

Everyday LIFE 3.0 thus represents a transitional phase: a world of pervasive connectivity without fully autonomous intelligence. It established the infrastructural, cultural, and behavioral foundations upon which LIFE 4.0 would later build. As users became accustomed to constant assistance, personalization, and responsiveness, expectations shifted toward systems that could not only execute commands but also anticipate needs, learn preferences, and make decisions, a shift realized more fully with the emergence of advanced AI in LIFE 4.0.

### 12.2.4 Everyday LIFE 4.0: Intelligent Assistance, Smart Living, and Predictive Routines

Everyday LIFE 4.0 represents a qualitative shift from digital convenience to AI-empowered automation and prediction. In this phase, systems no longer merely respond to user commands; they learn from behavior, infer preferences, anticipate needs, and act proactively. Every day, environments, homes, vehicles, marketplaces, and personal devices become adaptive systems that continuously optimize routines. This marks a transition from task execution (LIFE 3.0) to the orchestration of daily life, in which algorithms coordinate activities across time, space, and domains.

#### 12.2.4.1 Home Automation and Smart Living Ecosystems

At the center of Everyday LIFE 4.0 is the smart home ecosystem, enabled by the Internet of Things (IoT), cloud computing, and machine learning. Contemporary smart homes integrate sensors (motion, temperature, humidity, cameras), actuators (bright lights, locks, thermostats), intelligent appliances (AI refrigerators, robot vacuums), and a central hub or cloud service that coordinates decision-making. Residents can monitor and control environments remotely via smartphones, but increasingly the system operates autonomously.

Devices such as Nest and Ecobee thermostats learn occupancy patterns and automatically adjust heating and cooling, reducing energy consumption while maintaining comfort. AI-enabled refrigerators track contents and expiration dates, recommend recipes, and generate shopping lists. Security systems use computer vision to distinguish household members from strangers, reducing false alarms while increasing safety. By 2024, the global smart home market reached approximately $122 billion and is projected to exceed $630 billion by 2032, reflecting compound annual growth rates above 20 percent [25].

Beyond convenience, smart homes deliver efficiency and sustainability. AI-driven Home Energy Management Systems (HEMS) forecast energy demand and renewable generation, scheduling high-load activities such as electric vehicle charging or laundry during off-peak or high-solar periods. Surveys indicate that over 40 percent of homeowners view innovative technologies as a primary tool for reducing energy use, linking intelligent automation to climate and cost objectives [26].

Entertainment systems are similarly personalized. Streaming platforms and smart speakers tailor content recommendations using behavioral data, making leisure time algorithmically curated. This personalization reflects a broader LIFE 4.0 pattern: environments adapt continuously to individual preferences with minimal conscious input.

#### 12.2.4.2 Commuting, Mobility, and Autonomous Transport

Everyday LIFE 4.0 also reshapes mobility. Advances in autonomous driving, advanced driver-assistance systems (ADAS), and AI-optimized transport networks are redefining how people move through cities. While fully autonomous vehicle (AV) fleets are not yet universal, early deployments such as Waymo's driverless taxis and Tesla's Autopilot and Full Self-Driving systems are already reducing the cognitive and physical burden of driving.

Economic forecasts suggest that by 2035, autonomous mobility could generate $300–400 billion in annual economic value, transforming logistics, ride-hailing, and freight transport [2]. More importantly for daily life, AVs change how commute time is used: hours once devoted to driving can be reallocated to work, communication, entertainment, or rest. For individuals with long commutes, disabilities, or advanced age, autonomous transport promises greater independence and reduced stress.

Cities are also adopting AI for traffic and transit optimization. Real-time sensors and predictive algorithms dynamically adjust traffic signals to reduce congestion. Public transit agencies use AI forecasting to align bus and train schedules with demand patterns, improving reliability and reducing wait times. Logistics firms apply route-optimization algorithms to cut delivery times and fuel use. Studies in Europe suggest that widespread adoption of ADAS could reduce traffic accidents by up to 15 percent by 2030, highlighting safety as a key LIFE 4.0 benefit (European Transport Safety Council 2022).

#### 12.2.4.3 Personal Productivity and Smart Assistants

In Everyday LIFE 4.0, digital assistants evolve from passive tools into proactive coordinators. Voice- and text-based assistants embedded in phones, speakers, and computers manage calendars, reminders, information retrieval, and communication. Systems such as Google Assistant, Siri, and Alexa respond conversationally, while newer generative-AI assistants add planning, summarization, and creative support.

Surveys show that around 61 percent of users value AI assistants primarily for speed and convenience, using them for quick answers, reminders, and task management. According to a global Kantar study, 81 percent of AI users engaged with voice or chat assistants, with younger cohorts adopting them more frequently: approximately 83 percent of Gen Z and 81 percent of Millennials report regular use, compared with just over half of Boomers. Regional variation is significant, with daily use reported by 67 percent of respondents in India and 50 percent in China, compared with under 20 percent in France and Germany [27].

Beyond efficiency, users increasingly rely on assistants for creative and cognitive augmentation, brainstorming ideas, drafting messages, planning travel, or learning new skills. This reflects a shift from automating execution to augmenting thinking, a defining feature of LIFE 4.0.

#### 12.2.4.4 AI-Driven Shopping and Digital Marketplaces

Shopping in Everyday LIFE 4.0 becomes conversational, predictive, and deeply personalized. Retailers deploy AI recommendation engines that analyze browsing behavior, purchase history, and contextual signals to suggest products in real time. Industry reports indicate that over 70 percent of consumers expect personalized shopping experiences, making algorithmic curation a baseline expectation rather than a premium feature [21].

Chatbots and virtual shopping assistants guide customers through product selection, sizing, and price comparison. Generative-AI tools, such as conversational agents embedded in search and commerce platforms, are beginning to influence purchasing decisions directly. Surveys in 2025 found that over half of U.S. consumers planned to use AI chatbots to compare prices or find deals, while a substantial minority intended to generate shopping lists or product ideas with AI assistance.

Seasonal data illustrate the scale of this shift. During the 2025 holiday season, analysts reported a year-over-year increase of over 700 percent in AI-driven referral traffic, with grocery- and recipe-based AI assistants driving dramatic spikes in brand exposure. Shopping thus transitions from keyword search to AI-mediated discovery, where algorithms anticipate needs before consumers articulate them.

#### 12.2.4.5 Personalized Lifestyle, Health, and Financial Services

LIFE 4.0 extends beyond convenience into continuous lifestyle management. Wearables and health applications use AI to monitor vital signs, detect anomalies, and suggest behavior changes. Smartwatches track heart rate, sleep patterns, and physical activity, while algorithms flag irregularities such as arrhythmias before clinical diagnosis. Surveys indicate that over one-third of older adults believe AI will meaningfully support health management, reflecting growing trust in algorithmic health assistance [28].

In nutrition, AI meal planners tailor recommendations to dietary goals, medical conditions, and preferences. Telemedicine platforms use symptom-checking chatbots to triage patients, increasing access while reducing system strain. In personal finance, AI budgeting tools analyze spending patterns and suggest savings strategies, while robo-advisors manage investment portfolios using algorithmic risk models. Though usage statistics vary, industry forecasts consistently project rapid growth in AI-driven personal finance services [29].

#### 12.2.4.6 Family and Home Care

Perhaps the most socially significant transformation occurs in family care and domestic support. AI and robotics increasingly supplement caregiving roles, particularly for aging populations. Companion robots such as ElliQ converse with seniors, provide reminders, and monitor safety. Robotic vacuum cleaners and kitchen assistants reduce physical strain, enabling older adults to live independently longer. Smart baby monitors analyze breathing patterns and movement, alerting parents to potential issues, while AI-enabled cameras can notify caregivers if an elderly family member leaves the home unexpectedly.

These systems extend the reach of caregiving without entirely replacing human care, reflecting the dual-support model characteristic of LIFE 4.0 households.

Taken together, these developments constitute an "Internet of Daily Things," a living environment in which learning algorithms continuously optimize routines. In a mature LIFE 4.0 household, residents may wake to blinds opening automatically, coffee brewing to a preferred strength, calendars populating with suggested schedules, and homes adjusting energy use without conscious input.

The defining feature of Everyday LIFE 4.0 is not automation alone, but anticipation. Systems act before users ask, coordinating tasks across domains and time horizons. While this promises unprecedented convenience, efficiency, and personalization, it also raises critical questions about autonomy, transparency, and dependence, which are addressed in subsequent sections of the chapter.

## 12.3 OPPORTUNITIES AND TRANSFORMATIONAL POTENTIAL

AI and intelligent systems hold transformative potential for everyday life, extending far beyond incremental convenience. Much as electrification and mechanization reshaped domestic labor in the 20th century, LIFE 4.0 technologies promise structural improvements in time use, safety, personalization, accessibility, sustainability, and economic productivity. When integrated thoughtfully, these systems can reconfigure how individuals allocate attention, energy, and resources across daily routines.

### 12.3.1 Time Savings and Productivity

One of the most immediate and measurable benefits of Everyday LIFE 4.0 is time liberation. By automating routine chores, errands, and administrative tasks, AI reduces the cognitive and physical burden of everyday management. Historical precedents illustrate the scale of such effects: electrified washing machines, irons, and refrigerators dramatically reduced household labor in the 20th century, enabling greater workforce participation, especially among women [12, 30].

LIFE 4.0 accelerates this trajectory. AVs can convert commute time into productive or restorative time, while AI assistants handle scheduling, bill payment, travel planning, and routine decision-making. McKinsey estimates that widespread AI adoption could increase global productivity by 30–40 percent by 2035, primarily driven by time savings and task automation [2]. At the individual level, this translates into shorter workweeks, more flexible schedules, and expanded discretionary time if productivity gains are equitably distributed.

### 12.3.2 Improved Safety and Well-Being

AI systems significantly enhance safety in domestic and public environments. Intelligent security systems using facial recognition and anomaly detection can deter intrusions and automatically alert emergency services. In transportation, ADAS and autonomous driving technologies reduce human error, which accounts for most traffic accidents. European transport studies estimate that current ADAS trends alone could reduce road crashes by 15 percent by 2030 [31].

In health and well-being, AI-powered monitoring systems detect early signs of illness, falls, or cognitive decline. Wearables and home sensors can identify irregular heart rhythms, sleep disturbances, or changes in mobility before symptoms escalate into emergencies. For seniors and people with disabilities, assistive robotics, such as mobility exoskeletons and fall-detection systems, offer greater autonomy and safety, potentially reducing healthcare costs while improving quality of life [32].

### 12.3.3 Personalization and Convenience

LIFE 4.0 enables a shift from standardized services to deep personalization. AI systems tailor recommendations, environments, and services to individual preferences, habits, and goals. From nutrition planning and wardrobe styling to financial advice and learning pathways, AI functions as a personal assistant at scale.

Consumer expectations reflect this shift. Surveys consistently show that around 70 percent of consumers demand personalized experiences, a demand increasingly met through AI-driven analytics in retail, banking, and healthcare [21]. Personalized systems not only improve user satisfaction but also increase efficiency by reducing information overload and decision fatigue. In finance, for example, AI advisors dynamically rebalance portfolios in response to life events, while in healthcare, personalized risk assessments enable preventive rather than reactive care.

### 12.3.4 Accessibility and Inclusion

A particularly significant opportunity lies in accessibility and social inclusion. AI lowers barriers for individuals with physical, sensory, or cognitive limitations. Voice-controlled interfaces and smart home technologies enable hands-free living for people with mobility impairments. Real-time language translation delivered via smartphones or augmented reality (AR) glasses facilitates communication across linguistic boundaries, supporting migrants, travelers, and multicultural communities.

Assistive robots and adaptive interfaces can perform tasks for individuals with limited mobility or cognitive capacity, thereby extending independence for older people and people with disabilities. When deployed in low-resource contexts, AI tools can also compensate for shortages of skilled labor, such as healthcare workers or educators. In this sense, LIFE 4.0 technologies have the potential to democratize access to services, provided infrastructure and affordability gaps are addressed [33].

### 12.3.5 Sustainability and Resource Efficiency

AI-driven systems play an increasingly central role in sustainability and environmental efficiency. Smart homes and buildings optimize energy use through predictive control of heating, cooling, lighting, and water consumption. HEMS coordinate appliances, solar panels, and battery storage to minimize peak demand and reduce carbon emissions. Surveys indicate that nearly half of homeowners rank energy efficiency as a top concern, making AI-enabled optimization both an environmental and consumer priority [26].

At the urban scale, AI-driven traffic management reduces congestion, fuel consumption, and air pollution by dynamically adjusting signal timing and routing. Smart grids balance supply and demand more efficiently, integrating renewable energy sources at scale. Together, these systems contribute to lower emissions, reduced waste, and more resilient infrastructure, aligning LIFE 4.0 with global climate and sustainability goals [34].

### 12.3.6 Economic Growth and New Forms of Work

Although AI automates many routine tasks, it also generates new economic opportunities and job categories. Roles in data science, AI system design, robotics maintenance, cybersecurity, and AI ethics are expanding rapidly. McKinsey estimates that AV development alone could generate hundreds of billions of dollars in economic value, with spillover effects across logistics, manufacturing, and services [2].

At a national level, early investment in digital infrastructure and AI ecosystems correlates with higher adoption and productivity. Singapore, for example, achieved approximately 59 percent AI usage through proactive investment in connectivity, education, and public–private partnerships, while similarly developed nations have slower investment lags. This suggests that LIFE 4.0 opportunities are not automatic; they depend on strategic governance and long-term planning.

In summary, Everyday LIFE 4.0 offers transformative opportunities across convenience, safety, productivity, inclusion, sustainability, and economic growth. By reducing routine burdens and optimizing resource use, AI can improve both individual well-being and systemic efficiency. Evidence from transport, energy, and household automation already demonstrates measurable gains in safety and quality of life [35].

However, realizing these benefits requires thoughtful implementation. Without inclusive design, ethical safeguards, and equitable access, productivity gains may concentrate among already advantaged populations. When guided responsibly, LIFE 4.0 technologies can make households healthier, cities cleaner, work more flexible, and daily life less burdensome, reinforcing the central promise of intelligent assistance: not to replace human agency, but to amplify it.

## 12.4 CHALLENGES, RISKS, AND ETHICAL CONSIDERATIONS

Despite its promise of convenience, efficiency, and improved quality of life, LIFE 4.0 introduces significant risks and ethical dilemmas. As intelligent systems become embedded in homes, transport, healthcare, and everyday decision-making, failures or misuses can have direct and intimate consequences. Unlike earlier technological waves, LIFE 4.0 systems often operate continuously, invisibly, and autonomously, raising concerns about trust, power, equity, and human agency.

### 12.4.1 Privacy and Data Security

One of the most immediate risks of Everyday LIFE 4.0 lies in data collection and surveillance. Smart homes, wearables, voice assistants, and AI-driven services collect vast volumes of personal data, including behavioral patterns, biometric signals, location histories, and intimate household activities. Consumer surveys reflect growing unease: over 80 percent of users rank data security as a top concern in smart home adoption [26].

Always-on microphones, cameras, and sensors enable continuous monitoring. If compromised, smart homes can effectively become "spy-in-the-home" environments, exposing residents to stalking, burglary, or psychological manipulation. The IoT dramatically expands the attack surface, with vulnerabilities ranging from insecure door locks to hacked baby monitors [23]. While regulations such as the GDPR emphasize consent and data minimization, enforcement is uneven, and fragmented vendor ecosystems complicate accountability [36].

Ensuring robust encryption, user control, and transparency is therefore essential but technically and institutionally challenging in a globally distributed, multi-vendor environment.

### 12.4.2 Security, Reliability, and System Failure

LIFE 4.0 systems also pose physical safety risks when reliability fails. AVs, robotic assistants, and AI-controlled infrastructure depend on complex software stacks, cloud services, and machine-learning models that can malfunction or behave unpredictably. A software bug in an AV, or a cyberattack on a home energy system, could result in injury, property damage, or loss of essential services.

Because many LIFE 4.0 systems rely on cloud connectivity, outages or network failures can disable heating, mobility, medical monitoring, or communication. Moreover, AI systems remain vulnerable to adversarial inputs and subtle manipulations that cause misclassification or erroneous behavior [37]. Experts increasingly argue that widespread trust in LIFE 4.0 will require certifiable AI safety standards, redundancy, and rigorous testing analogous to those used in aviation or medical devices [38].

### 12.4.3 Ethical and Social Impact

Algorithmic systems inevitably reflect the biases embedded in their training data. Numerous studies show that facial recognition systems perform less accurately on women and people of color, raising serious concerns about discrimination and misidentification in security and policing applications [39]. In everyday contexts, biased recommendation systems may reinforce stereotypes, limit exposure to diverse viewpoints, or manipulate user choices.

There is also a risk of behavioral manipulation. Virtual assistants and recommendation engines are often designed to optimize engagement or commercial outcomes, not necessarily user well-being. The Imagining the Digital Future project warns that persuasive AI could erode autonomy by nudging users toward sponsor-friendly decisions, reviving the adage, if you are not paying for it, you are the product [40]. Balancing commercial incentives with ethical responsibility remains a central challenge of LIFE 4.0.

### 12.4.4 Employment Disruption and Economic Adjustment

Automation driven by AI and robotics threatens significant job displacement, particularly in transportation, retail, manufacturing, and administrative work. While new roles in AI development, maintenance, and oversight will emerge, displacement may occur faster than labor markets can adapt. Historical precedent from agricultural mechanization to industrial automation shows that transitions can be socially disruptive without retraining and safety nets [41].

LIFE 4.0 raises the stakes because automation now affects not only manual labor but also cognitive and service roles. Ensuring that productivity gains translate into shared prosperity rather than concentrated wealth will require proactive labor policy, education reform, and social protection systems.

### 12.4.5 Digital Divide and Equity

Access to LIFE 4.0 technologies is highly unequal. High-income households and digitally advanced nations benefit disproportionately from AI-enabled convenience and efficiency, while billions of people still lack reliable electricity, internet access, or digital literacy. Microsoft estimates that over 4 billion people remain digitally underserved, lacking the foundational infrastructure required to benefit from AI [42].

Language inequality further compounds this divide. Large language models perform far better in high-resource languages such as English than in many low-resource languages, where accuracy can fall below 55 percent. Without deliberate intervention, LIFE 4.0 risks amplifying global inequities rather than alleviating them [33].

### 12.4.6 Autonomy and Human Agency

As AI systems become more proactive, concerns grow about the erosion of human agency. Over-reliance on algorithmic recommendations can reduce critical thinking and decision-making skills, a phenomenon sometimes described as "automation complacency." Users may accept AI outputs uncritically, even when systems are wrong or biased.

Surveys indicate ambivalence: nearly half of consumers prefer not to let AI make household decisions independently, citing fears of loss of control [26]. Ethical design, therefore, requires human-in-the-loop mechanisms, explainability, and user override options, ensuring that AI remains assistive rather than authoritative [43].

### 12.4.7 Privacy of Vulnerable Users

Particularly sensitive are the risks faced by vulnerable populations, including children, the elderly, and people with disabilities. AI caregiving systems generate highly sensitive health and behavioral data that could be exploited if misused or inadequately protected. Children may lack informed consent, while older adults may not fully understand data practices associated with AI assistants.

Ensuring that such data are used strictly for care and not sold, repurposed, or weaponized for advertising is a critical ethical obligation. Advocacy groups emphasize that protecting vulnerable users should be a baseline requirement, not an optional feature of LIFE 4.0 systems [32].

In summary, LIFE 4.0 magnifies long-standing ethical challenges while introducing new ones tied to autonomy, intimacy, and scale. Privacy risks, system failures, bias, job displacement, and inequitable access threaten to undermine public trust if left unaddressed. The very features that make LIFE 4.0 attractive, automation, personalization, and invisibility, also heighten its potential harms. Addressing these challenges requires responsible AI design, grounded in transparency, fairness, accountability, and respect for human dignity. Governments must establish enforceable standards, industry must prioritize ethical engineering, and users must be educated about risks and controls. Without such safeguards, the conveniences of LIFE 4.0 may come at the cost of autonomy, equity, and trust. With them, intelligent assistance can remain a tool for empowerment rather than a source of social harm.

## 12.5 THE FUTURE OUTLOOK (LIFE 4.0 SCENARIO)

Looking ahead, researchers and futurists anticipate an era of deeper and more seamless integration of AI into everyday life, extending the trajectory established in LIFE 4.0. Rather than discrete tools or applications, AI is expected to function as a continuous, adaptive infrastructure, embedded across homes, cities, services, and personal routines. Competing scenarios exist. In one, AI remains largely assistive, highly capable but bounded, while in another, breakthroughs in general artificial intelligence (GAI) enable systems to perform complex, high-skill reasoning tasks once reserved for human professionals. Which pathway dominates will depend on progress in technical capability, trust, regulation, and social acceptance.

In optimistic projections, virtual assistants evolve into seamless companions, coordinating schedules, managing households, and offering contextual advice on health, finance, or career development. In more ambitious visions, AI systems approach human-level reasoning by the 2030s or 2040s, enabling certified deployment in domains such as law, medicine, engineering, and scientific research, provided robust safety, transparency, and accountability mechanisms are established [29, 44]. Across scenarios, the defining feature of LIFE 4.0's future is not automation alone, but anticipation: systems that act before users explicitly ask.

### 12.5.1 Ambient Intelligence as Background Infrastructure

A central trend shaping the future of LIFE 4.0 is the rise of ambient intelligence AI that fades into the background while continuously adapting environments to human needs. Smart sensors embedded in homes, offices, and cities will monitor contextual signals such as occupancy, weather, energy demand, and mobility flows. Lighting adjusts to mood and circadian rhythms; traffic signals respond dynamically to congestion; and climate systems optimize comfort and efficiency without manual intervention.

As the IoT expands, it may evolve into an "Internet of Everyday Things," in which microchips embedded in clothing, utensils, furniture, and packaging connect physical objects to cloud-based AI. This vision extends Mark Weiser's concept of ubiquitous computing, where technology becomes calm, invisible, and integrated into the fabric of life rather than a focal point of attention [45, 46]. In such environments, daily routines are orchestrated continuously, reducing cognitive load while increasing responsiveness.

### 12.5.2 Ubiquitous Personal AI Agents

Another defining feature of LIFE 4.0's future is the emergence of personalized AI agents acting on behalf of individuals. Rather than a single assistant responding to commands, each person may be represented by a persistent AI delegate that understands preferences, constraints, and goals. These agents could negotiate service contracts (such as electricity or internet pricing), coordinate travel across modes, manage subscriptions, and pre-authorize healthcare interactions, effectively acting as digital proxies.

In healthcare, AI agents might manage preventive care by scheduling screenings, monitoring biomarkers, and coordinating insurance claims. In commerce, agents could compare prices, negotiate discounts, and strategically time purchases. Such delegation promises efficiency but also raises governance questions about accountability, consent, and alignment, reinforcing the need for clear regulatory frameworks governing agent autonomy [38, 47].

### 12.5.3 Enhanced Reality and the Blurring of Physical–Digital Boundaries

The future of LIFE 4.0 also includes enhanced reality, as AR glasses and mixed-reality systems overlay digital information directly onto the physical world. Grocery shopping may display nutritional data, allergy warnings, or sustainability scores in real time. Tourists could access instant translations, historical context, or navigational cues through sight-based overlays. Combined with haptic feedback and advanced sensors, these systems blur the line between physical and digital experience.

This convergence suggests a shift from screen-based interaction to situated, embodied computing, in which information is delivered contextually rather than explicitly searched for. Researchers argue that such systems may fundamentally reshape learning, navigation, and perception, making everyday environments themselves interactive interfaces [48].

### 12.5.4 Evolving Social Norms and Governance

As AI becomes more deeply embedded, social norms will adapt. Privacy expectations may shift as individuals trade selective personal data for convenience or personalization, while education systems emphasize digital literacy and AI awareness from an early age. Understanding how algorithms work and when to question them may become a core civic skill.

Regulatory frameworks are also likely to evolve. Governments are already drafting AI-specific legislation, such as the EU AI Act, that classifies high-risk applications and mandates safeguards. Over time, these frameworks may expand to cover everyday AI systems, including household assistants and personal agents, particularly following high-profile failures or incidents [33, 47].

### 12.5.5 Potential Divergence: Incrementalism versus Transformation

Futures scholars caution that LIFE 4.0 may follow divergent paths. In a conservative scenario, AI development remains incremental, focused on narrow tasks and optimization. By mid-century, everyday life may look familiar, more convenient, but not fundamentally transformed. In contrast, breakthrough scenarios envision rapid progress toward AGI, enabling systems to perform a wide range of cognitive tasks with minimal human oversight.

Some experts, such as Ray Kurzweil, predict accelerated convergence toward AGI by the 2040s, while others emphasize persistent limitations in common-sense reasoning, value alignment, and contextual understanding [49, 50]. Under the latter view, AI's role remains supportive rather than authoritative, augmenting human decision-making rather than replacing it.

### 12.5.6 LIFE 4.0 at the Tipping Point

From today's vantage point (mid-2020s), LIFE 4.0 appears to be at a tipping point. Early adopters already schedule household tasks via voice assistants, shop through conversational interfaces, and rely on AI for navigation and coordination. Looking two decades ahead, it is plausible to imagine homes that autonomously manage supplies, vehicles that transport children without drivers, and health systems that continuously optimize family well-being through preventive analytics.

In such a future, routine living becomes effortless, while intelligent systems absorb complexity. The challenge will be ensuring that this effortlessness does not come at the expense of autonomy, equity, or meaning. As scholars emphasize, the success of LIFE 4.0 will depend less on technical capabilities than on human values that guide design, governance, and use [43].

## 12.6 CONCLUSION

Everyday LIFE 4.0 represents a decisive turning point in the long historical evolution of human routines, marking the shift from tools that merely assist physical or digital tasks to intelligent systems that anticipate, coordinate, and increasingly shape daily life itself. As traced throughout the chapter, everyday living has progressed from the labor-intensive manual dependence of LIFE 1.0, through the mechanical relief of LIFE 2.0 and the networked convenience of LIFE 3.0, toward a present in which AI actively mediates time, space, decisions, and care. What distinguishes LIFE 4.0 is not automation alone, but the embedding of learning, prediction, and adaptation into the fabric of ordinary activities, from household management and mobility to shopping, health monitoring, and family care.

The opportunities of this transformation are substantial. Intelligent assistance promises significant time savings, enhanced safety, personalized services, improved accessibility, and more sustainable resource use. For many households, AI already reduces routine burdens, supports independence for aging populations, optimizes energy consumption, and enables more flexible forms of work and care. At a societal scale, these gains translate into productivity growth, new economic roles, and the possibility of more inclusive service delivery. In this sense, LIFE 4.0 extends a long-standing human aspiration: to offload drudgery onto technology to reclaim time, attention, and well-being.

Yet the chapter also makes clear that these benefits are neither automatic nor evenly distributed. LIFE 4.0 amplifies long-standing challenges around inequality, privacy, and power, while introducing new risks tied to autonomy, surveillance, bias, and system dependence. Intelligent systems that operate invisibly and continuously raise profound ethical questions about control, consent, and accountability, especially for vulnerable users such as children, the elderly, and digitally marginalized populations. Without deliberate governance, inclusive design, and public literacy, the conveniences of intelligent assistance risk deepening digital divides and eroding human agency rather than enhancing it.

Ultimately, the significance of Everyday LIFE 4.0 lies not in technological sophistication alone, but in the values that guide its integration into daily routines. As AI becomes a routine companion in ordinary life, the central challenge is to ensure that intelligent systems remain tools for empowerment rather than instruments of dependency or exclusion. The future of LIFE 4.0 will be shaped less by what AI can do than by how societies choose to design, regulate, and live with it. When aligned with human dignity, transparency, and equity, intelligent assistance can fulfill its deepest promise: not to replace human life, but to support richer, safer, and more meaningful ways of living in the AI epoch.

## REFERENCES

1. Datar Portal, Digital 2025, G.O. Report, Editor, 2025.
2. McKinsey Global Institute, The Economic Potential of Generative AI, 2023.
3. World Bank Group, World Development Indicators: Digital Adoption, 2023.
4. Pew Research Center, Public Attitudes Toward Artificial Intelligence, 2023.
5. De Vries, J., The Industrial Revolution and the Industrious Revolution. *Journal of Economic History*, 1994. 54(2): p. 249–270.
6. Mokyr, J., *The Lever of Riches: Technological Creativity and Economic Progress*. New York: Oxford University Press, 1990.
7. Humphries, J., *Childhood and Child Labour in the British Industrial Revolution*. Cambridge: Cambridge University Press, 2010.
8. Goldin, C., *Understanding the Gender Gap*. New York: Oxford University Press, 1990.
9. Cunningham, H., *The Invention of Childhood*. London: Penguin, 2006.
10. Braudel, F., *The Structures of Everyday Life: The Limits of the Possible*. New York: Harper & Row, 1981.
11. Thompson, E.P., Time, Work-Discipline, and Industrial Capitalism. *Past & Present*, 1967. 38: p. 56–97.
12. Greenwood, J., A. Seshadri, and M. Yorukoglu, Engines of Liberation. *Review of Economic Studies*, 2005. 72(1): p. 109–133.
13. Goldin, C., The Quiet Revolution That Transformed Women's Employment, Education, and Family. *American Economic Review*, 2006. 96(2): p. 1–21.
14. De Vries, J., *The Industrious Revolution: Consumer Behavior and the Household Economy, 1650–1850*. Cambridge: Cambridge University Press, 2008.

15. Gartman, D., *Auto Slavery: The Labor Process in the American Automobile Industry.* New Brunswick, NJ: Rutgers University Press, 2004.
16. Allen, R.C., *The British Industrial Revolution in Global Perspective.* Cambridge: Cambridge University Press, 2009.
17. Smil, V., *Creating the Twentieth Century.* Oxford: Oxford University Press, 2005.
18. Castells, M., *The Rise of the Network Society*, 2nd ed. Oxford: Wiley-Blackwell, 2010.
19. GSMA, *The Mobile Economy*, 3rd ed. Oxford: Wiley-Blackwell, 2024.
20. Research, J., *Voice Assistants: Market Forecasts & Adoption.* London: Springer, 2023: p. 1–34.
21. McKinsey Global Institute, *The Value of Personalization at Scale.* Springer: London, UK, 2022. 4(0): p. 45–64.
22. Srnicek, N., *Platform Capitalism.* Cambridge: Polity Press, 2017.
23. Zuboff, S., *The Age of Surveillance Capitalism.* New York: PublicAffairs, 2019.
24. Gillespie, T., *Custodians of the Internet.* New Haven, CT: Yale University Press, 2018.
25. Statista, *Smart Home Market Size and Forecast*, 2024.
26. Schneider Electric, Smart *Homes and Energy Efficiency Report*, 2023.
27. Kantar, *Global AI Usage and Attitudes Survey*, 2024.
28. AARP, *Older Adults' Views on Artificial Intelligence and Health*, 2023.
29. OECD, *Artificial Intelligence in Financial Markets*, 2023.
30. Cowan, R.S., *More Work for Mother.* New York: Basic Books, 1983.
31. European Transport Safety Council, *The Role of Advanced Driver Assistance Systems in Road Safety*, 2022.
32. World Health Organization (EHO), *Ethics and Governance of AI for Health*, 2021.
33. UNESCO, *Recommendation on the Ethics of Artificial Intelligence*, 2021.
34. IEA, *Digitalization and Energy*, 2023.
35. Policy Group, *AI in Transit and Mobility: Measurable Benefits*, 2022.
36. European Commission, *General Data Protection Regulation (GDPR)*, 2018.
37. Goodfellow, I., J. Shlens, and C. Szegedy, Explaining and Harnessing Adversarial Examples. arXiv, 2015.
38. OECD, *Artificial Intelligence in Society*, 2019.
39. Buolamwini, J., and T. Gebru, Gender Shades. *Proceedings of Machine Learning Research*, 2018. 81: p. 1–15.
40. Rainie, L., and J. Anderson, *The Future of Well-Being in a Tech-Saturated World.* Pew Research Center, Washington DC, USA, 2018.
41. Autor, D.H., Why Are There Still So Many Jobs? *Journal of Economic Perspectives*, 2015. 229(3): p. 3–30.
42. Microsoft, *Bridging the Digital Divide*, 2022.
43. Floridi, L., et al., AI4People—An Ethical Framework. *Minds and Machines*, 2018. 28: p. 689–707.
44. Brynjolfsson, E., and A. McAfee, *Machine, Platform, Crowd.* New York: W. W. Norton, 2017.
45. Weiser, M., The Computer for the 21st Century. *Scientific American*, 1991. 265(3): p. 94–104.
46. Floridi, L., *The Fourth Revolution.* Oxford: Oxford University Press, 2014.
47. European Commission, *Proposal for a Regulation Laying Down Harmonised Rules on Artificial Intelligence (AI Act)*, 2021.
48. Milgram, P., and F. Kishino, A Taxonomy of Mixed Reality Visual Displays. *IEICE Transactions on Information and Systems*, 1994. 77(12): p. 1321–1329.
49. Kurzweil, R., *The Singularity Is Near.* New York: Viking, 2005.
50. Marcus, G., and E. Davis, *Rebooting AI.* New York: Pantheon, 2019.

# Index

Note: **Bold** page numbers refer to tables.

## D

## E

## F



## G

## N



## O



## P



## R

## V



## W



## X